# Radiation Effects on Solid Surfaces

**Manfred Kaminsky,** EDITOR

*Argonne National Laboratory*

A symposium sponsored by

the Division of Colloid and

Surface Chemistry at the

170th Meeting of the

American Chemical Society,

Chicago, Ill.,

August 25–26, 1975

ADVANCES IN CHEMISTRY SERIES **158**

AMERICAN CHEMICAL SOCIETY

WASHINGTON, D. C.      1976

Library of Congress CIP Data

Radiation effects on solid surfaces.
(Advances in chemistry series; 158 ISSN 0065-2393)

Includes bibliographical references and index.

1. Solids, Effect of radiation on—Congresses. 2. Surfaces (Technology)—Congresses.
I. Kaminsky, Manfred. II. American Chemical Society. Division of Colloid and Surface Chemistry. III. Series.

QD1.A355 no. 158 [QC176.8.R3] 540'.8s [530.4'1]
ISBN 0-8412-0331-8 ADCSAJ 158 1–397

# Advances in Chemistry Series

**Robert F. Gould,** *Editor*

# FOREWORD

ADVANCES IN CHEMISTRY SERIES was founded in 1949 by the American Chemical Society as an outlet for symposia and collections of data in special areas of topical interest that could not be accommodated in the Society's journals. It provides a medium for symposia that would otherwise be fragmented, their papers distributed among several journals or not published at all. Papers are refereed critically according to ACS editorial standards and receive the careful attention and processing characteristic of ACS publications. Papers published in ADVANCES IN CHEMISTRY SERIES are original contributions not published elsewhere in whole or major part and include reports of research as well as reviews since symposia may embrace both types of presentation.

# CONTENTS

# PREFACE

The symposium, "Radiation Effects on Solid Surfaces," was held to discuss various surface phenomena which can occur under energetic particle and photon irradiations of solid surfaces. The organization of chapters in this volume follows that of the conference program. First some of the basic surface processes occurring under surface irradiations with atoms, ions, neutrons, electrons, positrons, and photons (including x-rays) are discussed. Next, recently developed techniques based on particle and/or photon irradiations of surfaces are presented. Finally, surface irradiation effects in solar and nuclear energy applications are covered.

The editor is very grateful to his colleagues who contributed to this symposium and to the session chairmen: G. A. Somorjai (University of California, Berkeley), F. L. Vook (Sandia Laboratories, Albuquerque, N. M.), Monroe S. Wechsler (Iowa State University, Ames, Iowa), and Klaus M. Zwilsky (DMFE–USERDA). He would also like to acknowledge the efficient secretarial help of Debra Herman, Argonne National Laboratory. The editor would like to thank G. A. Somorjai, director of the Division of Colloid and Surface Chemistry, ACS for his encouragement to organize the symposium.

Argonne National Laboratory
Argonne, Ill. 60439
October, 1976

MANFRED KAMINSKY

# INTRODUCTION

Investigating the effect of high energy radiation on the structure and composition of surfaces is one of the important endeavors of modern surface science research. Energetic ions or neutral particles incident on the surface or particles generated internally in radioactive systems disrupt the atomic structure in the near-surface region. These induced defects in structure can cause mechanical damage or can be the primary targets of chemical attack—both destructive to the solid structure. Thus the importance of studies of radiation effects on solid surfaces to the nuclear technology cannot be overemphasized. Most of our difficulties in assembling safe nuclear reactors stem from materials problems caused by the high temperature, high energy radiation environment.

Most of the papers in this volume describe recent results of fundamental research in the field and discuss many of the modern techniques that have become available in recent years. The research results and their interpretation based on the underlying theory are presented in several well written papers that are easy to follow even for the uninitiated reader. The principles and applications of modern techniques of surface science are discussed equally well.

It is hoped that this volume will provide a valuable reference for those working in the fields of radiation damage or surface science. This book will also be of interest to those who would like to learn about this exciting field of surface chemistry.

University of California            G. A. SOMORJAI
Berkeley, Calif.

# Physical Sputtering: A Discussion of Experiment and Theory

HAROLD F. WINTERS

IBM Research Laboratory, San Jose, Calif. 95193

*The theory of sputtering of amorphous and polycrystalline materials has been further developed. Experimental results for the energy, mass, and angular dependence of sputtering yields and recent experiments on the sputtering of chemisorbed gas and sputtering by molecular ions illustrate several effects that are important in determining sputtering yields and, in particular, the relationship of these effects to Sigmund theory. There are several mechanisms which might lead to discrepancies between theory and experiment.*

The ejection of material from solid surfaces under bombardment by energetic ions (or neutrals) is known as sputtering. The sputtering yield, $S$, is defined as the number of target atoms ejected per incident ion. Review articles on physical sputtering have been published by Güntherschulze and Meyer (*1*), Wehner (*2*), Behrish (*3*), Kaminsky (*4*), Carter and Colligon (*5*), McDonald (*6*), and Tsong and Barber (*7*). McCracken (*8*) has published a paper recently on the interaction of ions with solid surfaces in which there is a long section on sputtering. In addition, Sigmund has published a series of review articles on the collision theory of displacement damage (*9*) which includes a long chapter on sputtering (*10*). All of these articles, and in particular, various articles by Sigmund and co-workers, have been of material benefit in preparing this paper. This is all the more true because the author, while very interested in the interaction of low energy ions ( < 1000 eV) with solid surfaces, is not a specialist in the field of physical sputtering. It is hoped that this chapter will benefit from his slightly detached point of view rather than suffering from a lack of intimacy with the subject.

The thrust of this paper is to outline concisely the current theoretical approaches to the sputtering of amorphous and polycrystalline targets and then to interpret some of the important experiments on the basis of the derived results. No attempt at comprehensive coverage will

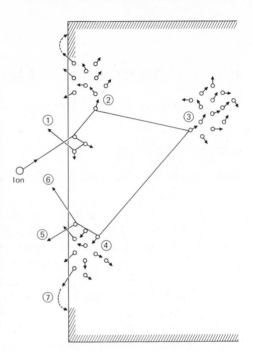

*Figure 1. Schematic of possible collision processes which occur under ion bombardment. (1) Surface atom receives energy and after several collisions is reflected away from the target; (2) the incoming ion creates a primary recoil which in turn produces a collision cascade that penetrates the surface; (3) collision cascade which does not penetrate the surface; (4) reflected ion creates a cascade which penetrates the surface; (5) reflected ion gives energy to a surface atom which is sputtered; (6) ion reflected into the vacuum with kinetic energy; and (7) atom with momentum component directed away from the surface returns because of attractive forces.*

be made because of the number of review articles already in the literature.

All modern approaches to the theory of physical sputtering are based on the binary collision model which assumes that energy is transferred from the impinging ion to the target atoms by a sequence of binary collisions; i.e., the ion only interacts with one target atom at a time. This process is illustrated schematically in Figure 1. The first collision does not, in general, lead directly to sputtering since the hit target atom has a momentum component in the direction away from the surface. Therefore, sputtering is a multiple collision process involving a cascade of moving target atoms. Whereas the concept of a collision cascade governing the sputtering process is a common feature in all recent sputtering theories (*11, 12, 13, 14*), there are some differences in the processes that various authors consider important and consequently in the approximations made to solve the problem (*see* Ref. *11*).

A yield calculation consists of a number of steps:

(1) To determine the differential cross section, $d\sigma$, for the transfer of energy between $T$ and $T + dT$ from the ion to the target atom and from one target atom to another (this step primarily involves the approximation of interaction potentials).

(2) To determine the amount of energy deposited near the surface.

(3) To convert this energy into the density of low energy recoil atoms.

(4) To determine the number of recoil atoms which reach the very surface.

(5) To select those atoms which are able to overcome surface binding forces and thus be emitted into the gas phase.

### The Formalism of Sputtering Theory

**Classical Mechanics.** Two-particle collisions are often adequately described using the ideas of classical dynamics, and all modern sputtering theories make this assumption. The quantity, $d\sigma(T)$, which is needed by sputtering theorists, can be formally obtained from the following well known equations (5, 15).

$$T = T_m \sin^2 \frac{\theta}{2} \; ; \quad T_m = \frac{4M_1 M_2}{(M_1 + M_2)^2} E \tag{1}$$

$$\theta = \pi - 2P \int_{R_0}^{\infty} \frac{dR/R}{\left[1 - \frac{U(R)}{E_R} - \frac{P^2}{R^2}\right]^{1/2}} \tag{2}$$

$$E_R = \frac{M_2 E}{M_1 + M_2}$$

$$d\sigma(\theta) = -2\pi P dP \tag{3}$$

where (*see* Figure 2) $M_1$ and $M_2$ are the masses of the colliding atoms, $E$ the initial energy, $T$ the energy transfer (the energy of the recoiling atom), $\theta$ the center of mass scattering angle, $P$ the impact parameter, $R$ the internuclear distance, $R_0$ the distance of closest approach, and $U(R)$ the interatomic potential. $R_0$ is the root of the equation (15):

$$\left(1 - \frac{U(R_0)}{E_R}\right) R_0{}^2 = P^2$$

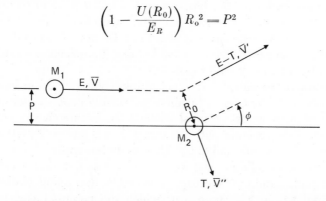

*Figure 2. Scattering of two particles viewed from the laboratory system. $\phi$ is the laboratory scattering angle.*

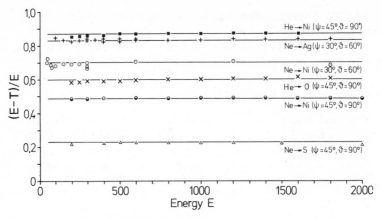

*Figure 3. Fractional energy loss* (E − T)/E *vs. primary energy at constant* φ *for ion scattering experiments. Data from Ref. 17.*

$d\sigma(T)$ is formally obtained by inverting Equation 2 to obtain $P$ as a function of $\theta$. $P$ and $dP$ are then substituted into Equation 3 yielding $d\sigma(\theta)$ which is subsequently changed to $d\sigma(T)$ using Equation 1. The input quantity needed to use this procedure is the interaction potential, $U(R)$.

The binary collision approximation is crucial to sputtering theorists, and hence we make a few comments at this time. The best experimental evidence supporting the model is from investigations conducted at many laboratories involving the scattering of ion beams from surfaces and the subsequent measurement of the angle and energy of the reflected ions (*16*). For binary collisions, the quantity $(E − T)/E$ (*see* Figure 2) is independent of $E$ for constant $\theta$ (*see* Equation 1). (Constant $\theta$ implies constant $\phi$.) Figure 3 shows data from Heiland and Taglauer (*17*) which demonstrate that the binary approximation is valid to quite low energies. This is consistent with computer-simulated results of Karpuzov and Yurasova (*18*) who investigated the reflection of 50–500 eV argon ions from a copper crystal and concluded that ion reflection is adequately described by the binary collision model.

**Interatomic Potentials.** It is not absolutely necessary to have accurate interatomic potentials to calculate reasonably good sputtering yields because the many collisions involved tend to obscure the details of the interaction. This, together with the fact that accurate potentials are only known for a few sysems makes the Thomas–Fermi approach quite attractive. The Thomas–Fermi statistical model assumes that $V(r)$ varies slowly enough within an electron wavelength that many electrons can be localized within a volume over which the potential changes by a fraction of itself. The electrons can then be treated by statistical mechanics and obey Fermi–Dirac statistics. In this approximation, the potential in the atom is given by:

$$V(r) = \frac{Ze^2}{r} \, \phi\left(\frac{r}{a}\right); \quad a = 0.885 \, a_0 Z^{-1/3} \qquad (4)$$

where $Z$ is the atomic number and $a_0$ the Bohr radius, 0.529 Å. For a derivation of Equation 4, *see* Ref. *19*. $\phi(r/a)$ is the Thomas–Fermi screening function shown in Figure 4.

It is convenient to describe the interatomic potential, $U(R)$, with the same functional form as Equation 4. This has been accomplished by Bohr (*21*) who estimated the interaction energy between two atoms by the formula:

$$U(R) \cong \frac{Z_1 Z_2 e^2}{R} \exp - \frac{R}{a}$$

where $R$ is the internuclear distance and $\exp(-R/a)$ the screening function. Subsequent authors usually represented their interaction potentials in the same form but with modified screening functions. Firsov (*22*) showed that, within the limits of accuracy of the Thomas–Fermi statisti-

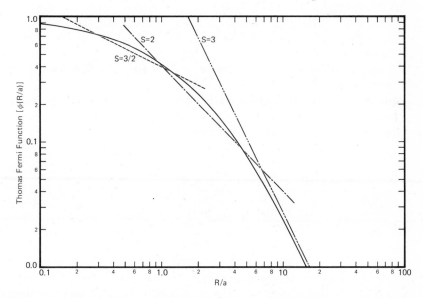

*Figure 4.   Thomas–Fermi screening function, $\phi(R/a)$, (see Equation 4) for neutral atoms (——) and power approximations (– – –) from Equation 7. Values of $\phi(R/a)$ are from Ref. 20. Constants used in Equation 7 are: $k_{1.5} = 0.591$, $k_2 = 0.833$, $k_3 = 2.75$. See Equation 7 for definition of s.*

cal model, the interaction between atoms at a distance less than $10^{-8}$ cm could be described by the potential:

$$U(R) = \frac{Z_1 Z_2 e^2}{R} \, \phi \left( \frac{R}{a_F} \right) \tag{5}$$

where $\phi(R/a_F)$ is the Thomas–Fermi screening function shown in Figure 4 and

$$a_F = 0.8853 \, a_0 \, (Z_1^{1/2} + Z_2^{1/2})^{-2/3}$$

Lindhard et al. (23) preferred a screening radius:

$$a_L = 0.8853 \, a_0 [Z_1^{2/3} + Z_2^{2/3}]^{-1/2}$$

for the same functional form of Equation 5. The two screening radii are numerically equal within the accuracy of the Thomas–Fermi approach. In subsequent sections we will use $a_L$ and refer to it as $a$.

Equation 5 is often used to describe the interaction between the incoming ion and the target atoms. The interaction between two target atoms generally occurs at low energy where the Thomas–Fermi potential overestimates the interaction. Under this situation a Born–Mayer potential is more appropriate (11), i.e.:

$$U(R) = A e^{-bR} \tag{6}$$

where $A$ and $b$ are constants. Typical values for $A$ and $b$ have been tabulated by Abrahamson (24).

An especially useful approximation for the Thomas–Fermi potential has been developed by Lindhard (23) and co-workers where the screening function is assumed to have the form:

$$\phi \left( \frac{R}{a} \right) = \frac{k_s}{a} \left( \frac{a}{R} \right)^{s-1} \tag{7}$$

where $k_s$ and $s$ are constants. $U(R)$ then becomes:

$$U(R) = \frac{k_s}{s} \frac{Z_1 Z_2 e^2 a^{(s-1)}}{R^s} \tag{8}$$

Figure 4 shows that Equation 7 reasonably approximates the screening function over limited energy ranges. The inverse power approximation made in Equation 7 is quite attractive since it allows Equation 2 to be integrated in closed form for several values of $s$ (25).

The substitution of Equation 8 into Equation 2 followed by approximations (*see* Ref. 23) and integration leads to:

$$d\sigma(T) = CE^{-m}T^{-1-m}dT; \quad m = \frac{1}{s} \tag{9}$$

where

$$C = \frac{1}{2}\pi\lambda_m a^2 \left(\frac{M_1}{M_2}\right)^m \left(\frac{2Z_1Z_2e^2}{a}\right)^{2m}$$

and $\lambda_{m=1} = 0.5$, $\lambda_{m=1/2} = 0.327$, $\lambda_{m=1/3} = 1.309$. $m = 1$ corresponds to Rutherford scattering, and Sigmund (*11*) has shown that $m = 0$ approximates scattering from a Born–Mayer potential. For $m = 0$, $C_0 = (1/2)\pi\lambda_0 a^2$ where $\lambda_0 = 24$ and $a = 0.219$ Å. For $m = 0$ $a$ is assumed to be independent of $z$. Sigmund also suggests that for medium mass ions and atoms over most of the keV range, $m = 1/2$ is a fair approximation while in the lower keV and upper eV region, $m = 1/3$ should be adequate.

Equation 9 is an extremely useful description for the differential cross section and has been used extensively in a variety of applications. With the use of Equation 9, the nuclear stopping power becomes:

$$s_n(E) = \int_0^{T_m} T d\sigma = \frac{C}{1-m} \left[\frac{4M_1M_2}{(M_1+M_2)^2}\right]^{1-m} E^{1-2m} \tag{10}$$

This is one of the basic input quantities needed in Sigmund's sputtering theory.

**Basic Equations—Sigmund's Theory (*11*).** There have been three important theories in recent years on the sputtering of amorphous and polycrystalline solids. They are attributed to Sigmund (*11*), Thompson (*12*), and Brandt and Laubert (*13*). The predictions about various aspects of sputtering often agree. Sigmund's theory, however, is the most general and latest and therefore is outlined briefly.

Suppose an atom starts its motion with an arbitrary velocity vector $\bar{v}$ at $t = 0$ in a plane $x = 0$ in an infinite medium (Figure 5). The basic quantity of interest is the function $G(x,\bar{v}_0,\bar{v},t)d^3v_0 dx$ which is the average number of atoms moving at time $t$ in a layer $(x,dx)$ with a velocity $(\bar{v}_0,d^3v_0)$. The number of atoms with a velocity $(\bar{v}_0,d^3v_0)$ penetrating the plane $x$ in the time interval $dt$ is given by:

$$G(x,\bar{v}_0,\bar{v},t)d^3v_0|v_{0x}|dt$$

wher $v_{0x}$ is the $x$ component of $\bar{v}_0$. The backward sputtering yield for a surface at $x = 0$ is then:

$$S = \int |v_{0x}| d^3 v_0 \int_0^\infty dt\, G(0, \overline{v}_0, \overline{v}, t) \tag{11}$$

Based on Equation 11, $S$ is formally obtained by (1) writing a differential equation for $G(x, \overline{v}, \overline{v}_0, t)$ using standard techniques, (2) integrating the resulting equation over $\overline{v}_0$ and $t$ which yields an equation involving $H(\overline{v}, x)$, and (3) solving the resulting integral equation for $H(\overline{v}, 0)$ using the input quantities $d\sigma(T)$ (Equation 10), the heat of sublimation and, where necessary, an expression for electronic stopping.

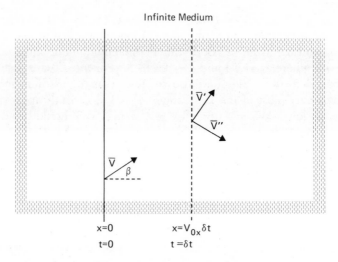

Figure 5.    *Geometry of the sputtering events considered in the Sigmund (1969) theory. The initial velocity of the ion is $\overline{v}$, and after one collision, the ion velocity is $\overline{v}'$; the velocity of the primary recoil is $v''$.*

We will now derive the equation used to determine $G(x, \overline{v}_0, \overline{v}, t)$. Consider a particle moving at $t = 0$ in $x = 0$ (Figure 5). After a time $\delta t$, it may or may not have collided. In any case, the final distribution cannot have changed. Therefore an expression for $G(x, \overline{v}, \overline{v}_0, t)$ based on initial conditions is equated to an expression for $G$ based on the situation as it exists after a time $\delta t$. The equation:

$$G(x, \overline{v}, \overline{v}_0, t)_{\text{initial}} = G(x, \overline{v}, \overline{v}_0, t)_{\text{after } \Delta x, \Delta t}$$

reads

$$G(x, \overline{v}, \overline{v}_0, t) = Nv\delta t \int [G(x, \overline{v}_0, \overline{v}', t) + G(x, \overline{v}_0, \overline{v}'', t)] d\sigma(v, v', v'')$$

$$+ [1 - Nv\delta t \int d\sigma(\overline{v}, \overline{v}', \overline{v}'')] G(x + \eta v\delta t, \overline{v}_0, \overline{v}, t + \delta t) \tag{12}$$

where $N$ is the atom density and $\eta = v_x/v$, where $v_x$ is the $x$ component of $v$, i.e., $\eta$ is the cosine of the angle of incidence. The first term on the right hand side expresses the collision probability specified by $\overline{v}',\overline{v}''$ and multiplied by the sum of contributions to $G$ by the two collision partners and then integrated over all collisions. The second term is the probability for not making a collision, multiplied by the contribution to $G$ by a particle with unchanged velocity but changed initial position and starting time. Expansion of $G(x + \eta v \delta t, \overline{v}_0, \overline{v}, t + \delta t)$ in terms of $\delta t$ followed by manipulation yields first order terms:

$$-\frac{1}{v}\frac{\delta}{\delta t}\, G(x,\overline{v}_0,\overline{v},t) - \eta\,\frac{\delta}{\delta x}\, G(x,\overline{v}_0,\overline{v},t) =$$
$$N\int_{v',v''} d\sigma[G(x,\overline{v}_0,\overline{v},t) - G(x,\overline{v}_0,\overline{v}'t) - G(x,\overline{v}_0,\overline{v}'',t)]  \qquad (13)$$

Equation 13 applies to an ion of the same species as the target, but similar arguments lead to an analogous equation for an arbitrary ion. For simplicity, electronic stopping has also been neglected but could easily be included as was done in the original treatment by Sigmund.

The function:

$$F(x,\overline{v}_0,\overline{v}) = \int_0^\infty G(x,\overline{v}_0,\overline{v},t)\,dt$$

is the total number of atoms which penetrate the plane $x$ with a velocity $(\overline{v}_0, d\overline{v}_0^3)$ during the development of the collision cascade. $F(x,\overline{v}_0,\overline{v})$ satisfies an equation that follows from integration of Equation 13 over $t$, i.e.:

$$\frac{1}{v}\,\delta(x)\,\delta(\overline{v} - \overline{v}_0) - \eta\,\frac{\partial}{\partial x}\, F(x,\overline{v}_0,\overline{v}) =$$
$$N\int d\sigma[F(x,\overline{v}_0,\overline{v}) - F(x,\overline{v}_0,\overline{v}') - F(x,\overline{v}_0,\overline{v}'')]  \qquad (14)$$

The function $H(x,\overline{v})$ represents the backward sputtering yield of a target atom for the case of a source at $x = 0$ and the sputtered surface in the plane $x$ where:

$$H(x,\overline{v}) = \int d^3 v_0 |v_{0x}| F(x,\overline{v}_0,\overline{v})$$

The integrations over $\overline{v}_0$ obey the conditions:

$$\eta_0 = \frac{v_{0x}}{v} \leq 0$$

$$E_0 = \frac{1}{2} M v_0{}^2 \geq U(\eta_0)$$

where $U(\eta_0)$ is a surface binding energy. Multiplying both sides of Equation 14 by $v_{0x}$ and integrating over $\bar{v}_0$ and also changing the velocity variables to energy variables yields:

$$-\delta(x)\eta\theta(-\eta)\theta[E - U(\eta)] - \eta\frac{\partial H(x,E,\eta)}{\partial x}$$
$$= N \int d\sigma[H(x,E,\eta) - H(x,E',\eta') - H(x,E'',\eta'')] \qquad (15)$$

Here, $\theta(\xi)$ is the Heaviside step function; $\theta(\xi) = 0$ for $\xi < 0$ and $\theta(\xi) = 1$ for $\xi \geq 1$. The backward sputtering yield is:

$$S(E,\eta) = H(x = 0,E,\eta) \qquad (16)$$

The fundamental problem in determining yield is to solve Equation 15 for $H$. Sigmund solves the problem by using a standard technique involving expansion in Legendre polynomials. The final result is given by:

$$H(x,E,\eta) = \frac{3}{4\pi^2} \frac{F(x,E,\eta)}{NC_0U_0} \qquad (17)$$

where $F(x,E,\eta)$ is now defined as the amount of energy deposited in a layer $(x,dx)$ by an ion of energy $E$ starting at $x = 0$ and by all recoil atoms. See Ref. 11 for the mathematical details used in obtaining Equation 17 from Equation 15.

For power scattering, it can be shown that:

$$F(0,E,\eta) = \alpha N s_n(E) \qquad (18)$$

where $s_n(E)$ is the nuclear stopping power and $\alpha$ is a dimensionless quantity depending on the relative masses and angle of incidence. $\alpha$ is shown as a function of $M_2/M_1$ in Figure 6 for perpendicular incidence. Direct energy dependence in $\alpha$ drops out for power scattering. Using Equations 17 and 18, the sputtering yield becomes:

$$S(E,\eta) = \frac{3}{4\pi^2} \frac{\alpha s_n(E)}{C_0U_0} = 0.042\,\alpha\,\frac{s_n(E)}{U_0} \qquad (19)$$

Although Equation 19 has been derived for the case where the ion and target are of the same substance, it is also valid for an arbitrary ion (see Ref. 11).

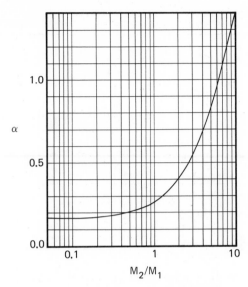

Physical Review

*Figure 6. Factor α in sputtering yield formula (Equation 21) calculated for power scattering and averaged between* $m = 1/3$ *and* $m = 1/2$ (11)

For low energies ($m = 0$; $E \lesssim 1$ keV), the yield becomes:

$$S(E,\eta) = \frac{3}{4\pi^2} \alpha \frac{4M_1 M_2}{(M_1 + M_2)^2} \frac{E}{U_0} \tag{20}$$

where the value for $s_n(E)$ was obtained from Equation 10. For keV energies and heavy-to-medium mass ions, the expression for the nuclear stopping power calculated by Lindhard (23), assuming a Thomas–Fermi interaction, is used, i.e.:

$$s_n(E) = 4\pi Z_1 Z_2 e^2 a \left[ \frac{M_1}{M_1 + M_2} \right] s_n(\epsilon)$$

where $s_n(\epsilon)$ is the reduced stopping power plotted in Figure 7. The sputtering yield from Equation 19 is therefore:

$$S(E,\eta) = 3.56 \, \alpha \frac{Z_1 Z_2}{[Z_1^{2/3} + Z_2^{2/3}]^{1/2}} \frac{M_1}{[M_1 + M_2]} \frac{s_n(\epsilon)}{U_0} \tag{21}$$

where $U_0$ is in eV. The relationship between the ion energy $E$ and reduced energy $\epsilon$ is given by:

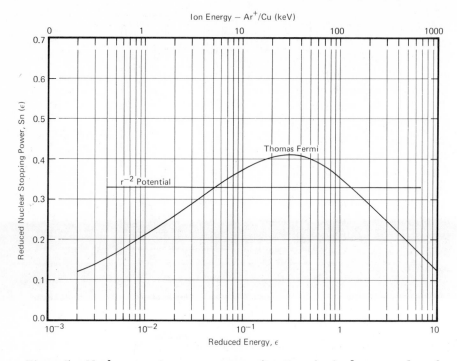

*Figure 7. Nuclear stopping power, $s_n$, as a function of $\varepsilon$ for bottom scale and of E for Ar⁺–Cu top scale. Horizontal line is for an $R^{-2}$ potential. Data from Ref. 23.*

$$\epsilon = \frac{aM_2E}{[Z_1Z_2e^2(M_1 + M_2)]} \tag{22}$$

Equations 20 and 21 along with Figures 6 and 7 are used as a framework to interpret most of the experimental data presented in this paper.

### Comparison of Experiment with Theory

To interpret experimentation results in terms of Equations 20 and 21, we are interested in the sputtering yield expressed as a function of (1) ion energy; (2) the angle of incidence; (3) the masses of the incident ion and target material; (4) the surface binding forces, i.e., $U_0$, and (5) energy $E_0$ of the sputtered atoms.

There have been numerous sputtering yield measurements on amorphous and polycrystalline targets during the past 20 years. Unfortunately many of the measurements are difficult to interpret because of uncontrolled experimental conditions. For example, the presence of chemisorbed gas often reduces the sputtering rate (26). This phenomenon is well known to people involved in the growth of thin films (*27*). An example of the influence of an adsorbed layer is seen in Figure 8 where the variation in sputtering yield with ion dose is shown for 600-eV $N_2^+$ ions on gold. The initial rapid increase was attributed by Colligon et al. (*28*) to the removal of an adsorbed layer. It is, therefore, quite clear that atomically clean surfaces should be used for accurate yield measurements. This requires good vacuum conditions.

It is becoming clear that many of the reported yields are dose dependent. Almén and Bruce (*29*) published some examples of 45-keV $V^+$ ion irradiation of copper and tantalum targets where they found a marked decrease in yield as a function of dose. For 45-keV $Ca^+$ ions on the same target, they found a weight gain. In the same article, the authors reported measurements of the $Z_1$ variation of the 45-keV sputtering yield on copper, silver, and tantalum targets. In the $Z_1$ variation they found peculiar oscillations following the chemical properties of the incident ions and ascribed these variations to a change in target material caused by the accumulation of projectile atoms. This assumption has recently been shown to be correct (*30, 31*) since the periodicity disappears for smaller doses.

A dose dependence is expected under situations where a change of chemical composition occurs because the development of the collision cascade must depend to some extent on the masses of the target atoms.

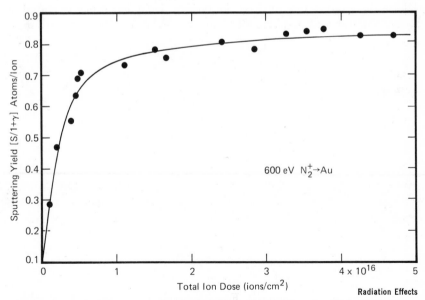

*Figure 8.  Variation of sputtering yield with ion dose for 600-eV $N_2^+$ ions on gold (28)*

Accumulation of significant amounts of projectile material near the surface is expected under many situations. Therefore, a definite need for yield measurements made under low dose conditions exists.

Figure 9 shows the concentration of nitrogen near the surface of a polycrystalline tungsten sample as a function of the number of bombarding $N_2^+$ ions. The saturation coverages are $8 \times 10^{15}$ and $9 \times 10^{15}$ atoms/ cm$^2$ at 300 and 450 eV, respectively. Based on this information, one would expect the sputtering yield for an initially clean surface to change until the surface had been bombarded with $4 \times 10^{16}$ $N_2^+$/cm$^2$ and then to remain relatively constant.

Noble gases are attracted to surfaces by weak van der Waals forces, and therefore when they reach the surface they are almost immediately desorbed into the vacuum. Consequently, ion bambardment effectively releases noble gas atoms trapped near the surface (33), and therefore the saturation concentration does not become nearly so large as for other projectiles (34) (compare Figures 9 and 10). We think that yield measurements involving noble gas ions are in general more reliable than those involving other ions, but even for noble gases, there appears to be a dose dependence (35).

**Variation of Sputtering Yield with Ion Energy.** For scattering using an inverse power potential (Equation 8), the quantity $\alpha$ (*see* Equation 21) is independent of energy and is only a function of the angle of incidence and relative masses. The energy dependence of the sputtering yield is therefore determined by the energy dependence of the stopping power. Brandt and Laubert (13) make similar predictions, and their

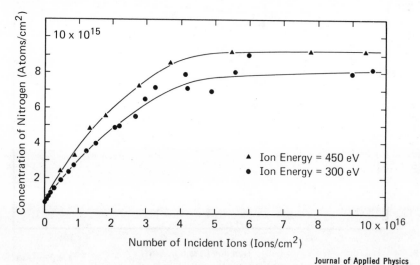

Journal of Applied Physics

*Figure 9.   Number of nitrogen atoms trapped near a tungsten surface vs. number of incident $N_2^+$ ions (32)*

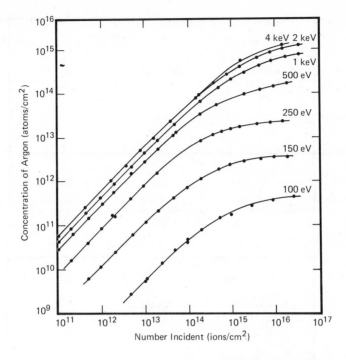

Canadian Journal of Physics

*Figure 10.    The number of argon atoms trapped near a tung-
sten surface as a function of the number of incident argon
ions of various energies (34)*

calculated results agree reasonably with those based on the Sigmund model.

Yields for many ion–target combinations have been calculated using Equations 20 and 21, and the agreement is remarkably good considering that there are no adjustable parameters in the theory. Figure 11 shows data for inert gas ion bombardment of copper where agreement is very good over the entire energy range. Table I shows calculated and measured yields for bombardment of various metals with 45-keV noble-gas ions. The agreement, in general, is somewhat worse than for copper. However, judging from the work of Andersen and Bay (35), we would expect the experimental yields of Almén and Bruce (29) to be significantly higher (possibly by a factor of two or more) if they had been taken under low dose conditions. This would make the agreement between experiment and theory much better.

Wehner (36) has amassed a large amount of data in the low energy range (0–600 eV) which, in our opinion, is quite reliable and relatively free of dose effects. These data are not corrected for secondary electron emission. However, for Kr⁺ and Xe⁺, where Sigmund's theory is most

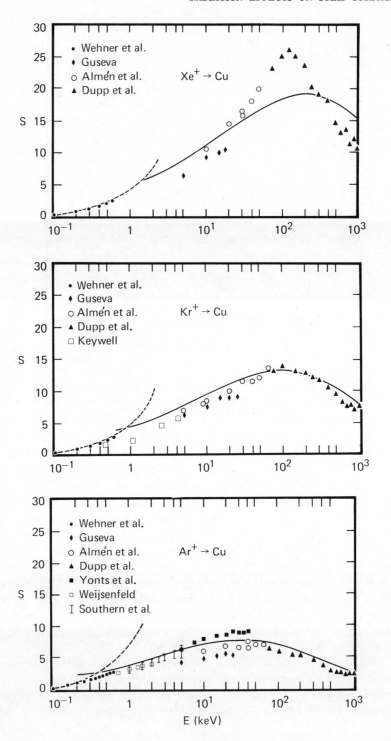

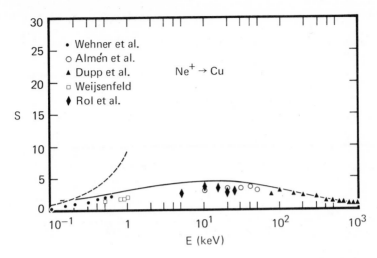

S

• Wehner et al.
○ Almén et al.
▲ Dupp et al.          Ne$^+$ → Cu
□ Weijsenfeld
♦ Rol et al.

E (keV)

Physical Review

*Figure 11. (left and above) Sputtering yields for Cu calculated from Equation 21 (———) and Equation 20 (— — —), compared with experimental results of Refs. 11, 29, 36–43 (11)*

applicable, the correction would, in general, change the yield by less than 10%. Calculated yields using Equation 20 agree excellently with Wehner's data for many target materials, but for others the measured values are smaller than the calculated ones by a factor of two.

Whereas the sputtering yield is generally proportional to the nuclear stopping power as predicted by Sigmund's theory, there are systematic variations for large and small mass ratio ($M_2/M_1$) as well as in the energy dependence. These discrepancies are discussed in later sections.

**Variation of Sputtering Yield with Mass of the Incident Ion.** Dose effects make it inconvenient in some instances to use absolute yield data when comparing experiment with theory. Therefore, Andersen and Bay have presented their data as normalized yields, i.e., the ratio of the sputtering yield for an ion of atomic number $Z_1$ to the self sputtering yield or to the argon sputtering yield. The normalization tends to eliminate or at least minimize dose effects.

Data for Si, Cu, Ag, and Au from Andersen and Bay (35) are shown in Figure 12. Absolute yields from Almén and Bruce (29) along with calculated values from Equation 21 are shown in Table I. Both sets of data and also theory indicate that the yield increases with $Z_1$. The quantitative agreement between theory and experiment (Figure 12) is good for amorphous silicon but gets progressively worse as the mass increases. The sputtering yield for heavy ions on heavy targets shows a more pronounced maximum in the energy dependence than does the stopping power (*see* Figure 11—Xe$^+$ on Cu). Such a discrepancy between theory

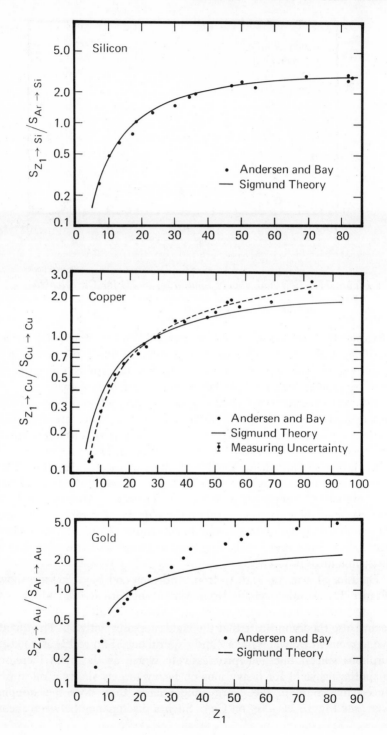

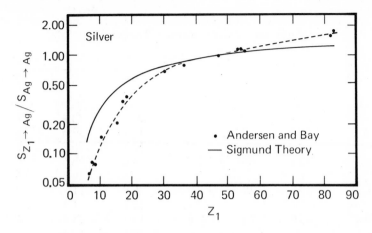

*Figure 12.    (left and above) Normalized sputtering yields for Si, Cu, Ag, and Au for 45-keV ions. Data from Refs. 35, 44.*

and experiment may result from thermal spikes in the dense collision cascades for heavy ion irradiation (45).

Sigmund's sputtering theory is based on the assumption that all interactions occur in binary collisions between a moving and a fixed atom. In very dense cascades (e.g., if cascades designated 2 and 3 in Figure 1 overlapped), the collision may occur between moving atoms, and therefore the theory breaks down. Andersen and Bay (35) suggest that both the deviations in energy dependence for heavy ions on heavy targets and the discrepancy between experiment and theory shown in Figure 12 are caused by this thermal spike effect, i.e., overlapping cascades. Further evidence for thermal spikes is found in the appearance of a low energy peak in the energy spectra of sputtered gold atoms (46, 47) as well as in the temperature dependence of the sputtering yield measured by Nelson (48).

Andersen and Bay (35) have experimentally demonstrated that anomalies in the yield can arise because of nonlinear effects in the collision cascade. This was accomplished through irradiation with molecular ions and subsequent comparison of the yield per atom from the molecular ion with the yield of single atomic ions at the same energy per atom. Overlapping of the cascades created by each individual atom will occur for the case of the molecular ion. This in turn leads to an increased energy density within the cascade. Table II shows that the yield ratio is greater than unity, indicating that nonlinear effects can cause an increase in the sputtering yield.

**Variation of the Sputtering Yield with Angle of Incidence.** The angular dependence of the sputtering yield is contained in the quantity

**Table I.**   **Sputtering Yield for Ne⁺, Ar⁺, Kr⁺, Xe⁺ Ions at 45 keV on Different Polycrystalline Target Materials**[a]

| Target Material | $U_0(eV)$ | $Z_2$ | Sputtering Ratios (atoms/ion) $Ne^+$ | $Ar^+$ | $Kr^+$ | $Xe^+$ |
|---|---|---|---|---|---|---|
| Pb | 2.01 | 82 | 3.6 | 10.5 | 24.0 | 44.5 |
|    |      |    |     | (29.7) | (44.4) | (74.6) |
| Ag | 2.94 | 47 | 4.5 | 10.8 | 23.5 | 36.2 |
|    |      |    |     | (12.7) | (20.1) | (24) |
| Sn | 3.11 | 50 | 1.8 | 4.3 | 8.5 | 11.8 |
|    |      |    |     | (12.5) | (19.5) | (24.9) |
| Cu | 3.46 | 29 | 3.2 | 6.8 | 11.8 | 19.0 |
|    |      |    | (3.7) | (6.7) | (11.9) | (15.5) |
| Au | 3.79 | 79 | 3.6 | 10.2 | 24.5 | 39.0 |
|    |      |    |     | (16.1) | (23.9) | (27.8) |
| Pd | 3.87 | 46 | 2.5 | 5.3 | 10.5 | 14.4 |
|    |      |    |     | (9.24) | (14.1) | (18.1) |
| Fe | 4.29 | 26 | 1.3 | 2.3 | 4.0 | 4.9 |
|    |      |    |     | (5.15) | (8.92) | (11.7) |
| Ni | 4.43 | 28 | 1.4 | 3.5 | 5.6 | 7.6 |
|    |      |    |     | (5.35) | (9.01) | (11.8) |
| V  | 5.3  | 23 | 0.3 | 1.0 | 1.7 | 1.9 |
|    |      |    |     | (3.85) | (6.21) | (8.9) |
| Pt | 5.82 | 78 | 1.9 | 5.3 | 11.3 | 16.0 |
|    |      |    |     | (9.35) | (15.3) | (19.2) |
| Mo | 6.82 | 42 | 0.6 | 1.5 | 2.7 | 3.8 |
|    |      |    |     | (5.14) | (7.70) | (10.1) |
| Ta | 8.06 | 73 | 0.7 | 1.6 | 3.1 | 4.0 |
|    |      |    |     | (6.62) | (10.0) | (12.3) |
| W  | 8.70 | 74 | 1.0 | 2.3 | 4.7 | 6.4 |
|    |      |    |     | (5.07) | (9.31) | (11.9) |

[a] Data from Ref. 29. Theoretical values, in parentheses, were calculated from Equation 21. The values for $U_0$ are from Ref. 49. The values for $\alpha$ and $s_n(\epsilon)$ were estimated from Figures 6 and 7.

**Table II.**   **Ratios between the Sputtering Yield per Atom for Irradiation with Molecular and Atomic Ions for Different Ion Target Combinations**[a]

| Projectiles | Targets $Si$ | $Ag$ | $Au$ |
|---|---|---|---|
| C–Cl₂ | — | 1.09 | — |
| Se–Se₂ | 1.15 | 1.44 | 1.44 |
| Te–Te₂ | 1.30 | 1.67 | 2.15 |

[a] Data from Ref. 35.

$\alpha$ which is independent of energy for power scattering and only weakly dependent on the power $m$ in the cross section formula (Equation 9). Even for inelastic collisions, Sigmund finds that $\alpha$ depends only weakly on energy but is somewhat sensitive to the electronic stopping constant, i.e., $\alpha$ decreases with increasing electronic loss. Since $\alpha$ is almost independent of ion energy, the angular dependence of the sputtering is also predicted to be relatively independent of energy.

For not too oblique angles, Sigmund (*11*) suggests that the angular dependence is given by:

$$\frac{S(\eta)}{S(1)} = \eta^{-f} = (\cos \beta)^{-f} \tag{23}$$

where $S(1)$ is the yield at perpendicular incidence, and $f$ is a calculable constant. For $M_2/M_1 < 1, f \sim 1.7$. For $M_2/M_1 > 1$, the factor $f$ slowly

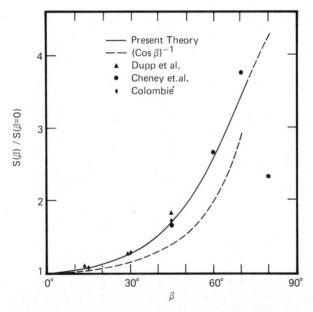

*Figure 13. Variation of sputtering yield with angle of incidence for Ar⁺ ions on polycrystalline copper. (———), evaluated from Sigmund theory for* m = 1/2. *(– – –), for (cos β)⁻¹. Data from Refs. 50, 51, 52.*

decreases until it reaches a value somewhat less than 1. Theory is compared with experiment in Figure 13, and the agreement is quite good. The $1/\cos \theta$ dependence of the sputtering yield which has been suggested by many investigators (*12, 13, 53*) does not agree with experiment (*see* Figures 13 and 14).

Oechsner (*54*) has recently reported an extensive investigation of the sputtering yield of various materials as a function of the angle of incidence. His data for copper are shown in Figure 14. The yield ratio $[S(\beta)/S(0)]$ increases rather slowly in the beginning, then more rapidly, and finally passes through a maximum at near grazing incidence after which it falls toward zero at $\beta = 90°$. Similar behavior has been observed for a variety of target materials (*51, 54, 55, 56*).

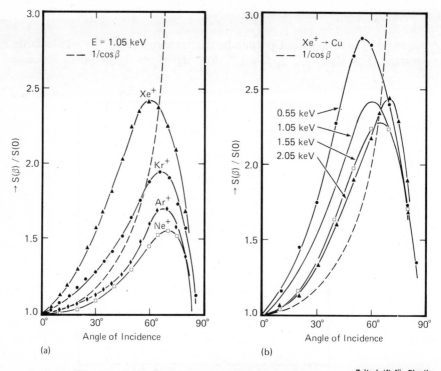

Zeitschrift für Physik

*Figure 14.* (a) (left) Sputtering yield for polycrystalline copper as a function of the angle of incidence. The incident energy was 1.05 keV. S(0) is the yield at normal incidence. The dashed curve represents a 1/cos β dependence. (b) (right) Sputtering yield for polycrystalline copper as a function of angle of incidence and incident energy for irradiation with Xe⁺ ions (54).

Oechsner's data taken at relatively low ion energies do not appear to agree with theory as well as data taken at higher energies. The angular dependence generally increases faster than cos $\beta$ as predicted by the Sigmund theory. However, there are deviations between experiment and theory which Oechsner (*54*) states lie beyond the limits of experimental error.

According to Figure 14a, the initial slope of the measured curves $S(\beta)/S(0)$ changes considerably while in the corresponding range of $M_2/M_1$, the quantity $f$ remains nearly constant (*11*). It is not clear how to explain this behavior theoretically. Moreover, Figure 14b indicates that the energy dependence is somewhat greater than expected.

The original Sigmund theory does not account for the decrease in yield at grazing incidence, but it appears that application of a surface correction may eliminate this deficiency. A sputtering theory should include multiple collisions of the ion except for $M_1 \gg M_2$. Sigmund found it most convenient to satisfy this criteria by assuming an infinite medium. However, yields calculated in this manner should be corrected for the fact that atoms can be reflected back through the intersecting surface only once. For example, at low bombardment energies many ions are reflected back into the vacuum (process 6—Figure 1) with a large fraction of their original kinetic energy (*57*). Theories based on an infinite medium do not take this into account. When a surface correction, however, is applied to Sigmund's theory, it tends to behave more like the experimental data (*58*).

**The Energy Distribution of Sputtered Atoms.** Random cascade theory predicts that, under appropriate conditions, the energy distribution of sputtered atoms should have a $1/E_0^2$ dependence if there were no surface binding forces (*11, 12, 59*). The effect of the surface binding energy $U_0$ is to modify the distribution at low energies. According to Thompson (*12*), the energy spectrum of sputtered atoms varies like $E_0^{-2}$ at high energies, passes through a maximum in the region of $U_0$, and then falls linearly to zero at zero energy.

Energy spectra have been measured by Stuart et al. (*60*), Oechsner (*61*), Politick and Kistemaker (*62*), and Chapman et al. (*46*). Figures 15 and 16 show energy spectra for $Ar^+$ bombardment of gold. The general agreement between theory and experiment is quite good. The curves peak at about the binding energy, and the rest of the spectrum has an approximatly $E_0^{-2}$ dependence. Increasing the target temperature (Figure 16) causes the peak to move to smaller energies. The low temperature contribution is almost surely caused by evaporation from the region containing the thermal spike. In very dense cascades, such as $Xe^+$ on Au, the low energy peak occurs even at room temperature (*46*), again suggesting that the thermal spike is making some contribution to the sputtering yield.

**The Yield for Light Ions.** Weismann and Behrisch (*63*) demonstrated that backscattered ions make a large contribution to the sputtering yield for the irradiation of a heavy target with a light ion (*see* process

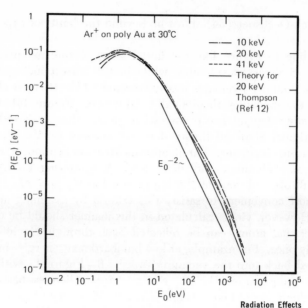

*Figure 15. Energy distribution of sputtered gold atoms under bombardment with 10-keV, 20-keV, and 41-keV Ar⁺. Theory is from model in Ref. 12 (46).*

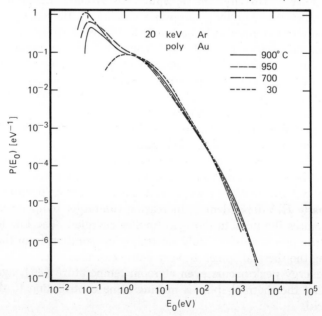

*Figure 16. Energy distribution of sputtered gold atoms under irradiation by 20-keV Ar⁺ for several different target temperatures (46)*

4—Figure 1). They evaporated a thin film of copper onto substrates of Be, V, Nb, and Ta and subsequently measured the yield of copper as a function of substrate material. The various materials produced different intensities of backscattered ions, and therefore the contributions to the sputtering yield of incident and backscattered ions could be separated. Their results indicate that possibly 50% of the sputtering events result from backscattering.

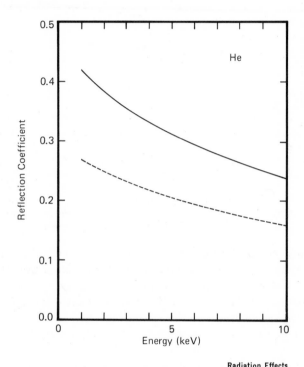

Radiation Effects

*Figure 17.    Theoretical calculations of the reflection coefficients as a function of energy for ⁴He⁺ incident on Nb. (——), with surface correction; (– – –), without surface correction (64).*

Calculations by Bøttiger and Winterbon (*64*) and others (*65, 66*) have shown that a large fraction of the incident ions are reflected away from the surface (*see* process 6—Figure 1). Their calculations for the reflection coefficient of ⁴He⁺ are shown in Figure 17. These results indicate that a significant fraction of the initial kinetic energy is not deposited in the target material. Furthermore, the results depend sensitively on a surface correction. This is because in the theoretical model, an ion can pass through the hypothetical surface several times while in an actual experiment, it can pass back through the surface only once.

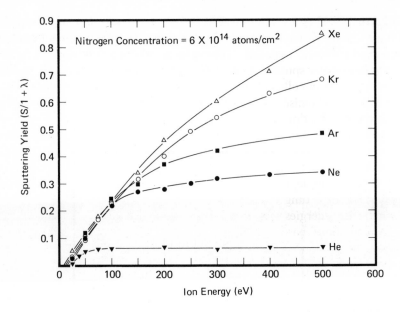

*Figure 18. Sputtering yield for nitrogen chemisorbed on tungsten as a function of ion energy. λ is the secondary electron coefficient (67).*

A theory of sputtering for light ions on heavy substrates should consider the effects of (1) electronic stopping, (2) large angle scattering, and (3) a surface correction. When processes 1 and 2 are included (*66*), α (*see* Equation 21), contrary to the case for heavy-keV ions, depends sensitively on ion energy, i.e., the sputtering yield is no longer directly proportional to the nuclear stopping power. Weismann (*66*) suggests that the rather large values of α at low and intermediate ε-energies indicate a considerable contribution to the yield from backscattered ions. Comparing experiment with theory shows the measured values to be about one half the calculated ones. The general shape of the yield curves, however, agree well with Weismann's predictions.

**The Sputtering Yield for Chemisorbed Gas.** Winters and Sigmund (*67*) have measured the sputtering yield for nitrogen chemisorbed on tungsten for ion energies up to 600 eV (*see* Figure 18). When this data is contrasted with yields for elemental materials (*36*), some interesting comparisons can be made. Elemental materials with large sublimation energies generally have small sputtering yields and relatively large apparent threshold energies. However, nitrogen chemisorbed on tungsten has a large desortpion energy, ~6.7 eV/atom, a very large sputtering yield (*see* Table III), and also a low apparent threshold energy. This is just the opposite of what one would intuitively expect. Moreover, the yield for helium reaches a maximum at ~100 eV and then remains constant to 600 eV. Neon shows a tendency in the same direction. By comparison, elemental yields never reach a maximum in this energy range.

Whereas the sputtering of elemental materials is dominated by the development of a collision cascade (process 2—Figure 1), the sputtering of a light gas chemisorbed on a target of large atomic weight appears to be dominated by direct collisions between the incoming and/or reflected ion and the adsorbed species (67) (processes 1 and 5—Figure 1). This difference in behavior occurs because energy is not effectively transferred between the nitrogen and tungsten (i.e., $4M_1M_2/(M_1 + M_2)^2 \simeq 0.25$). For example, a sputtering event involving transfer of energy from an $He^+$ ion to a tungsten atom and then to a nitrogen atom could not occur for ion energies less than several hundred eV.

Calculations based on direct collisions between the incoming ion and the adsorbed gas, including sputtering events resulting from development of a collision cascade, agree with experiment within approximately a factor of two. The difference in the shape of the yield curves between He, Ne, and the heavier noble gases results from the fact that a Thomas–Fermi interaction is appropriate for He,Ne in this energy range (*see* Equation 5) while a Born–Mayer potential (*see* Equation 6) is more appropriate for the heavier gases. The large yields and low thresholds occur because the energy initially given to the nitrogen is not easily transferred to the tungsten, and therefore the nitrogen atoms are reflected away from the surface with a large fraction of their initial kinetic energy.

Winters and Sigmund predict that the cascade mechanism would begin to dominate at higher energies because of the increasing yield of sputtered tungsten and also because of the decreasing cross section for ion–nitrogen collisions. Under this situation the adsorbed nitrogen should have an exceptionally low yield because of ineffective energy transfer

### Table III.   Ratio of Sputtering Yields[a]

| | $\dfrac{S\ (nitrogen\ on\ tungsten)}{S\ (silver)}$ | | $\dfrac{S\ (nitrogen\ on\ tungsten)}{S\ (tungsten)}$ | |
|---|---|---|---|---|
| | *100 eV* | *500 eV* | *100 eV* | *500 eV* |
| He | 2.9 | 0.62 | > 100 | > 10 |
| Ne | 1.6 | 0.39 | 11.5 | 2.4 |
| Ar | 0.77 | 0.31 | 8.0 | 1.6 |
| Kr | 1.2 | 0.41 | > 6 | 1.4 |
| Xe | 1.1 | 0.51 | > 6 | 1.7 |

[a] Data from Ref. *67*. The nitrogen yields have been extrapolated to a coverage of $1.2 \times 10^{15}$ atoms/cm$^2$ (~ 1 monolayer) for comparison purposes. The tungsten and silver yields are from Ref. *36*. Silver was chosen because it has one of the largest elemental sputtering yields.

from the moving tungsten to the adsorbed gas.   As of now there is no
experimental evidence to verify this prediction.

## Literature Cited

1. Güntherschulze, A., Meyer, K., *Vacuum* (1953) **3**, 360.
2. Wehner, G. K., *Adv. Electron. Electron Phys.* (1955) **7**, 239.
3. Behrisch, R., *Ergeb. Exakten. Naturwiss.* (1964) **35**, 295.
4. Kaminsky, M., "Atomic and Ionic Impact Phenomena on Metal Surfaces,"
   Springer-Verlag, Berlin-Heidelberg, 1965.
5. Carter, G., Colligon, J. S., "Ion Bombardment of Solids," Elsevier, 1968.
6. MacDonald, R. J., *Adv. Phys.* (1970) **19**, 457.
7. Tsong, I. S. T., Barber, D. J., *J. Mater. Sci.* (1973) **8**, 123.
8. McCracken, G. M., *Rep. Prog. Phys.* (1975) **38**, 241.
9. Sigmund, P., *Rev. Roum. Phys.* (1972) **17**, 823 and 969.
10. *Ibid.*, 1079.
11. Sigmund, P., *Phys. Rev.* (1969) **184**, 383.
12. Thompson, M. W., *Philos. Mag.* (1968) **18**, 377.
13. Brandt, W., Laubert, R., *Nucl. Instrum. Methods* (1967) **47**, 201.
14. Koichi, K., Hojou, K., Koga, K., Toki, K., *J. Appl. Phys. Jpn.* (1973) **12**,
    1297.
15. Hasted, J. B., "Physics of Atomic Collisions," Butterworths, 1964.
16. Smith, D., *Surf. Sci.* (1970) **25**, 171.
17. Heiland, W., Taglauer, F., private communication.
18. Karpuzov, P. S., Yurasova, V. E., *Phys. Status Solidi* (1971) **B47**, 41.
19. Schiff, L. I., "Quantum Mechanics," p. 281, McGraw-Hill, 1955.
20. Gombas, P., *Handb. Physik* (1956) **36**, 109.
21. Bohr, N., *Mat. Fys. Medd. Dan. Vid. Selsk.* (1948) **18** (8).
22. Firsov, O. B., *J. Exp. Theor. Phys.* (1957) **33**, 696 [English Trans.: *Sov.
    Phys.-JETP* (1958) **6**, 534.]
23. Lindhard, J., Nielsen, V., Scharff, M., *Mat. Fys. Medd. Dan. Vid. Selsk.*
    (1968) **36** (10).
24. Abrahamson, A. A., *Phys. Rev.* (1969) **178**, 76.
25. Goldstein, H., "Classical Mechanics," p. 73, Addison–Wesley, 1959.
26. Yonts, O. E., Harrison, D. E., *J. Appl. Phys.* (1960) **31**, 1583.
27. Jones, R. E., Winters, H. F., Maissel, L. I., *J. Vac. Sci. Technol.* (1968)
    **5**, 84.
28. Colligon, J. S., Hicks, C. M., Neokleous, A. P., *Radiat. Eff.* (1973) **18**, 119.
29. Almén, O., Bruce, G., *Nucl. Instrum. Methods* (1961) **11**, 257 and 279.
30. Andersen, H. H., Bay, H., *Radiat. Eff.* (1972) **13**, 67.
31. Andersen, H. H., *Radiat. Eff.* (1973) **19**, 139.
32. Winters, H. F., *J. Appl. Phys.* (1972) **43**, 4809.
33. Carter, G., Colligon, J. S., "Ion Bombardment of Solids," p. 385, Elsevier,
    1968.
34. Kornelsen, E. V., *Can. J. Phys.* (1964) **42**, 364.
35. Andersen, H. H., Bay, H. L., *J. Appl. Phys.* (1975) **46**, 2416.
36. Wehner, G. K., General Mills Rept. No. **2309**, July 1962.
37. Guseva, M. I., *Fiz. Tverd. Tela.* (1959) **1**, 1540 [English Trans.: *Sov.
    Phys.–Solid State* (1960) **1**, 1410].
38. Dupp, G., Scharman, A., *Z. Physik* (1966) **192**, 284.
39. Keywell, F., *Phys. Rev.* (1955) **97**, 1611.
40. Yonts, O. C., Normand, C. E., Harrison, D. E., *J. Appl. Phys.* (1960) **31**,
    447.
41. Weijsenfeld, C. H., Thesis, University of Utrecht, 1966.
42. Southern, A. L., Willis, W. R., Robinson, M. T., *J. Appl. Phys.* (1963) **34**,
    153.
43. Rol, P. K., Fluit, J. M., Kistemaker, J., *Physica* (1960) **26**, 1000.

44. Andersen, H. H., Bay, H. L., "Ion Surface Interaction, Sputtering, and Related Phenomena," p. 63, Gorden and Breach, 1973.
45. Sigmund, P., unpublished data.
46. Chapman, G. E., Farmery, B. W., Thompson, M. W., Wilson, I. H., *Radiat. Eff.* (1972) **13**, 121.
47. Thompson, M. W., Nelson, R. S., *Philos. Mag.* (1962) **7**, 2015.
48. Nelson, R. S., *Philos. Mag.* (1965) **11**, 291.
49. Honig, R. E., *RCA Rev.* (1962) **23**, 567.
50. Dupp, G., Scharman, A., *Z. Physik* (1966) **194**, 448.
51. Cheney, K. B., Pitkin, E. T., *J. Appl. Phys.* (1965) **36**, 3542.
52. Colombie, N., University of Toulouse, 1964, unpublished data.
53. Rol, P. K., Fluit, J. M., Kistemaker, J., *Physica* (1960) **26**, 1009.
54. Oechsner, H., *Z. Physik* (1973) **261**, 37.
55. Bach, H., *J. Noncryst. Solids* (1970) **3**, 1.
56. Molchanov, V. A., Tel'Kovskii, V. G., *Dokl. Akad. Nauk SSSR* (1961) **136**, 801 [English Trans.: *Sov. Phys.-Dokl.* (1961) **6**, 137].
57. Winters, H. F., *Phys. Rev.* (1974) **10**, 55.
58. Sigmund, P., unpublished data.
59. Robinson, M. T., *Philos. Mag.* (1965) **12**, 145.
60. Stuart, R. V., Wehner, G. K., Anderson, G. S., *J. Appl. Phys.* (1969) **40**, 803.
61. Oechsner, H., *Z. Phys.* (1970) **238**, 433.
62. Politiek, J., Kistemaker, J., *Radiat. Eff.* (1969) **2**, 129.
63. Weissmann, R., Behrisch, R., *Radiat. Eff.* (1973) **19**, 69.
64. Bøttiger, J., Winterbon, K. B., *Radiat. Eff.* (1973) **20**, 65.
65. Robinson, J. E., *Radiat. Eff.* (1974) **23**, 29.
66. Weissmann, R., Sigmund, P., *Radiat. Eff.* (1973) **19**, 7.
67. Winters, H. F., Sigmund, P., *J. Appl. Phys.* (1974) **45**, 4760.

RECEIVED January 5, 1976.

# 2

# Aluminum Oxide Sputtering: A New Approach to Understanding the Sputtering Process for Binary Targets

PATRICIA A. FINN, DIETER M. GRUEN, and DENNIS L. PAGE[1]

Chemistry Division, Argonne National Laboratory, Argonne, Ill. 60439

*The relative abundances of the products Al, $Al_2O$, and AlO sputtered in 15- and 40-kV $Ar^+$ and 15-kV $H^+$ bombardments of aluminum oxide targets (anodized film, polycrystalline disk, sapphire) are functions of the target material and of the nature, flux, and fluence of the ion beam. This suggests that in collisional sputtering, the material's sensitive parameters are the surface binding energies of the sputtered species. These energies are functions of the surface composition present at the moment of a particular sputtering event and should be identified with the partial molar enthalpy of vaporization of a particular species. The aluminum oxide species—Al, $Al_2O$, AlO, $Al_2O_2$, $AlO_2$, $Al(O_2)_2$, and $AlO_3$—are characterized by matrix isolation spectroscopy aided by $O^{18}$ isotopic substitution experiments.*

Matrix isolation spectroscopy provides a powerful tool supplementing other more traditional methods, such as mass spectrometry, for elucidating the sputtering process (*1, 2*). Those studies described the development of matrix isolation spectroscopy into a sensitive method for measuring sputtering yields of metals bombarded with noble gas projectiles. The principal sputtered species, because of the interaction of energetic noble gas ions with metal surfaces, are metal atoms resulting from physical sputtering—a process which is reasonably well described by Sigmund's theory of collisional sputtering (*3*). The number of sputtered atoms can be counted by performing an atomic absorption measurement on the matrix isolated atoms. Dividing this number by the number of bombarding ions striking a unit surface area gives the sputtering yield.

[1] CSUI Argonne semester participant from Virginia Wesleyan College.

The unique definition of sputtering yield ($S$ = atoms/ion), which is applicable to monoatomic solids such as metals, fails for binary or more complex compounds. Sputtering studies on metal oxides (4), for example, have shown that their behavior, not unexpectedly, is considerably more complex that that of metals. Kelly and Lam (4) proposed three general categories for metal oxides according to whether the dominant sputtering processes are—collisional, collisional and thermal, or collisional and oxygen sputtering. In their work, the surface binding energies were identified with the heats of atomization of the oxides for collisional and oxygen sputtering for which $Al_2O_3$ is given as an example. An assumption underlying the interpretation of the results of Kelly and Lam in terms of different sputtering processes, therefore, is that the sputtered products are atomic in nature. However, recent work by Coburn et al. (5) shows that for oxides, molecular products can constitute large (up to 50%) fractions of the neutral sputtered products from binary targets such as oxides. Work on identifying sputtered products from binary products is still fragmentary. One may expect a variety of sputtered products including not only metal and oxygen atoms but also metal oxide molecules in various proportions depending on the particular metal oxide under investigation. In view of the important role played by molecular products in sputtering from certain binary targets, we state at the outset that close attention must be paid to the thermodynamics of vaporization processes and that the important material-dependent quantities controlling the sputtering coefficient may well turn out to be the partial molal enthalpies of vaporization of the various sputtered species. To develop a unified and comprehensive theory applicable to sputtering of metal oxides, and indeed of any binary target, it appears that it is necessary to determine the sputtering yields for each sputtered species. Only when data of this kind are available will it be possible to test the reliability of a given theory and to provide guidance for future theoretical refinements.

Determining the sputtering yields for each sputtered species from a metal oxide is a more challenging experimental problem than yield determinations on metals where target weight loss measurements or quantitative analysis of the amount of sputtered material on a collector are the usual methods. The experimental problem lies in devising techniques that allow simultaneous determination of both the identity and the amount of each sputtered species.

Current methods for identifying sputtered products either characterize only atomic species as in the case of atomic absorption or emission measurements on sputtered atoms in flight (6, 7) or, as with secondary ion mass spectroscopy, detect only the ionic fraction of the sputtered material which, in general, represents less than 1% of the total (8, 9).

Mass spectrometric detection of sputtered neutrals by special techniques is relatively recent (5, 10). Broadly speaking, studies of sputtering yield and species identification have been conducted as separate experiments, and none of the methods used hitherto for species identification seem to lend themselves readily to quantitative yield measurements for individual species.

Matrix isolation spectroscopy offers a possible technique for the quantitative determination of each sputtered species. Both neutral and ionic species are collected and immobilized in a noble gas matrix. In general, ions are neutralized on deposition. Species collection can be carried out over a long enough time to provide sufficient amounts of sputtered material for spectroscopic analysis. A variety of spectroscopies (uv–visible, ir, Raman, ESR) can be carried out on the matrix-isolated atoms or molecules. In the case of atoms, matrix isolation spectroscopy can serve both to identify the sputtered atoms by means of their characteristic resonance transitions and to determine the sputtering yield from oscillator strength data coupled with absorption intensity measurements (1, 2). For metal oxide molecules, matrix isolation spectroscopy is a powerful tool not only for species identification, but also for obtaining molecular structural information from an analysis of vibrational spectra (11, 12). Unfortunately, because of the very limited quantity of experimental data available on oscillator strengths of vibronic transitions or transition moments in metal oxide spectra, it is possible at present to determine amounts of sputtered oxide molecules in only a few cases. In principle, however, such data can be obtained for every molecule of interest so that matrix isolation spectroscopy can lend itself both to characterizing the sputtered species and to measuring each sputtering channel quantitatively.

The present study was undertaken to explore the application of matrix isolation spectroscopy to quantitative molecular sputtering by investigating a particular oxide, aluminum oxide. A subsidiary aim was to study the effect of reactive projectiles ($H^+$) vs. nonreactive projectiles ($Ar^+$) on the nature, distribution, and yield of sputtered products.

The sputtering yield of aluminum oxide was determined by target weight loss measurements (13, 14) and by measuring the collapsing interference fringes of oxide films (13, 15). Sputtered aluminum on collectors was determined by a neutron activation method (16, 17) from both aluminum metal (16) and aluminum oxide targets (17). However, since the method of measurement determined only aluminum, no values of total sputtering yields for aluminum oxide can be taken from these data.

The nature of sputtered products from aluminum and aluminum oxide surfaces has been characterized using atomic absorption (6, 18) or emission (7, 18) spectra as diagnostic tools. Both neutral Al (6, 18) and $Al^+$ (7) ions have been observed in this way. Spectral features caused by "molecules" in sputtered products have also been observed (6). Another technique which has also been used to characterize the nature of sputtered products from $Al_2O_3$ is secondary ion mass spectrometry (8, 9, 19). The energy distribution of $Al^+$ from Al metal targets has been determined (19). Sputtering an aluminum metal target (9) and following the time evolution of sputtered products every 0.2 sec for a total sputtering time of 4.6 sec showed that early in the sputtering run, $Al^+$, $Al^{2+}$, $Al_2^+$, $AlO^+$, $Al_2O^+$, and $AlO_2^+$ were among the sputtered products. In another mass spectrometric study (8) of the sputtering of Al metal in which the background pressure of oxygen was deliberately varied, $Al^+$, $O^-$, $Al_2^+$, $AlO^-$, $AlO_2^-$, and $Al^{2+}$ were found among the sputtered products. Finally, and not unexpectedly, the sputtering of oxide-coated Al metal or of bulk $Al_2O_3$ yields a rather complex distribution of sputtered products, since the vapor in equilibrium with $Al_2O_3$ contains the species Al, O, AlO, $Al_2O$, $AlO_2$, and $Al_2O_2$ (20).

In the present study, the sputtered products, which normally exist only during their time of flight, were isolated in solid argon matrices at 12–14°K and then examined spectroscopically. Various forms of aluminum oxide (anodic films, polycrystalline disk, single crystal sapphire) were bombarded with 15-kV or 50-kV argon ions or with 15-kV protons. The physical state of the oxide surface, the flux as well as the total beam fluence, and the nature of the bombarding ions were important parameters determining the composition and relative concentrations of the sputtered species.

The identification of the species sputtered from aluminum oxide was aided in part by information available in the literature. However, because not all species observed in the present work had previously been identified, detailed studies including oxygen isotopic substitution experiments were undertaken on matrix-isolated species. These studies and the results on the distribution of sputtered species are described below.

### Experimental

**Ion Bombardment Experiments.** Mass and energy-selected ion beams were generated using either a laboratory-type electromagnetic isotope separator (21) (50 kV, 3–6 $\mu$A) or a Radiation Dynamics Duoplasmatron source (15 kV, 10–800 $\mu$A). Because of the insulating properties of aluminum oxide, integrated currents could only be measured approximately. No attempt was made to neutralize surface charge on the target.

A brass target assembly was used. The main portion consisted of a flat plate containing 16 equally spaced holes (each 0.079 cm in diameter located on a 1.746-cm bolt circle). The plate was silver soldered to a larger threaded brass piece forming an enclosed cavity 0.635 cm deep by

1.905 cm in diameter). A 0.3175-cm steel tube silver soldered to the rear of the cavity served as the inlet for the argon matrix gas. The 1.27-cm targets were centered on the flat plate and held in place by a threaded brass cap with a 0.9525-cm diameter hole in its center. Sixteen equally spaced holes in the cap matched those on the flat plate.

A closed-cycle helium refrigerator (Cryogenics Technology Inc.) with a rotatable cold station encased in a copper heat shield with two 2.54-cm in-line apertures was mated with a four-window Dewar tail. Two in-line optical windows (NaCl) were at right angles to the path of the ion beam and the target. For most bombardments, the target, in line with the ion beam, was located ~ 2 cm from the deposition plate (NaCl) which had a 0.635-cm hole drilled through its center to allow passage of the ion beam. The ion beam entered the Dewar chamber through a 0.476-cm limiting aperture. A gate valve attached to the Dewar assembly allowed it to be detached from the ion source.

For two 5-hr bombardments of sapphire with 15-kV $Ar^+$ or $H^+$, the limiting aperture was increased to 0.9525 cm. The deposition plate had a 1.27-cm hole drilled through its center, and a target assembly equipped to handle 1.905-cm targets was used. This geometry resulted in an increased flux to the target.

In each run, the Dewar assembly was attached to the ion source and evacuated overnight. The matrix gas flow was started 2–5 min prior to bombardment and varied from 4–7 mmole/hr. Bombardment times ranged from 30 min to 5 hr. The background pressure during a run was typically $7–11 \times 10^{-4}$ Pa. Beam currents measured during $Ar^+$ bombardments ranged from 3–6 $\mu A$ on the isotope separator to 40 $\mu A$ on the Duoplasmatron. For $H^+$ bombardments, beam currents were ~ 150–800 $\mu A$. Because of the low sputtering yields, matrix-isolated sputtered products could only be detected in the uv–visible by their electronic spectra.

**Hollow Cathode Experiments.** In the hollow cathode experiments, the gate valve and limiting aperture were removed from the Dewar tail and replaced with a hollow cathode sputtering device described previously (22). A quartz window replaced the target assembly, and KCl windows were used in place of NaCl windows—the deposition plate now being solid, i.e., without a central hole. A 1-hr sputtered cleaning treatment with 99.9995% argon was given to aluminum screws used for the first time. A screw typically lasted 14–15 hr.

Matrix deposition experiments began by passing pure argon through the sputtering device. Oxygen–argon mixtures (0.015–5% $O_2$) were then introduced either directly through the sputtering device or via a second flow line whose tip was situated near the deposition plate. For the former method, designated uniflow, ozone (23, 24) was the main product at oxygen concentrations greater than 0.15%. Large amounts of aluminum oxide were concurrently being formed in the screw which limited the length of time that a discharge could be maintained. In the latter method, designated duoflow, pure argon entered through the sputtering device. Ozone was again detected, but the various aluminum oxide species were also present even at the highest oxygen concentrations.

The oxygen–argon mixtures were diluted by the argon entering via the sputtering device. Ratios of total oxygen to argon varied from 14,000:1 to 55:1. Each flow rate varied from 3–9 mmole/hr, and deposition times were 1–2 hr. In a typical hollow cathode experiment, the discharge current was 40–60 mA, and the voltage was 450–720 V.

Temperatures at the deposition plate were measured with a hydrogen vapor pressure thermometer and with a 7% iron-doped gold vs. chromel thermocouple. Evanohm heating wires wound around the cold station and connected to an Adams temperature controller were used to maintain a constant heat load during annealing studies ( $\pm 2°$ ). Temperatures were raised in successive 5°K increments from 12° to 45°K.

Isotopic substitution studies were done with pure $^{16}O_2$, pure $^{18}O_2$, and $^{16}O_2$–$^{18}O_2$ mixtures. For all experiments, ir band intensities were quite low ( $\% T > 0.95$ ).

Gases—99.9995% pure—(Ar, Kr, Xe, $H_2$, $^{16}O_2$) were provided by Air Products Co. and were used without further purification. The $^{18}O_2$ (99%) was provided by Miles Lab., Inc. Single crystal, unoriented sapphire disks were used as obtained from the Linde Division of Union Carbide. The aluminum oxide polycrystalline disks were made by compressing reagent grade alumina at $7 \times 10^8$ Pa in a conventional pellet press. The $2.4 \times 10^{-7}$-m thick aluminum oxide anodixed film was prepared from reagent grade aluminum electrolyzed in a phosphate solution. The aluminum disk and aluminum screws used were 99.5% pure. Spectra were recorded on a Cary 17H spectrometer and/or a Beckman IR-12 at 12°K.

*Results and Discussion*

**Species Identification.** The ultimate goal of this work is to measure the sputtering yield for each atomic or molecular sputtering channel as a function of the nature, energy, and flux of the projectiles bombarding an aluminum oxide surface. A variety of sputtered products, including but not limited to, Al, O, $Al_2O$, AlO, $AlO_2$, and $Al_2O_2$, can be expected based on species known to occur in equilibrium vapors over condensed-phase aluminum oxide. Since sputtering is a nonequilibrium process in the usual thermodynamic sense, species other than those listed above could also occur. The application of matrix isolation spectroscopy to quantitative studies of the sputtering process clearly depends on detailed information concerning the electronic and vibrational spectra of each sputtered product. The problems requiring solution have three major aspects. First, insofar as matrix isolation studies are concerned, prior work has been done on the electronic spectra of Al atoms and AlO molecules, as well as on the vibrational spectra of AlO, $Al_2O$, and $Al_2O_2$. For several of these species however, spectral assignments are still under active discussion in the literature as pointed out below, and our work has provided additional spectral data leading to new vibrational assignments.

A second problem area has to do with the possible occurrence of aluminum–oxygen species among the sputtered products which are not produced as a result of vaporization from a Knudsen cell. All matrix isolation studies on the aluminum–oxygen system have hitherto relied on producing the high temperature molecules by the Knudsen cell vaporization technique. To prepare and to characterize the known as well as the new aluminum–oxygen species, the hollow-cathode sputtering technique was used in our work in both the uniflow and duoflow modes, as already described.

The third major set of data needed for quantitative sputtering yield measurements are the oscillator strengths or the transition moments of the electronic or vibrational absorption band characteristic of each species. Obtaining such data for matrix-isolated atoms has been the chief goal of earlier work. In the present study, this aspect has received relatively less attention, and it will require an additional series of experiments to obtain reliable numbers. We will consider each of the aluminum oxide species, both as previously characterized in the literature and as determined in our experiments.

AL. Most of the references for the gas phase transitions of aluminum atoms are cited by Wei et al. (25). Ammeter and Schlosnagle (26) have reported the spectra of aluminum atoms isolated in several noble gas matrices at 4.2°K. In their work, aluminum atoms were obtained from a resistance-heated tantalum cell at 1323°K. In the argon matrix, the band at 29500 cm$^{-1}$ was assigned to the $4s(^2S) \leftarrow 3p(^2P_{1/2})$ transition and the four bands at 34,220, 34,880, 37,240, and 37,860 cm$^{-1}$ to the $3d(^2D) \leftarrow 3p(^2P_{1/2})$ transition. They mention the possibility that not all the peaks on the high energy side of the spectrum belong to the $^2D \leftarrow ^2P$ transition.

Aluminum surfaces are readily coated with a thin adherent oxide layer (18). Because the sputtering device was dismantled after each run, an oxide coating was always present initially. After a short sputtering time of less than 5 min, the principal product was aluminum atoms.

Continued sputtering does not eliminate the small constant amount of suboxide molecules also produced. After a 30-min sputtering with argon, the uv–visible spectrum (Figure 1B) of the matrix-isolated products consists of a main band at 29,500 cm$^{-1}$, a doublet at 34,300 and 34,900 cm$^{-1}$, and two additional bands at 30,500 cm$^{-1}$ and 37,800 cm$^{-1}$. The intensities of the latter two bands increase markedly when higher currents (100–150 mA) at the same flow rates are used during the sputtering experiment. During annealing studies the two bands did not exhibit the same behavior as the 29,500 cm$^{-1}$ band nor did they disappear. The latter behavior often occurs for multiple sites. In mass spectroscopic studies (8, 9, 27) of the products formed during ion bombardment of aluminum metal, Al$_2$ is the most abundant cluster produced. Therefore, we assign the 30,500 cm$^{-1}$ and 37,800 cm$^{-1}$ bands to this species. The possibility exists however, that one or both of these bands is caused by the species Al$_x$ where $x \geq 2$. A low intensity band at 26,500 cm$^{-1}$ may also be caused by polymeric Al$_x$ where $x \geq 2$.

The mechanism by which these polymeric species are formed is not known. Brewer (28) noted that $Pb_2$ could be isolated either from the vapor or by synthesis from lead atoms by a surface diffusion mechanism. Huber et al. (29) found that the formation of $Mn_2$ dimer species is a function of the ratio of metal atoms to noble gas molecules. As the metal atom concentration increases, the dimer concentration also increases.

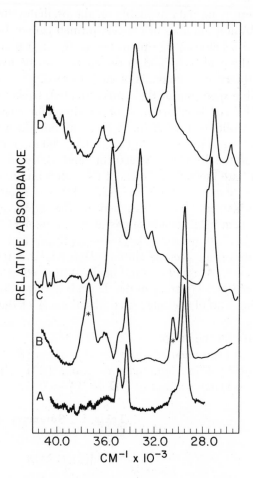

*Figure 1. Matrix spectra of aluminum atoms and aluminum dimer molecules. A, 50-kV Ar+ ion bombardment of aluminum, isolated in an argon matrix; B, hollow cathode sputtering of aluminum isolated in an argon matrix; C, hollow cathode sputtering of aluminum isolated in a krypton matrix; D, hollow cathode sputtering of aluminum isolated in a xenon matrix.*

In some of the present experiments the discharge current in the sputtering device was increased to 100–150 $\mu$A without increasing the noble gas flow rate. In effect, this increased the concentration of aluminum atoms in the matrix gas and led to an increase in intensity of bands ascribed to the $Al_2$ dimer in the deposit. One concludes that decreasing the Ar/Al ratio leads to an increase in $Al_2$ formation because of enhanced surface diffusion.

Because there is no data available on the oscillator strengths of $Al_x$ ($x \geq 2$), the absolute number of clusters formed cannot be determined. However, the ratio between the intensities of the 29,500 cm$^{-1}$ aluminum atom peak and the peaks assigned to the cluster can be used to monitor the changes in the relative amounts of each formed.

The uv-visible spectrum of the matrix-isolated, back-sputtered products from a 50-kV Ar$^+$ bombardment of an aluminum disk is compared with similar data obtained using the hollow cathode device in Figure 1. The former spectrum is presented in Figure 1A. The intense band located at 29,500 cm$^{-1}$ and the doublet at 34,300 cm$^{-1}$ and 34,900 cm$^{-1}$ with its center at 34,600 cm$^{-1}$ are the main features of the spectrum. A small shoulder signals the presence of the 30,500 cm$^{-1}$ peak while the 37,800 cm$^{-1}$ peak is absent. The 29,500 cm$^{-1}$ band assigned to the $4s(^2S_{1/2}) \leftarrow 3p(^2P_{1/2})$ transition, a 4152 cm$^{-1}$ shift from the gas phase value of 25,348 cm$^{-1}$. The doublet is assigned to the $3d(^2D_{3/2}) \leftarrow 3p(^2P_{1/2})$ and $3d(^2D_{5/2}) \leftarrow 3p(^2P_{1/2})$ transitions, an average shift of 2165 cm$^{-1}$ from the gas phase value of 32,435 cm$^{-1}$. A pseudo-crystal field splitting effect could account for the removal of the degeneracy of the $^2D$ peak, giving a doublet splitting of 600 cm$^{-1}$. The ratio of the integrated absorbances of the bands in the doublet is 1.41.

The $gf$ values for the gas phase ($^2S \leftarrow {}^2P$) and ($^2D \leftarrow {}^2P$) transitions have been derived experimentally by several workers (30, 31, 32). Previous work is discussed in Cunningham's paper (32) where values of $gf$ ($^2S \leftarrow {}^2P$) $= 0.111 \pm .01$ and $gf$ ($^2D \leftarrow {}^2P$) $= 0.173 \pm .01$ are cited.

### Table I.  Summary of Aluminum

| | $3p \to 4s$ (cm$^{-1}$) | Matrix Shift $\Delta\bar{\nu}(cm^{-1})$ | (cm$^{-1}$) |
|---|---|---|---|
| Ar matrix, 12°K | 29500 | 4152 | 34300 34900 |
| Kr matrix, 12°K | 27100 | 1752 | 33000 33550 |
| Xe matrix, 12°K | 27070 | 1722 | 30750 31375 |
| Gas phase | 25348 | | 32435 |

The oscillator strengths ($f = g_u f/g_l$) derived from these $gf$ values are 0.0555 and 0.0865, respectively, giving 0.642 for the ratio of the gas phase oscillator strengths of the ($^2S \leftarrow {}^2P$) and ($^2D \leftarrow {}^2P$) transitions. In an argon matrix, (Figure 1A) the ratio of the integrated absorbances of the two transitions ($A = nf$, where $n$ equals number of atoms) is 1.24, nearly twice as large as the gas phase value. Previous experiments with matrix-isolated gold (1) and niobium (2) atoms have shown that the oscillator strength of a particular transition can be perturbed by the matrix environment.

For aluminum atoms, matrix perturbations of the energy of the resonance transition are exceptionally large. The ground state of aluminum is $3p(^2P_{1/2})$ while the first excited state is $4s(^2S_{1/2})$. The observed matrix shift reflects the expected, strongly repulsive matrix interaction of the excited $4s$ state with a single electron in an outer $s$ orbit which substantially penetrates the valence shells of the rare gas cage. The matrix shift, 4152 cm$^{-1}$ in an argon matrix, is much larger than the 1500–2500 cm$^{-1}$ shifts usually observed for transitions within the same orbital (2165 cm$^{-1}$ for the $3p \to 3d$ transition of Al). However, it is the same order of magnitude as the 5550 cm$^{-1}$ shift observed by Brewer and Chang (28) for the analogous $6p \to 7s$ transition in lead atoms. The oscillator strength of the $3p \to 4s$ transition in Al can therefore be expected to be strongly perturbed. Using the previously determined ratios of the oscillator strengths in the gas phase and in an argon matrix (0.642 and 1.24) and making the assumption that the $3p \to 3d$ oscillator strength is unchanged, one would conclude that the oscillator strength of the $3p \to 4s$ transition has increased from 0.055 in the gas phase to 0.107 in an argon matrix.

The spectra of aluminum atoms trapped in krypton and xenon matrices (Figure 1C and D) have also been studied. The spectra are similar to those reported by Ammeter and Schlosnagle (26) insofar as band positions are concerned. However, there are major differences in the relative intensities of the bands and the assignments. Table I lists

**Atom Matrix UV–Visible Spectra**

| $3p \to 3d$ Splitting, $\Delta\bar{v}(cm^{-1})$ | Ratio–Integ. Absorbances | Av. Matrix Shift $\Delta\bar{v}(cm^{-1})$ | Ratio–Integ. Absorbances $\dfrac{3p \to 4s}{3p \to 3d}$ |
|---|---|---|---|
| 600 | 1.41 | 2165 | 1.24 |
| 550 | 1.72 | 840 | 0.485 |
| 625 | 1.59 | −1373 | 0.205 |
| | | | 0.642 |

the band assignments of the $3p \rightarrow 4s$ and the $3p \rightarrow 3d$ transitions in argon, krypton, and xenon matrices. The matrix shifts, splitting of the $^2D$ state and the ratio of the integrated absorbances of the lines of interest are reported.

To verify the assignment of bands in the krypton and xenon matrices, several experiments were done in which the back-sputtered products from 15-kV, $4.1 \times 10^{17}$ ions/cm$^2$ Ar$^+$ bombardments of an aluminum disk were isolated either in a krypton or a xenon matrix. The uv-visible spectra of these matrices were then compared with those obtained from the hollow cathode experiments. For the krypton matrix containing back-sputtered species, both the band at 27,700 cm$^{-1}$ and the band at 35,500 cm$^{-1}$ were absent. For the xenon matrix, the bands at 25,750 cm$^{-1}$ and 33,600 cm$^{-1}$ were greatly reduced in size compared with the peaks assigned to the $(3p \rightarrow 4s)$ and $(3p \rightarrow 3d)$ transitions of aluminum. (Both the 27,700 cm$^{-1}$ and 35,500 cm$^{-1}$ bands in krypton and the 25,750 and 33,600 cm$^{-1}$ bands in xenon are assigned to Al$_x$ species where $x \geq 2$).

The matrix shifts for the $3p \rightarrow 4s$ transition in krypton and xenon are approximately of the same magnitude but much less than in argon. The average shift for the $3p \rightarrow 3d$ transition is to the blue in Ar (2165 cm$^{-1}$) and in Kr (840 cm$^{-1}$) but to the red in Xe (1373 cm$^{-1}$). The splitting of the $^2D$ state which varies from Ar (600) $\rightarrow$ Kr (550) $\rightarrow$ Xe (625) cm$^{-1}$ in the respective matrices indicates that the crystal field splitting varies only slightly in the three matrices. The ratios of the integrated absorbances of the doublet peaks varies from 1.41 $\rightarrow$ 1.72 $\rightarrow$ 1.59, the average being 1.57.

When the ratios $(R)$ of the integrated absorbances (i.e., oscillator strengths) of the $(3p \rightarrow 4s)$ transition vs. the $(3p \rightarrow 3d)$ transition are compared, the large ratio observed for Ar ($R = 1.24$), attributed primarily to an increase in oscillator strength of the $3p \rightarrow 4s$ transition, is reduced in a Kr matrix ($R = 0.485$) and is still lower in a Xe matrix ($R = 0.205$).

Assuming the oscillator strength of the $3p \rightarrow 3d$ transition to be unchanged in going from the gaseous to matrix-isolated Al, the experimental results can be rationalized on the basis of the $3p \rightarrow 4s$ oscillator strength being nearly the same in a Kr matrix but lower in a Xe matrix compared with the gaseous atom value.

The other bands in the krypton and xenon matrices are assigned to Al$_x$ ($x \geq 2$) species. To make definitive assignments, additional experiments at variable gas flows and also at different current settings are required.

ALO. In the gas phase, a number of transitions in the AlO molecular spectrum have been well characterized $(33, 34, 35, 36)$. For example, the $B^2\Sigma \leftarrow X^2\Sigma$ transition occurs at 20,635 cm$^{-1}$ $(0, 0)$, 21,500 cm$^{-1}$ $(1, 0)$, and 22,450 cm$^{-1}$ $(2, 0)$ in the gas phase $(34)$. This transition has been observed in noble gas matrices by Knight and Weltner (36) by isolating AlO molecules vaporizing at 2500°C from resistively heated tungsten or tungsten–rhenium cells containing a mixture of $Al_2O_3$ and aluminum metal. Bands were observed at 21,859 $cm^{-1}$, 22,726 $cm^{-1}$, and 23,556 $cm^{-1}$ in an argon matrix and assigned to the (0,0), (1,0), and (2,0) transitions, respectively.

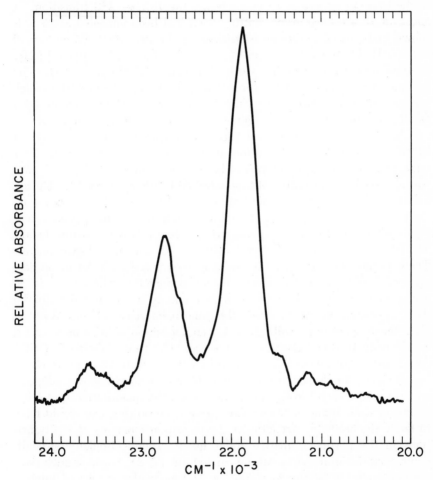

Figure 2.  *Visible spectrum of AlO isolated in an argon matrix*

In the ir, a band at 917 $cm^{-1}$ was assigned to the Al–O stretch (36). Two other broad bands at 974 $cm^{-1}$ and 944 $cm^{-1}$ were also observed by Knight and Weltner (36) but not assigned. The AlO assignment supposes a matrix shift of 60–85 $cm^{-1}$ to lower frequencies compared with the gas phase value of 979.2 $cm^{-1}$ deduced by Lagerquist (37) from a rotational analysis of the $B^2\Sigma \leftarrow {}^2\Sigma$ system. A calculated value of 989.2 cm$^{-1}$ was obtained by Yoshimine et al (38) using limited single configuration SCF wave functions. A value of 1003.8 cm$^{-1}$ was calculated by Das et al. (39) using Dunham's analysis.

Using the hollow cathode sputtering device and either a uniflow or duoflow system, AlO molecules were prepared, matrix isolated, and studied spectroscopically to characterize the $B^2\Sigma \leftarrow X^2\Sigma$ transition. A spectrum of a typical argon matrix which had been annealed at 25°K is shown in Figure 2. Vibronic transitions are located at 21,865 cm$^{-1}$ ± 10 cm$^{-1}$ (0,0), 22,730 cm$^{-1}$ ± 10 cm$^{-1}$ (1,0), and 23,560 cm$^{-1}$ ± 10 cm$^{-1}$ (2,0) in good agreement with the work of Knight and Weltner (36). The integrated absorbances of the three peaks are 0.665, 0.28, and 0.05 which are intermediate between the Franck–Condon factors of Nicholls (40) (0.73, 0.22, and 0.04) and Yoshimine et al. (38) (0.56, 0.36, and 0.07).

Matrices in which intense vibronic transitions were observed at 21865 cm$^{-1}$ showed no absorption at 917 cm$^{-1}$, either at 15°K or after annealing to higher temperatures. Instead, a band at 946.5 ± 0.5 cm$^{-1}$ was observed concurrently with the vibronic bands in the visible. During annealing studies, spectral complexities caused by multiple sites, observed both in the vibronic and pure vibrational bands, usually disappeared near 25°K. The resultant spectra consisted of a few relatively narrow bands. The 946.5-cm$^{-1}$ band was shifted to 901.0 ± 0.5 cm$^{-1}$ in $^{18}$O experiments. The experimental ratio ($\nu_{16}/\nu_{18}$) is 1.0505 compared with the theoretical ratio of 1.0368 for AlO assuming no anharmonicity. The Al$^{18}$O visible spectrum was shifted 40 cm$^{-1}$ to higher energy as compared with the Al$^{16}$O spectrum, but it preserved the same band shape. Figure 3 represents the ir spectrum resulting from a co-deposition of Al atoms with a 1:1:2500 mixture of $^{16}$O$_2$, $^{18}$O$_2$, and argon. The peaks labeled C (946.5 cm$^{-1}$) and C' (901.0 cm$^{-1}$) are caused by Al$^{16}$O and Al$^{18}$O, respectively.

Both in the hollow cathode and in ion bombardment studies, it would be very useful to be able to determine the quantitative sputtering yields for each atomic or molecular species. To do this, one would have to know the oscillator strengths or the transition moments of the transitions in the particular matrix under investigation. Oscillator strength determinations in matrices can be carried out (1, 2); however such data are not available for the Al atomic or AlO molecular species of interest here. To obtain crude estimates of relative sputtering yields of Al atoms and AlO molecules, the gas phase oscillator strengths for both Al atoms and AlO molecules will be used.

The oscillator strengths of the three vibronic states of AlO for the $B^2\Sigma \leftarrow X^2\Sigma$ transition have been calculated by Yoshimine et al. (38) to be: $f(0,0) = 0.07$, $f(1,0) = 0.0485$, and $f(2,0) = 0.0111$. As in the case of Al atoms, the vibronic oscillator strengths can be expected to be per-

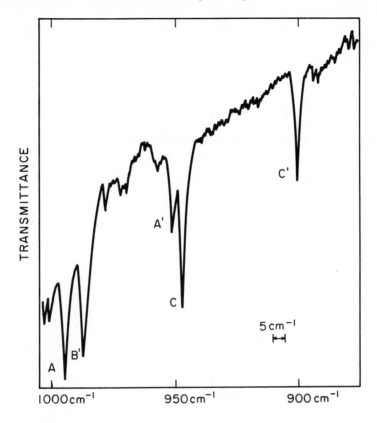

*Figure 3. IR spectrum of the products of aluminum atoms and a 1:1:2500 $^{16}O_2$:$^{18}O_2$:Ar gas mixture at 12°K. A:$Al_2^{16}O$; A':$Al_2^{18}O$; B':$^{18}O_3$; C:$Al^{16}O$; C':$Al^{18}O$.*

turbed in the matrix environment. To determine the relative numbers of Al atoms vs. AlO molecules sputtered under various conditions, Smakula's equation (41) is used:

$$n_x f_x = k \frac{n}{(n^2 + 2)^2} \alpha_x \omega_x$$

where $n_x$ is the number of atoms or molecules of species $x$, $f_x$ is its electronic oscillator strength, $k$ is a constant, $n$ is the index of refraction, $\alpha_x$ is the absorption coefficient of $x$, and $\omega_x$ is the full width at half height of the absorption peak. The ratio of aluminum atoms to AlO molecules is expressed as:

$$\frac{n_{Al}}{n_{AlO}} = \left(\frac{A_{Al}}{f_{Al}}\right)\left(\frac{f_{AlO}}{A_{AlO}}\right)$$

Table II.   Ratios of Relative Concentrations

| Description | Energy (KeV) | Ion |
|---|---|---|
| 1. Uniflow–hollow cathode, 99.995% Ar | | |
| 2. Uniflow–hollow cathode, 0.015% $O_2$ in Ar | | |
| 3. Uniflow–hollow cathode, 0.15% $O_2$ in Ar | | |
| 4. Uniflow–hollow cathode, 0.5% $O_2$ in Ar | | |
| 5. Uniflow–hollow cathode, 1.0% $O_2$ in Ar | | |
| 6. $2.0 \times 10^{-9}$m Oxide coating on Al disk | 50 | $Ar^+$ |
| 7. $2.4 \times 10^{-7}$m Anodic oxide film on Al disk | 50 | $Ar^+$ |
| 8. Polycrystalline $Al_2O_3$ disk | 50 | $Ar^+$ |
| 9. Single crystal sapphire | 50 | $Ar^+$ |
| 10. Polycrystalline $Al_2O_3$ disk | 15 | $Ar^+$ |
| 11. Single crystal sapphire | 15 | $Ar^+$ |
| 12. Single crystal sapphire | 15 | $Ar^+$ |
| 13. Single crystal sapphire | 15 | $H^+$ |
| 14. Single crystal sapphire | 15 | $H^+$ |

$^a$ Presence of large amounts of $Al_2$ not included in $^nAl$.

where $A_{Al}$ is the integrated absorbance of the 29,500 $cm^{-1}$ band and $A_{AlO}$ is the integrated absorbance of each vibronic level of the $B^2\Sigma \leftarrow X^2\Sigma$ transition. The ratios of $n_{Al}/n_{AlO}$ deduced by making use of the above relationship for some hollow cathode experiments and for some of the ion bombardment experiments are given in Table II.

As the oxygen concentration in the argon sputtering gas was increased, the $n_{Al}/n_{AlO}$ ratio decreased until a limiting value was reached. During annealing studies the intensities of the aluminum atomic absorption bands decreased more rapidly than those caused by AlO, indicating more rapid diffusion of the atomic species in the argon matrix.

$AL_2O$. For $Al_2O$, a molecule whose matrix ir spectrum has been studied extensively (42, 43, 44, 45, 46), only one mode at 994 $cm^{-1}$, the $\nu_3$ antisymmetric stretch, has been definitively assigned. Both Linevsky et al. (42) and Makowiecki et al. (43) reported the $\nu_1$ symmetric stretch at 715 $cm^{-1}$, but this assignment has been questioned on the basis of the radically different annealing behavior of the 994 $cm^{-1}$ and 715 $cm^{-1}$ bands (36, 46). Makowiecki et al. (43) assigned the $\nu_2$ bending mode to an absorption at 503 $cm^{-1}$, but this assignment has also been thrown into question (44, 45, 46). The $\nu_2$ mode is most likely to be found below 250 $cm^{-1}$ (36, 42, 44). In the course of the present study on matrix-isolated sputtered aluminum oxide molecules, no bands were observed which could be assigned to either the $\nu_1$ or $\nu_2$ mode of $Al_2O$. The $\nu_3$ mode was seen at 993.5 $cm^{-1}$ $\pm$ 0.5 for $Al_2O$ and at 951.5 $cm^{-1}$ $\pm$ 0.5 for $Al_2^{18}O$ in agreement with earlier work (42). Both bands, labeled A and A', are shown in Figure 3. The $(\nu_{16}/\nu_{18})$ ratio is 1.044.

## of Aluminum Atoms to AlO Molecules

| Fluence (ions/cm²) | $\dfrac{n_{Al}}{n_{AlO}(0,0)}$ | $\dfrac{n_{Al}}{n_{AlO}(1,0)}$ | $\dfrac{n_{Al}}{n_{AlO}(2,0)}$ | Figure No. |
|---|---|---|---|---|
| | $10^4$ | — | — | |
| | $10^2$ | — | — | |
| | 29.0 | 25.3 | — | |
| | 4.4 | 4.1 | — | |
| | 5.5 | — | — | |
| $2 \times 10^{17}$ | 8.2 | — | — | 8A |
| $2 \times 10^{17}$ | 13.3 | — | — | 8B |
| $4.1 \times 10^{17}$ | 2.7 | — | — | 8C |
| $2.0 \times 10^{17}$ | 0.7 | — | — | 8D |
| $6.7 \times 10^{17}$ | 11.0 | — | — | 9A |
| $6.7 \times 10^{17}$ | 11.0 | — | — | 9B |
| $6.0 \times 10^{18}$ | 4.4[a] | 6.5[a] | 8.2[a] | 9C |
| $2.0 \times 10^{19}$ | 0.7 | — | — | 10A |
| $1.2 \times 10^{20}$ | 10.6 | 13.7 | — | 10B |

Because only one mode is observed for $Al_2O$, the $C_2v$ symmetry hitherto assumed for this molecule in matrices may be questioned. For gaseous $Al_2O$, electron deflection measurements (47) were interpreted as favoring a $D_{\infty h}$ linear structure. Ir studies (48) on gaseous $Al_2O$ have also been interpreted in terms of a linear structure while the equilibrium value of the valence angle calculated from electron diffraction measurements varies from 141°–150°C depending on the choice of vibrational frequencies adopted from different literature sources (12). Wagner's (49) ab initio calculations on $Al_2O$ predict a linear symmetrical structure with the ir active modes $\nu_3$ and $\nu_2$ at 1057 cm⁻¹ and 102 cm⁻¹, respectively. The $\nu_2$ bending mode can be expected to be weak and difficult to observe. Raman spectra on matrix-isolated $Al_2O$ could be extremely helpful to provide information on the symmetric stretch. In view of the above discussion, the linear structure for $Al_2O$ must be seriously considered for the matrix-isolated molecule.

Linevsky et al. (42) assigned bands in the uv region to electronic transitions of $Al_2O$. McDonald and Innes (35) later showed that these bands were caused by the $^2\Delta_i \to A^2\pi_i$ transition of AlO. There seems therefore to be no previous record of an electronic transition of $Al_2O$.

In matrix deposits displaying the 993.5 cm⁻¹ absorption of $Al_2O$, subsequent examination of the visible spectrum revealed, in addition to bands at 29,500 cm⁻¹ (Al atoms), 30,500 cm⁻¹ ($Al_2$ dimer), and 21,865 cm⁻¹ (AlO molecules), a previously unreported band at 25,000 cm⁻¹. This band is tentatively assigned to an electronic transition of $Al_2O$. Using a Dupont curve resolver, the band was deconvoluted into four Gaussian constituents, with maxima at 24,400 cm$^{-1}$, 24,700 cm$^{-1}$, 25,000 cm$^{-1}$, and 25,300 cm$^{-1}$. The normalized integrated absorbances of the four Gaussians were 0.14, 0.28, 0.36, and 0.21, respectively, and the average separation between maxima was 300 cm$^{-1}$. The spectrum of the Al$_2$O transition is shown in Figure 4A. The (Al$_2$$^{18}$O) visible spectrum in an Ar matrix had the same band shape as the (Al$_2$$^{16}$O) spectrum, however the maxima of the Gaussian constituents are shifted $\sim$ 100 cm$^{-1}$ to higher energy.

The electronic transition of Al$_2$O in a Kr matrix (Figure 4B) is similar in appearance to that in an Ar matrix. Four bands located at 24,380 cm$^{-1}$, 24,590 cm$^{-1}$, 24,800 cm$^{-1}$, and 25,010 cm$^{-1}$ with normalized integrated absorbances of 0.16, 0.26, 0.40, and 0.18 and an average separation of 210 cm$^{-1}$ were observed. In a Xe matrix, the spectrum was partially obscured by the appearance of other features. However, it was

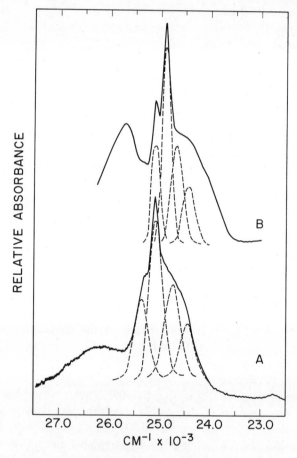

*Figure 4. Visible spectrum of matrix-isolated Al$_2$O.*
*A, argon matrix; B, krypton matrix.*

Table III.   Ratio of Relative Concentration of $Al_2O$
Molecules to AlO Molecules

| Description | $I_{Al}{}^a$ | $I_{AlO}{}^b$ | $I_{Al_2O}{}^c$ | $\dfrac{I_{Al_2O}}{I_{AlO}}$ |
|---|---|---|---|---|
| 1. Uniflow, 99.9995% Ar | 1.6 | 0.02 | 0.4 | 20 |
| 2. Uniflow, 0.015% $O_2$ in Ar | — | 0.13 | 1.1 | 10 |
| 3. Uniflow, 0.15% $O_2$ in Ar | 0.4 | 0.03 | 0.05 | 2 |
| 4. Uniflow, 0.5% $O_2$ in Ar | 0.3 | 0.15 | 0.10 | 0.6 |
| 5. Uniflow, 1.0% $O_2$ in Ar | 0.3 | 0.30 | 0.10 | 0.3 |

$^a$ $I_{Al}$, intensity of 29,500 cm$^{-1}$ band of aluminum.
$^b$ $I_{AlO}$, intensity of 21,865 cm$^{-1}$ band of AlO.
$^c$ $I_{Al_2O}$, intensity of 25,000 cm$^{-1}$ band of $Al_2O$.

possible to estimate positions of band maxima at 24,310 cm$^{-1}$, 24,540 cm$^{-1}$ 24,750 cm$^{-1}$, and 24,970 cm$^{-1}$ with normalized integrated absorbances of 0.26, 0.27, 0.19, and 0.28 and an average separation of 220 cm$^{-1}$.

During annealing studies the band shapes were retained, indicating the absence of multiple sites. The band shapes were independent of oxygen concentration in the matrix gas, but the $Al_2O$ band intensities changed dramatically when the oxygen concentration was varied, relative to the intensities of bands caused by Al atoms (29,500 cm$^{-1}$) and AlO molecules (21,865 cm$^{-1}$). A series of studies of this kind was performed using the hollow cathode device with various oxygen–argon mixtures used as the sputtering gas. The flow rate, the applied voltage, and the current were not varied in this series of experiments, the results of which are shown in Table III. As the oxygen concentration in the sputtering gas is increased, there is a steady decrease in the amount of $Al_2O$ sputtered relative to AlO. This behavior is in accord with chemical intuition.

$AL_2O_2$. Marino and White (45) observed the presence of a band at 496 cm$^{-1}$ in an experiment in which a 5% $^{16}O_2$ in Ar mixture was co-deposited with the effluent from a Knudsen cell containing Al and $Al_2O_3$. The band shifted to 481 cm$^{-1}$ in a 5% $^{18}O_2$ Ar mixture. In a 5% $^{16}O_2$: $^{16}O^{18}O$:$^{18}O_2$ in Ar mixture, three bands were observed at 496 cm$^{-1}$, 489 cm$^{-1}$, and 481 cm$^{-1}$. It is likely on the basis of previous studies of $Si_2O_2$ (50), $Sn_2O_2$ (51), $Li_2O_2$ (52, 53), $Na_2O_2$ (54), and $K_2O_2$ (55), all of which have been assigned a rhombic structure, that the assignment of the 496 cm$^{-1}$ band to the $B_{2u}$ mode of an $Al_2O_2$ dimer molecule is correct.

In our studies with $^{16}O_2$, the presence of two bands at 496.5 cm$^{-1}$ and 490 cm$^{-1}$ was noted. In $^{18}O_2$ work, the bands were shifted to 481 cm$^{-1}$ and 474 cm$^{-1}$, respectively, which represent isotopic frequency ratios ($\nu^{16}O_2/ \nu^{18}O_2$) of 1.032 and 1.034. The annealing behavior of these bands is shown in Figure 5 which represents results on aluminum oxide species isolated in Ar matrices containing 0.04% $O_2$ ($^{16}O_2/^{18}O_2 = 1.5$). The 496.5 cm$^{-1}$ and 490 cm$^{-1}$ bands grow in with increasing temperature and maintain a

3:1 intensity ratio relative to each other. The 490 cm⁻¹ band disappeared at 40°K while the 496.5 cm⁻¹ band was still detectable at this temperature. At 45°K, both bands had disappeared.

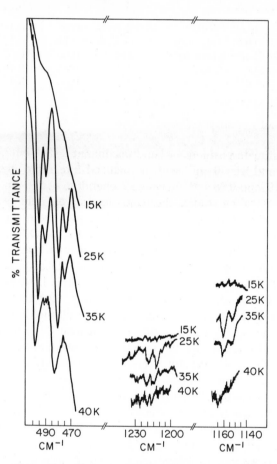

*Figure 5.   Al₂O₂—annealing study in a 1:1:2500 ¹⁶O₂:¹⁸O₂:Ar matrix.*

Because the 496.5 cm⁻¹ and the 490 cm⁻¹ bands have similar annealing behavior, it is unlikely that they are caused by multiple site occupation. We suggest that the two bands arise from two different species having the same stoichiometry but different structures. Assuming that both species have the molecular formula, $Al_2O_2$, we assign the 496.5 cm⁻¹ band to the $Al_2O_2$ dimer with a rhombic structure. The most likely assignment for the 490 cm⁻¹ band is to a ring structure with $C_{2v}$ symmetry. Such a structure might be expected to have an ir-active O–O stretch frequency. A broad doublet at 1213–1217 cm⁻¹ was observed when the 490 cm⁻¹ band was present. Because the intensity of the doublet was approximately 1/20th of the 490 cm$^{-1}$ band, only its gross behavior upon annealing could be observed. The doublet decreased in intensity at approximately the same rate as the 490 cm$^{-1}$ band. With $^{18}O_2$, the doublet shifted to 1154–1159 cm$^{-1}$, an isotopic frequency ratio of 1.050. (The isotopic frequency ratio for $^{16}O_2/^{18}O_2$ is 1.061.) Assigning the 1213–1217 cm$^{-1}$ doublet to an O–O stretch in the puckered ring structure of the $Al_2O_2$ species seems likely,

The $Al_2O_2$ species is the main reaction product at $O_2$ concentration larger than 0.2%. Possible reaction mechanisms for the production of $Al_2O_2$ at lower $O_2$ concentrations as a result of isochronal annealing are: $2\ AlO \rightarrow Al_2O_2$; $Al_2O + O \rightarrow Al_2O_2$; $Al + AlO_2 \rightarrow Al_2O_2$; and $Al_2 + O_2 \rightarrow Al_2O_2$. Because of the large number of matrix reactions leading to $Al_2O_2$, it is perhaps not surprising that this species appears so permanently in the spectra.

$AlO_2$ AND $Al(O_2)_2$. Evidence for the existence of several binary dioxygen species has been given by Andrews et al. ($LiO_2$ ($53, 56$), $NaO_2$ ($54, 57$), $KO_2$ ($55, 57$), $RbO_2$ ($55$)), Huber et al. ($58$) ($NiO_2$, $PdO_2$, $PtO_2$), and Bos et al. ($11$) ($SnO_2$). Except in the latter paper where a linear $D_{\infty h}$ structure was suggested, the $MO_2$ species have been observed to have sideways bonding of the oxygen molecule to the metal atom. When an additional oxygen molecule reacts, forming $M(O_2)_2$ ($53–58$), the structure has been determined to be $D_{2d}$, i.e., each oxygen molecule sideways bonded to the metal atom and twisted into a spiro configuration. Neither the $AlO_2$ nor the $Al(O_2)_2$ species have been observed previously spectroscopically, although the $AlO_2$ species has been noted in several mass spectrometric studies ($20$).

In duoflow experiments in which a low concentration of oxygen was used, a narrow band at 1168 cm$^{-1}$ is usually observed initially. As the sample was annealed, the band gradually diminished being replaced by another narrow band at 1177.5 cm$^{-1}$. At 35°K the latter was the only band present. It disappeared at 45°K. In Figure 6, the annealing study of the 0.007% $^{16}O_2$:Ar matrix is shown. Using $^{18}O_2$, the same behavior was noted for two bands located at 1137 cm$^{-1}$ and 1143 cm$^{-1}$. (The isotopic frequency ratios are 1.030 and 1.027, respectively.) In a 60:40 mixture of $^{16}O_2$ and $^{18}O_2$, both pairs of bands were present and displayed the same annealing behavior. No additional bands were detected which implies that the species to which each of these bands corresponds, contains only one dioxygen molecule. Because it seems unlikely that there are two species with distinct geometrics formed in the matrix, both bands (1168 cm$^{-1}$ and 1177.5 cm$^{-1}$) are assigned to the $Al(O_2)$ species. The presence of the two bands is attributed to multiple site occupation within the matrix environment.

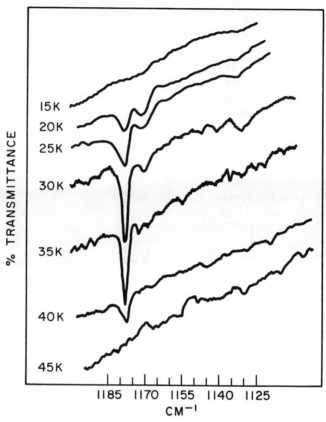

*Figure 6.   Al(O₂)—annealing study in a 1:14,300 ¹⁶O₂:Ar
matrix*

The mixed $^{16}O_2/^{18}O_2$ experiment did not settle the question of metal-to-oxygen bonding in $AlO_2$. An experiment with $^{16}O_2$:$^{16}O^{18}O$:$^{18}O_2$:Ar which would differentiate between the two possible bonding situations—side or end bonded—was not done because the low intensities of the lines would make it impossible to differentiate them from the noise level. On the basis of studies with other metal–oxygen system (53–58) we assign the sideways bonded configuration ($C_{2v}$) to $AlO_2$. The band observed at 1177.5 cm$^{-1}$ is therefore assigned to the $v_1$ mode. If our assignment of $C_{2v}$ symmetry to $AlO_2$ is correct, then the $v_2$ and $v_3$ modes should also be observed. After studying our spectra, two other bands were noted whose behavior was similar to the 1177.5 cm$^{-1}$ band. In $^{16}O_2$–Ar mixtures, these bands were located at 717 cm$^{-1}$ and 525 cm$^{-1}$; in $^{18}O_2$–Ar mixtures, they were located at 693 cm$^{-1}$ and 510 cm$^{-1}$. (The isotopic frequency ratios are 1.034 and 1.029, respectively.) These two bands are tentatively assigned to the $v_2$ and $v_3$ modes. Because the baselines of our spectra changed drastically in different spectral regions, the relative intensities of the $\nu_1$, $\nu_2$, and $\nu_3$ mode could not be compared.

If the oxygen concentration was increased in the duoflow experiments, a new band appeared at 1130 cm$^{-1}$ in addition to the bands at 1168 cm$^{-1}$ and 1177.5 cm$^{-1}$. During annealing, this band steadily increased in intensity and broadened at temperatures greater than 30°K. At 45°K, the band was still present. Figure 7 is a display of the spectra obtained after annealing a 0.25% $^{16}O_2$:Ar matrix. In an $^{18}O_2$–Ar experiment, the 1130 cm$^{-1}$ band shifted to 1101 cm$^{-1}$, an isotopic frequency shift of 1.026.

In a 60:40 mixture of $^{16}O_2$:$^{18}O_2$, four bands were observed at 1130 cm$^{-1}$, 1117 cm$^{-1}$, 1107 cm$^{-1}$, and 1101.5 cm$^{-1}$. This band pattern indicates that the absorbing species contains two equivalent dioxygen molecules.

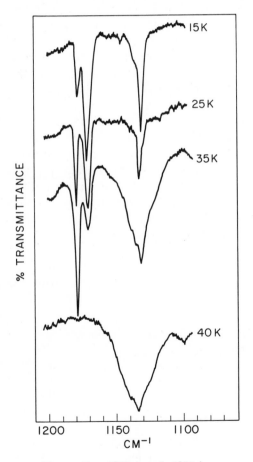

*Figure 7. Al(O$_2$) and Al(O$_2$)$_2$—annealing study in a 1:400 $^{16}O_2$: Ar matrix*

Therefore the bands are assigned, respectively, to the $^{16}O-^{16}O$ stretch in $(^{16}O_2)Al(^{16}O_2)$, the $^{16}O-^{16}O$ and $^{18}O-^{18}O$ stretches in $(^{16}O_2)Al(^{18}O_2)$, and the $^{18}O-^{18}O$ stretch in $(^{18}O_2)Al(^{18}O_2)$. Because an $^{16}O_2:^{16}O^{18}O:^{18}O_2$: Ar experiment was not done because of low signal quality, we again cannot differentiate definitely between side- or end-bonded oxygen. On the basis of our previous assignment of $AlO_2$, we assume sideways-bonded oxygen and therefore assign $D_{2d}$ symmetry, i.e., a spiro configuration to $Al(O_2)_2$.

An additional band was observed whose annealing behavior was similar to the 130 cm$^{-1}$ band. The band was located at 687 cm$^{-1}$ in $^{16}O_2$ and at 671 cm$^{-1}$ in $^{18}O_2$, an isotopic frequency shift of 1.024. This band is also assigned to the species $Al(O_2)_2$.

$O_3$. Ozone was detected in some of our experiments using both uniflow and duoflow modes of operation. The presence of ozone is evidenced by a peak near 1040 cm$^{-1}$ (23, 24). Concurrently, in the uv an intense, broad band centered at 39,200 cm$^{-1}$ was observed, which is most reasonably assigned to the Hartley band of ozone. In the gas phase, this transition is characterized by an intense and almost symmetric peak near 39,200 cm$^{-1}$ (59). Using a hollow cathode sputtering device in the duo-flow mode with an aluminum screw as cathode produces ozone by several possible mecahnisms. Argon gas emerging from the sputtering device either as activated atoms or ions can interact with oxygen molecules in the ambient oxygen–argon mixture producing either oxygen atoms or oxygen ions. These can react with another oxygen molecule to form ozone. Furthermore, sputtered aluminum atoms can react with oxygen molecules to produce AlO and oxygen atoms. The latter can then react with oxygen molecules to form ozone.

$AlO_3$. Because ozone was present in some matrices in relatively large quantities, the possibility of forming $AlO_3$ existed although it had not been reported previously. Jacox and Milligan (60) isolated the $O_3^-$ species in argon matrix and observed the $\nu_3$ mode at 802 cm$^{-1}$ in $^{16}O_2$ and 757.8 cm$^{-1}$ in $^{18}O_2$. This corresponds to an isotopic frequency ratio of 1.058. For various alkali ozonides the $\nu_3$ mode is observed near 800 cm$^{-1}$ (61) and near 820 cm$^{-1}$ for the alkaline ozonides (62). One would expect for an aluminum ozonide that the $\nu_3$ mode would also be observed in the 800–850 cm$^{-1}$ spectral region.

As the oxygen concentration increased in our matrices, the intensity of the ozone peak at 1040 cm$^{-1}$ increased. Also, as the matrices were annealed, the ozone intensity initially increased and then decreased while a band located at 853 cm$^{-1}$ increased in intensity. Using $^{18}O_2$, this band was shifted to 809 cm$^{-1}$ which corresponds to an isotopic frequency ratio of 1.052. In mixed $^{16}O_2:^{18}O_2$:Ar experiments, a series of bands were observed in the 800–850 $cm^{-1}$ region. No definitive assignment of all bands could be made.

Because of the behavior of the 853 $cm^{-1}$ band during annealing and its close proximity to known $O_3^-$ bands, we assign it to the $\nu_3$ mode of $O_3^-$ for the species $AlO_3$. A summary of the ir bands observed in the reaction of aluminum atoms and oxygen molecules is given in Table IV.

Table IV. Summary of the Matrix IR Data
for the Aluminum–Oxygen System

| Species | Symmetry | $^{16}O_2{:}Ar$ | $^{18}O_2{:}Ar$ | $^{16}O_2{:}-$ $^{18}O_2{:}Ar$ | $\nu^{16}O_2$ $\overline{\nu^{18}O_2}$ | Assignment |
|---|---|---|---|---|---|---|
| AlO | $C_{\infty v}$ | 946.5 | 901.0 | 946.5 | 1.0505 | Al–O |
| | | | | 901.0 | | |
| $Al_2O$ | $D_{\infty h}$ | 994.0 | 951.5 | 994.0 | 1.044 | $\nu_3$ |
| or | | | | 951.5 | | |
| $Al_2O_2$ | $C_{2v}$ | 496.5 | 481.0 | 496.5 | 1.032 | |
| | | | | 481.0 | | |
| | | 1217 | 1159 | 1217 | 1.050 | O–O |
| | | | | 1159 | | |
| | | 1213 | 1154 | 1213 | 1.050 | O–O |
| | | | | 1154 | | |
| $Al_2O_2$ | | 490.0 | 474.0 | 490.0 | 1.034 | |
| | | | | 474.0 | | |
| $Al(O_2)$ | $C_{2v}$ | 1177.5 | 1143.0 | 1177.5 | 1.027 | $\nu_1$ |
| | | | | 1143.0 | | |
| | | 1168.0 | 1137.0 | 1168.0 | 1.030 | $\nu_1$ |
| | | | | 1137.0 | | |
| | | 717 | 693 | 717 | 1.034 | $\nu_2$ |
| | | | | 693 | | |
| | | 525 | 510 | 525 | 1.029 | $\nu_3$ |
| | | | | 510 | | |
| $Al(O_2)_2$ | $D_{2d}$ | 1130.0 | 1101.5 | 1130.0 | 1.026 | O–O |
| | | | | 1117.0 | | |
| | | | | 1107.0 | | |
| | | | | 1101.5 | | |
| | | 687 | 671 | 687 | 1.024 | |
| | | | | 671 | | |
| $Al(O_3)$ | $C_{2v}$ | 853 | 809 | 853 | 1.052 | $\nu_3$ |
| | | | | 809 | | |

**Ion Bombardment Studies of Aluminum Oxide Surfaces.** The ion bombardment studies are conveniently grouped into five sets of experiments. The first deals with small bombardments ($2$–$4 \times 10^{17}$ ions/$cm^2$ of $Ar^+$) on several types of aluminum oxide targets (Figure 8; numbers 6–9 in Table II). In the second and third sets, the $Ar^+$ bombardments were increased to $6.7 \times 10^{17}$–$6 \times 10^{18}$ ions/$cm^2$ to determine the effect, if any, on the nature of sputtered products (Figure 9; numbers 10–12 in

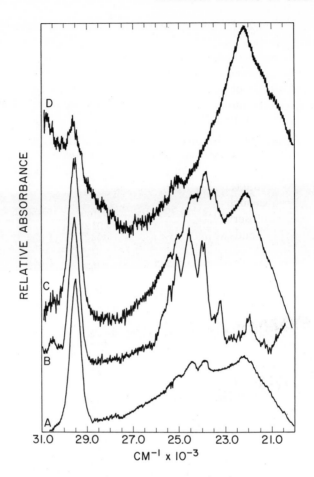

*Figure 8. Visible spectra of the sputtered products from 50-kV Ar⁺ ion bombardment of various targets, isolated in argon matrices. A, aluminum disk, 2 × 10¹⁷ ions/cm²; B, 2.4 × 10⁻⁷m anodized aluminum oxide film on an aluminum backing 2 × 10¹⁷ ions/cm²; C, polycrystalline disk of γ-alumina, 4.1 × 10¹⁷ ions/cm²; D, single-crystal sapphire-α-alumina 2 × 10¹⁷ ions/cm².*

Table II). The fourth and fifth sets of experiments deal with changes in sputtered products as a result of bombardments with the chemically reactive particle, H⁺ (Figure 10; numbers 13–14 in Table II).

Targets were not cooled during bombardments. Heat conduction to the brass target assembly was assumed to be sufficient to dissipate the power deposition (3 W maximum) to the targets and to keep the bombarded surfaces near room temperature.

AR⁺ BOMBARDMENTS. In argon–ion bombardments both the physical form of the aluminum oxide targets and beam fluence were important parameters determining the nature of the sputtered species. The oxide samples chosen as targets were: a piece of aluminum with a $2 \times 10^{-9}$ m thick surface coating of oxide, a $2.4 \times 10^{-7}$lm thick anodic $Al_2O_3$ film attached to its aluminum backing, a polycrystalline disk of $\gamma$-alumina compressed at $7 \times 10^8$ Pa, and an unoriented sapphire single crystal disk. Each was bombarded for 30–60 min with a 50-keV Ar⁺, 6-$\mu$A, 0.635-cm beam.

Because the yields of products were low, they were detected by means of their characteristic electronic transitions. Consequently, any sputtered molecules such as $Al_2O_2$, $AlO_2$ for which electronic transitions have not yet been observed would have remained undetected. The presence of uv–visible bands which were sometimes observed but not identified in the hollow cathode studies could have been caused by such species. The discussion here will deal only with those spectral features which were

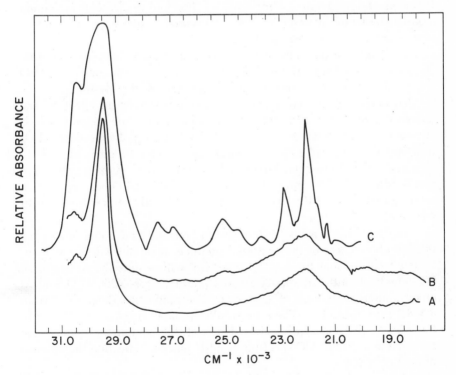

*Figure 9. Visible spectra of the sputtered products from 15-kV Ar⁺ ion bombardment of various targets, all isolated in argon matrices. A, polycrystalline disks of $\gamma$-alumina, $6.7 \times 10^{17}$ ions/cm²; B, single-crystal sapphire-$\alpha$-alumina, $6.7 \times 10^{17}$ ions/cm²; C, single-crystal sapphire-$\alpha$-alumina, 0.9525-cm aperture beam, $6.0 \times 10^{18}$ ions/cm².*

identified in earlier work with specific aluminum or aluminum oxide species.

In Figure 8, the visible spectrum of the matrix-isolated products obtained from each oxide sample bombarded with low fluences of $Ar^+$ is shown. Aluminum atoms as well as $Al_2O$ and $AlO$ molecules are back-sputtered and matrix-isolated in each bombardment.

The aluminum disk with a $2 \times 10^{-9}$-m oxide coating (Figure 8A) yielded primarily Al atoms ($29,500 \text{ cm}^{-1}$). In the spectral region where either $AlO$ or $Al_2O$ transitions are located, a broad band is found with a series of subsidiary maxima. The band at $\sim 22,000 \text{ cm}^{-1}$ is assigned to $AlO$; that at $25,000 \text{ cm}^{-1}$ to $Al_2O$. The ratio $n_{Al}/n_{AlO}$ is calculated to be 8.2. The anodized film, (Figure 8B), yielded the largest ratio of $n_{Al}/n_{AlO}$ $= 13.3$. Relative to $AlO$, the amounts of $AlO_2$ are increased as are the amounts of unidentified species.

The electronic spectrum of products sputtered from the polycrystal-line $Al_2O_3$ disk, (Figure 8C) is similar to that from the aluminum disk. However, the ratio $n_{Al}/n_{AlO}$ represents a threefold decrease of aluminum atoms. The bands at $24,400 \text{ cm}^{-1}$ and $23,900 \text{ cm}^{-1}$ caused by unidentified species are quite prominent in this spectrum.

The sapphire disk (Figure 8D) yielded quite a different spectrum. The main product was $AlO$ with $n_{Al}/n_{AlO} = 0.7$. A small amount of $Al_2O$ was produced, and the unidentified species were negligibly small.

To determine if longer bombardment times have an effect on the differences observed with the various aluminum oxide targets, $6.7 \times 10^{17}$-ions/cm² bombardments were carried out on the polycrystalline $Al_2O_3$ disk and the sapphire disk. The electronic spectra (Figure 9A and B) obtained by collecting backsputtered material from both targets, were nearly identical. The ratio of $n_{Al}/n_{AlO}$ was 11.0. $Al_2O$ was present in small quantities, and the two bands associated with unidentified species were absent.

To determine if a further increase in bombardment times affects the relative abundances of the sputtered products, a sapphire disk was bombarded with $Ar^+$ ions for $6 \times 10^{18}$ ions/cm² with results shown in Figure 9C. The backsputtered, matrix-isolated products were $AlO$, $Al_2O$, and Al atoms and large amounts of $Al_2$ dimer. The ratio of $n_{Al}/n_{AlO}$ varies from 4.4–8.2 (see Table II). The ratio would probably be closer to 11.0 as in the $6.7 \times 10^{17}$ ions/cm² bombardment if the aluminum dimer were included in $n_{Al}$. The $AlO$ spectrum is fairly narrow indicating the absence of multiple sites.

These observations of $Ar^+$ bombardments of aluminum oxide targets lead to the conclusion that for low bombardment times, $2$–$4 \times 10^{17}$ ions/ cm², the nature of the target is an important variable in determining the abundances of the various sputtered products. For longer bombardments,

$6.7 \times 10^{17}$ ions/cm², oxygen seems to be preferentially sputtered initially because the ratio of aluminum atoms to AlO molecules increases by a factor of four, implying that the surface becomes oxygen deficient. However, the nature of the target seems to have less effect on the sputtered products than at lower bombardment times. For still higher bombardments ($6 \times 10^{18}$ ions/cm²) the principal effect seems to be on the amount of $Al_2$ dimer produced which increased drastically when compared with Al atoms.

H⁺ BOMBARDMENTS. The reducing properties of a proton beam can be expected to enhance the production of Al atoms or $Al_2O$ at the expense

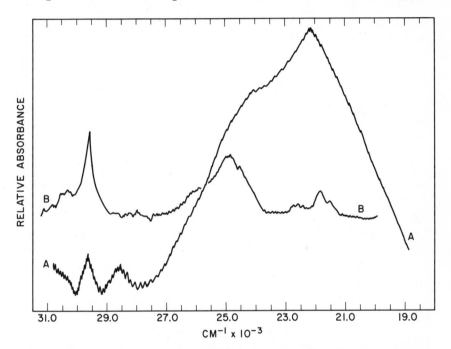

Figure 10.   *Visible spectra of the sputtered products from 15-kV H⁺ bombardment of sapphire, isolated in argon matrices. A, 0.476-cm aperature beam, 2.0 × 10¹⁹ ions/cm²; B, 0.9525-cm aperture beam- 1.2 × 10²⁰ ions/cm².*

of AlO. The formation of AlH or AlOH also are possible. In the gas phase (*63, 64*), the $A'\pi \leftarrow X'\Sigma^+$ transition of AlH occurs at 23,471 cm⁻¹ (*63*) and the $b^3\Sigma^- \leftarrow X'\Sigma^+$ transition at 26,217 cm⁻¹. A shift of 2000 cm⁻¹ in an argon matrix environment would cause a superposition of the AlH and $Al_2O$ bands.

To determine if alterations in sputtered products occur as a result of exposure to 15-kV H⁺, sapphire disks were given 2.0 × 10¹⁹ ions/cm² bombardments. The lower sputtering yield caused by H⁺ compared with

Ar$^+$ bombardment reduces the product concentrations in the matrix. However, it was still possible to detect and identify sputtered species by electronic spectroscopy.

From the intensity of the 22,000 cm$^{-1}$ band (Figure 10A), the main product is seen to be AlO with $n_{Al}/n_{AlO} = 0.7$, a result virtually identical to that for a $2 \times 10^{17}$-ions/cm$^2$, 50-kV Ar$^+$ bombardment of sapphire (Figure 8D). However, comparison of Figure 10A with Figure 8D shows that the intensity of the band in the 25,000 cm$^{-1}$ region which has been assigned to Al$_2$O is markedly increased in the H$^+$ bombardment. Whether these bands are composed entirely of Al$_2$O or a combination of Al$_2$O and AlH is still an open question. In any event the much-enhanced production of Al$_2$O (or possibly AlH) over AlO is evidence for the reducing effect of the proton beam and for a specific chemical sputtering channel.

Again, as in the case of Ar$^+$ bombardments, the effect of increasing the bombardment time on the distribution of sputtered products in a H$^+$ bombardment was investigated. A sapphire disk was bombarded for 1.2 $\times 10^{20}$ ions/cm$^2$ (Figure 10B). The $n_{Al}/n_{AlO}$ ratio was 12.2–16 times larger than that observed for the $2.0 \times 10^{19}$ ions/cm$^2$ bombardment. The very marked enhancement of the $n_{Al}/n_{AlO}$ ratio in going from $2.0 \times 10^{19}$-ions/cm$^2$ to a $1.2 \times 10^{20}$-ions/cm$^2$ H$^+$ bombardment is probably caused by two factors. The first is the oxygen depletion of sputtered surface layers noted previously for the highest fluence Ar$^+$ bombardments. The second factor is related to the first and represents the specific reducing action of the H$^+$ beam which still further enhances the preferential sputtering of Al atoms. The reducing influence of the H$^+$ beam is also evidenced by the greatly enhanced intensity of the Al$_2$O peak (25,000 cm$^{-1}$). The intensity now is approximately the same as the aluminum peak and 12 times higher than the AlO peak.

The changes in sputtered products which occur as a result of long bombardments of sapphire by Ar$^+$ or H$^+$ are illustrated in Figure 9C and Figure 10B. One can speculate that for still higher proton bombardments, Al atom and Al$_2$ dimer formation would be enhanced still further and that the Al$_2$O concentration would increase relative to that of AlO.

ION BOMBARDMENT RESULTS OF ALUMINUM OXIDE. The work reported here has demonstrated that the relative abundances of Al atoms, Al$_2$ dimers, and Al$_2$O and AlO molecules sputtered from aluminum oxide surfaces depend on (1) the nature of the target (single crystal, polycrystalline, anodic film); (2) bombardment times which are related to ion fluences as well as to (3) the nature of the bombarding particle (Ar$^+$ vs. H$^+$). There probably are additional, as yet undiscovered, parameters that affect the nature of the sputtered products. The sputtering process involving aluminum oxide is very complicated indeed, and sputtering of binary targets generally may turn out to be a very complex phenomenon.

To apply collision theory to sputtering of binary substances, it appears that the surface binding energies for each sputtered species have to be known, and these will in general be functions of surface composition. The surface composition of a binary target changes during ion bombardment because of either preferential loss of one of the components or because of implantation and subsequent reaction (in the case of reactive energetic projectiles) with the lattice (65, 66, 67). The chief material-dependent quantity controlling S, the sputtering yield, is $E_b$, the surface binding energy which enters collision theory (3) in the forms $\approx 1/E_b$. It has been proposed (4) to identify $E_b$ with the heats of atomization of the oxides. However, in view of the results of Coburn et al. (5) and those presented here which show not only that major fractions of the sputtered product are molecular in nature but also that the abundances among the products depend on target and bombardment conditions, a more realistic approach is to associate a distinct $E_b$ with each sputtered species to be used in turn in calculating the sputtering yield for that particular species from collision theory.

We propose that the surface binding energy of a given sputtered species be identified with the partial molar enthalpy of vaporization of the species from a freely vaporizing surface whose composition is to be taken at that prevailing at the time of a particular sputtering event. Since vaporization enthalpies are commonly obtained from Knudsen cell vaporization data, the experimental values have to be corrected for application to a freely vaporizing surface. It must also be kept in mind that the data which refer to bulk compositions are not necessarily characteristic of surface and near-surface region compositions. Furthermore, vaporization data, on oxides for example, are usually obtained at high temperatures (1000°–3000°K) while the temperature of a surface undergoing sputtering is a quantity which is very difficult to specify under the highly nonequilibrium conditions associated with a sputtering event. Nonetheless, it seems that in view of the rudimentary state in which the application of collision theory to binary targets finds itself, the introduction of the concept of partial molar enthalpies of vaporization associated with individual sputtered species could well be a useful point of departure.

To explore the implications of this concept for understanding the sputtering of oxides, a brief qualitative discussion of vaporization processes in binary oxides is given first. Finally, it is shown how the concept of partial molar enthalpies of vaporization can help to rationalize the sputtering results obtained with aluminum oxide.

Sublimation processes in oxides have been under intensive study for many years (68, 69, 70), and a great deal of thermodynamic data on enthalpies and free energies of vaporization is available. Trends in the total vapor pressures and in the vaporization of refractory oxides to the gaseous elements and to metal oxide molecules are reasonably well established as a function of atomic number (69). In most of the work that has been done however, it has been assumed that the subliming composition is stoichiometric. However, the situation with most oxides is that compositions, even if they start near stoichiometry, become non-stoichiometric as a result of vacuum sublimation. To illustrate the effect which non-stoichiometry has on the partial pressures and therefore on the partial molar enthalpies of vaporization of various gaseous species in equilibrium with a solid phase, let us consider, with Ackermann and Thorn (71), a system composed of the gaseous species M, MO, and $MO_2$ in equilibrium with a condensed phase, $MO_{2-x}$. If the stoichiometric oxide $MO_2$ is heated in a Knudsen cell in a vacuum to a temperature sufficiently high that sublimation occurs, a preferential loss of oxygen and a decrease in total effusion rate is observed in most cases. Quite small changes in composition ($x = 0.01$) can lead to changes in the free energy of formation of the oxide phase of several kJ/g–atom at high temperatures which in turn can sensitively affect the abundances of the various gaseous species in equilibrium with the solid. An attempt to sketch the variation of total and partial pressures with changing composition of the solid phase has been made for the hypothetical oxide $MO_2$ (see Figure 11) which qualitatively describes the behavior observed, for example, for zirconium dioxide (71). Relative to the partial pressure of the triatomic species $MO_2$, which remains fairly constant over the composition range $MO_2$–$MO_{1.94}$, the partial pressures of diatomic MO and atomic M increase drastically with decreasing oxygen content of the solid phase. By comparison with the changes in O, M, and MO partial pressures, the changes in total pressures as a function of the composition of the solid phase are relatively minor.

Molecular species are found in equilibrium with most refractory oxides, and so the minimum in the free energy curve of the solid phase as well as the congruently subliming composition are determined principally by the stabilities of the solid phase and the gaseous monoxide. These factors are such as to result in a minimum vaporizing composition close to the stoichiometric compound. For oxides such as $TiO_2$, $ZrO_2$, $WO_3$, and the actinide oxides $UO_2$, $NpO_2$, and $PuO_2$, the departures from stoichiometry of the minimum vaporizing compositions are large enough to be measured by relatively straightforward techniques, amounting to between one and a few percent of the total oxygen content. For other oxides, often referred to as line compounds, departures from stoichiometry are very small, amounting in the case of aluminum oxide, for example, to probably less than 0.1%. It will therefore be very difficult to obtain data on the partial molar enthalpies of vaporization of the various species in equilibrium with $Al_2O_3$ as a function of the O/Al ratio of the condensed phase. It is instructive nonetheless to see how the partial pressures of the various gaseous species can be expected to vary with composition. Although the actual minimum O/Al ratio is not known, it is expected to occur at the phase boundary between solid $Al_2O_3$ and liquid Al metal.

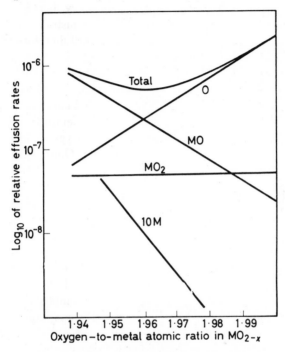

*Figure 11.   Schematic of the vaporization rates of various species in equilibrium with an $MO_2$ phase of variable composition*

Using the JANAF (72) thermochemical tables, one can construct the diagram shown in Figure 12 which graphs the partial pressures of O, Al, AlO, and $Al_2O$ in equilibrium with stoichiometric $Al_2O_3$ (right-hand side of Figure 12) and in equilibrium with a mixture of liquid Al metal and solid $Al_2O_{2-x}$ at 1800°K (left-hand side of Figure 12). The increase by a factor of $10^7$ in the partial pressure of atomic Al is of course caused by the presence of liquid Al and represents the vapor pressure of Al metal at 1800°K. Even more striking is the increase in the partial pressure of $Al_2O$ relative to AlO. While the $Al_2O/AlO$ pressures are in the ratio 1:10 for stoichiometric $Al_2O_3$, the ratio is $10^6:1$ at the liquid phase boundary. A very small change in stoichiometry therefore results in a remarkable enhancement of $Al_2O$ over AlO molecules in the equilibrium vapor.

The results of the thermodynamic calculations discussed above can be correlated qualitatively with the sputtering results on aluminum oxide.

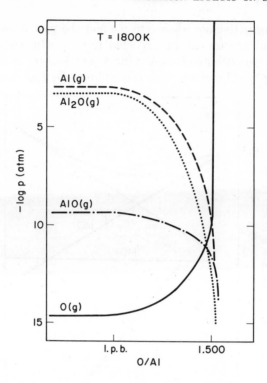

*Figure 12.   Partial pressures of Al, O, Al₂O, and AlO in equilibrium with solid Al₂O₃ and with liquid Al– solid Al₂O₃ at 1800°K*

High fluence Ar⁺ bombardment and particularly bombardments with H⁺ enhance the yield of Al atoms and $Al_2O$ molecules relative to the AlO molecules as shown by numerous experiments done in the course of the present work. Ion bombardment of $Al_2O_3$ and, in particular, energetic H⁺ bombardment appears to deplete surface and near-surface regions of oxygen leading to a slightly hypostoichiometric surface composition which may approximate the limiting phase boundary composition characteristic of $Al_2O_3$ in equilibrium with liquid Al metal. The change in the partial molar enthalpies of vaporization of Al, $Al_2O$, and AlO accompanying subtle but distinct departures of aluminum oxide from stoichiometry could then be the key parameters leading to the observed changes in the ratios of sputtered products.

*Acknowledgment*

We thank R. J. Ackermann for making the thermodynamic calculations on the $Al$–$Al_2O_3$ system.

## Literature Cited

1. Gruen, D. M., Gaudioso, S. L., McBeth, R. L., Lerner, J. L., *J. Chem. Phys.* (1974) **60**, 89.
2. Green, D. W., Gruen, D. M., Schreiner, F., Lerner, J. L., *Appl. Spectrosc.* (1974) **28**, 34.
3. Sigmund, P., *Phys. Rev.* (1969) **184**, 383.
4. Kelly, R., Lam, N. Q., *Radiat. Eff.* (1973) **19**, 39.
5. Coburn, J. W., Taglauer, E., Kay, E., *Proc. Int. Vacuum Congr. Japan, 6th* (1974), *J. Appl. Phys. Suppl.* **2** Part 1.
6. Stirling, A. J., Westwood, W. D., *J. Appl. Phys.* (1970) **41**, 742.
7. Jensen, K., Veje, E., *Z. Phys.* (1974) **269**, 293.
8. Castaing, R., Hennequin, J.-F., *Adv. Mass. Spectrom.* (1971) **5**, 419.
9. Hernandez, R., Lanusse, P., Slodzian, G., Vidal, G., *Method. Phys. Anal.* (1970) **6**, 411; *Chem. Abs.* (1971) **74**, 69195n.
10. Oechsner, H., Gerhard, W., *Surf. Sci.* (1974) **44**, 480.
11. Bos, A., Ogden, J. S., *J. Phys. Chem.* (1973) **77**, 1513.
12. Ivanov, A. A., Tolmachev, S. M., Ezhov, Yu, S., Spiridonov, V. P., Rambidi, N. G., *J. Struct. Chem.* (1973) **14**, 854.
13. Wehner, G. K., Kenknight, C., Rosenberg, D. L., *Planet. Space Sci.*, (1963) **11**, 885.
14. Davidse, P. D., Maissel, L. I., *J. Vac. Sci. Technol.* (1967) **4**, 33.
15. Kelly, R., *Can. J. Phys.* (1968) **46**, 473.
16. Andrews, A. E., Hasseltine, E. H., Olson, N. T., Smith, Jr., H. P., *J. Appl. Phys.* (1966) **37**, 3344.
17. Hasseltine, E. H., Hurlbut, F. C., Olson, N. T., Smith, Jr., H. P., *J. Appl. Phys.* (1967) **38**, 4313.
18. Stirling, A. J., Westwood, W. D., *Thin Solid Films* (1971) **7**, 1.
19. Hennequin, J.-F., *J. Phys. Paris* (1968) **29**, 655, 957.
20. Farber, M., Srivastava, R. D., Uy, O. M., *J. Chem. Soc. Faraday Trans.* (1972) **68**, 249.
21. Alvager T., Uhler, J., *Prog. Nucl. Tech. Instrum.* (1968) **3**.
22. Carstens, D. H. W., Kozlowski, J. F., Gruen, D. M., *High Temp. Sci.* (1972) **4**, 301.
23. Andrews, L., Spiker, Jr., R. C., *J. Phys. Chem.* (1972) **76**, 3208.
24. Brewer, L., Wang, J. L.-F., *J. Chem. Phys.* (1972) **56**, 759.
25. Wei, P. S. P., Tang, K. T., Hall, R. B., *J. Chem. Phys.* (1974) **61**, 3593.
26. Ammeter, J. H., Schlosnagle, D. C., *J. Chem. Phys.* (1973) **59**, 4784.
27. Herzog, R. F. K., Poschenrieder, W. P., Satkiewicz, F. G., *Radiat. Eff.* (1973) **18**, 199.
28. Brewer, L., Chang, C.-A., *J. Chem. Phys.* (1972) **56**, 1728.
29. Huber, H., Künding, E. PJ., Ozin, G. A., Poë, A. J., *J. Am. Chem. Soc.* (1975) **97**, 308.
30. Demtroder, W., *Z. Physik* (1962) **166**, 42.
31. Penkin, N. P., Shabanova, L. N., *Opt. Spectrosc.* (1963) **14**, 5; (1965) **18**, 504.
32. Cunningham, P. T., *J. Opt. Soc. Am.* (1968) **58**, 1507.
33. Tyte, D. C., Nicholls, R. W., "Iden. Atlas of Mol. Spectra 1: The AlO $A^2\Sigma - X^2\Sigma$ Blue Green System," Univ. of W. Ontario, London, Cnada, 1964.
34. Gole, J. L., Zare, R. N., *J. Chem. Phys.* (1972) **57**, 5331.
35. McDonald, J. K., Innes, K. K., *J. Mol. Spectrosc.* (1969) **32**, 501.
36. Knight, Jr., L. B., Weltner, Jr., W., *J. Chem. Phys.* (1971) **55**, 5066.
37. Lagerquist, A., Lennart Nilsson, N. E., Barrow, R. F., *Ark. Fys.* (1957) **12**, 543.
38. Yoshimine, M., McLean, A. D., Liu, B., *J. Chem. Phys.* (1973) **58**, 4412.
39. Das, G., Janis, T., Wahl, A. C., *J. Chem. Phys.* (1974) **61**, 1274.

64 RADIATION EFFECTS ON SOLID SURFACES

40. Nicholls, R. W., *J. Res. Natl. Bur. Stand.* (1962) **A66**, 227.
41. Smakula, A., Z. *Phys.* (1930) **59**, 603.
42. Linevsky, M. J., White, D., Mann, D. E., *J. Chem. Phys.* (1964) **41**, 542.
43. Makowiecki, D. M., Lynch, Jr., D. A., Carlson, K. D., *J. Phys. Chem.* (1971) **75**, 1963.
44. Snelson, A., *J. Phys. Chem.* (1970) **74**, 2574.
45. Marino, C. P., White, D., *J. Phys. Chem.* (1973) **77**, 2929.
46. Lynch, Jr., D. A., Zehe, M. J., Carlson, K. D., *J. Phys. Chem.* (1974) **78**, 236.
47. Buchler, A., Stauffer, J. L., Klemperer, W., Wharton, L., *J. Chem. Phys.* (1963) **39**, 2299.
48. Mal'tsev, A. A., Shevel'kov, V. F., *Vysokikh Temp.* (1964) **2**, 650.
49. Wagner, E. L., *Theor. Chim. Acta* (1974) **32**, 295.
50. Anderson, J. S., Ogden, J. S., *Chem. Phys.* (1969) **51**, 4189.
51. Ogden, J. S., Ricks, M. J., *J. Chem. Phys.* (1970) **53**, 896.
52. White, D., Seshadri, K. S., Dever, D. F., Mann, D. E., Linevsky, M. J., *J. Chem. Phys.* (1963) **39**, 2463.
53. Andrews, L., *J. Chem. Phys.* (1969) **50**, 4288.
54. *Ibid.* (1969) **73**, 3922.
55. *Ibid.* (1971) **54**, 4935.
56. Hatzenbuhler, D. A., Andrews, L., *J. Chem. Phys.* (1972) **56**, 3398.
57. Smardzewski, R. R., Andrews, L., *J. Chem. Phys.* (1972) **57**, 1327.
58. Huber, H., Klotzbücher, W., Ozin, G. A., Vander Voet, A., *Can. J. Chem.* (1973) **51**, 2722.
59. Herzberg, G., "Molecular Spectra and Molecular Structure," Vol. 3, p. 510, Van Nostrand Reinhold, New York, 1966.
60. Jacox, M. E., Milligan, D. E., *J. Mol. Spectrosc.* (1972) **43**, 148.
61. Spiker, Jr., R. C., Andrews, L., *J. Chem. Phys.* (1973) **59**, 1851.
62. Thomas, D. M., Andrews, L., *J. Mol. Spectrosc.* (1974) **50**, 220.
63. Grimaldi, F., LeCourt, A., Lefebvre–Brion, H., Moser, C. M., *J. Mol. Spectrosc.* (1966) **20**, 341.
64. Farkas, L., Z. *Phys.* (1931) **70**, 733.
65. Gruen, D. M., Wright, R. B., McBeth, R. L., Sheft, I., *J. Chem. Phys.* (1975) **62**, 1192.
66. Chatelet, J., Claassen, H. H., Gruen, D. M., Sheft, I., Wright, R. B., *Appl. Spectrosc.* (1975) **29**, 185.
67. Gruen, D. M., Siskind, B., Wright, R. B., unpublished data.
68. Gilles, P. W., "Vaporization Processes in Refractory Substances," *Thermodyn. Nucl. Mater. Proc. Symp.* (1962).
69. Ackermann, R. T., Thorn, R. J., Winslow, G. H., "Systematic Trends in Vaporization and Thermodynamic Properties," *Proc. Conf. Phys. Chem. Aerodyn. Space Flight* (1961) **12**.
70. Ackermann, R. J., Rauh, E. G., *High Temp. Sci.* (1973) **5**, 463.
71. Ackermann, R. J., Thorn, R. J., *High Temp. Technol.* (1964) (Supplement to *Pure Appl. Chem.*) IUPAC 141.
72. "JANAF Thermochemical Tables," Dow Chemical Co., Midland, Mich. Aug., 1965.

RECEIVED January 5, 1976. Work supported by U.S. Energy Research and Development Administration.

# Neutron Sputtering of Solids

P. DUSZA, S. K. DAS, and M. KAMINSKY

Argonne National Laboratory, Argonne, Ill. 60439

*Experiments on fast neutron (0.1–14.1 MeV) sputtering of solids have been performed using various neutron sources and techniques to determine sputtering yields. The yields for the same material irradiated with neutrons having comparable energy spectra vary greatly for several possible reasons. For 14.1-MeV neutron bombardment of cold-worked niobium with a coarse surface finish, some authors have observed deposits of chunks of the target material on collectors facing irradiated targets in addition to a fractional atomic layer while other authors have observed only the latter. The ejection of chunks appears to depend strongly on the stresses and the degree of roughness in the surface regions. Theoretical estimates of neturon sputtering yields and of a recent model of neutron impact-induced chunk emission are discussed.*

The phenomenon of "sputtering," i.e., ejection of material from solid surfaces under bombardment with energetic projectiles, has been studied widely for charged particle irradiation. For reviews on this subject *see* Refs. 1–7. Much less experimental and theoretical work has been done on sputtering of solid surfaces with neutrons. The cross section $\sigma_d$ for the production of displaced lattice atoms from collisions with neutrons is low compared with that typical for a charged particle such as a proton for the same energy. If 14.1-MeV neutrons interact, for example, with the nuclei of niobium lattice atoms, the displacement cross section (only the elastic scattering contribution is considered) has a value of approximately 2 barns (8). The corresponding displacement cross section for a 14.1-MeV proton is $\sim 1 \times 10^5$ barns. This value is calculated using Rutherford nuclear scattering cross section (9) for transfer of energies $> 50$ eV, which is sufficient for the displacement of lattice atoms. On the other hand, the average energy of a primary knock-on atom (PKA) of niobium under 14.1-MeV neutron irradiation is nearly three orders of magnitude larger than for 14.1-MeV protons. The release of sputtered particles is proportional to the number of displaced lattice atoms $v$ per primary knock-on atom and according to Kinchin and Pease, (10) $v \simeq E/2E_d$, where $E$ is the energy of the primary knock-on atom and $E_d$ is the displacement energy of a lattice atom. Therefore, one would expect that the sputtering yield, which is proportional to $\langle \sigma_d v(E) \rangle$, will be of the same order of magnitude for both 14.1-MeV proton and neutron sputtering. Since neutron sputtering yields have generally low values their measurements are relatively difficult.

Recently, there has been a great interest in measuring neutron sputtering yields for materials of importance to controlled thermonuclear fusion devices. In fusion reactors burning deuterium and tritium, large fluxes of 14.1-MeV neutrons will be produced. These neutrons will penetrate the vacuum vessel containing the plasma, commonly referred to as the "first wall." In addition, backstreaming neutrons from the blanket region [e.g. neutrons from ($n, 2n$) reactions, backscattered neutrons] with energies less than 14.1 MeV will also penetrate the first wall. Such high fluxes of energetic neutrons, especially the 14.1-MeV component, make this neutron energy spectrum substantially different from that of a fission reactor. This bombardment will contribute to first-wall erosion, and the material sputtered from the first wall may reach the plasma and contaminate it (11, 12, 13, 14).

In this paper we will first discuss aspects of the experimental techniques used to measure neutron sputtering yields, review and critically analyze some of the existing data on neutron sputtering yields, and discuss some of the theoretical models used to describe the release of particulate matter induced by neutron impact. The sputtering by fission fragments formed during thermal neutron irradiation of fissile materials will not be reviewed in detail.

## General Experimental Considerations

Many of the experimental techniques used for neutron sputtering are similar to those used for ion sputtering. Collectors are placed close to the target material being investigated so that the sputtered species will deposit on them. The range of energetic neutrons is generally much longer than the target dimensions commonly used. Therefore, for the case of a disc-shaped target, for example, sputtered species leave not only the large surface facing the neutron source (backward sputtering) but also the large surface on the opposite (exit) side (forward sputtering). In many neutron sputtering experiments collectors face the large front and back surfaces of target samples. Of course, the collectors themselves become sources of sputtered species. Since in many experiments (*15, 16, 17, 18*) the thickness of targets and collectors are small, the attenuation of the neutron flux is small for a given target–collector set. Several experimenters (*15, 16, 17, 18*) have irradiated a stack of several target–collector sets simultaneously. Of course in these types of experiments the flux decreases rapidly with distance away from the neutron source. For example, for the 14.1-MeV rotating target neutron source (RTNS) at Lawrence Livermore Laboratory (LLL) (*19, 20*) and for a point located on the axis of the deuteron beam striking the tritiated target, the neutron flux $\phi$ at a distance Z from the source is given by integration over the deuteron-irradiated (neutron) target spot (diameter $2r$) as (*19, 20*):

$$\phi = \frac{Y}{4\pi r^2} \ln \left[ 1 + \left(\frac{r}{Z}\right)^2 \right] \qquad (1)$$

where $Y$ is the total neutron emission rate over the $4\pi$ solid angle in neutrons/sec. For $r/Z < 1$ one can expand the last term in Equation 1 into a power series:

$$\phi = \frac{Y}{4\pi r^2} \left[ \left(\frac{r}{Z}\right)^2 - \frac{1}{2}\left(\frac{r}{Z}\right)^4 + \frac{1}{3}\left(\frac{r}{Z}\right)^6 \right] \qquad (2)$$

One can readily see that, at large distances where $Z \gg r$, higher order terms in the power series expansion can be neglected, and this expression reduces to the well known form $\phi = Y/4\pi Z^2$ for a point source.

Since the amount of material deposited on collector surfaces can be very small for practical fluence levels attainable in existing neutron sources, great care is required in handling targets and collectors and in analysis. The following seven factors appear to be important.

**Target and Collector Preparation and Characterization Prior to Neutron Irradiation.** In many of the earlier experiments (e.g., Refs. *21, 22, 23, 24, 25*) the characterization of the collector surface prior to irradiation was not reported. Only recently several experimental groups (*15–18, 26, 27, 28*) have reported the characterization of both target and collector surfaces prior to neutron irradiation by such techniques as ion microprobe analysis, scanning electron microscopy, and neutron activation analysis. Very small amounts of contamination of the collector surfaces by the target material to be collected can affect the results significantly. The use of ultrahigh purity collector materials, such as 99.999% pure monocrystalline silicon (111) (*15*) and the removal of surface impurities and inclusions by electropolishing and by chemical etching can alleviate the contamination problem to some extent.

**Assembly of Targets and Collectors.** This should be done under clean laboratory conditions prior to irradiation to avoid contamination, and the equipment should be shipped to the neutron source preferably under vacuum. This was done, for example, in the work described in Refs. *15, 16, 17.*

**The Neutron Irradiation Conditions.** Such parameters as target temperature, neutron energy spectrum, and dose need to be known to determine the sputtering yields. A variety of fast neutron sources [e.g., neutrons from fission reactors from the $(d, t)$ reaction and from the $(d, Be)$ reaction] can provide neutrons with different energy spectra and angular distributions. Different experimenters have used techniques and instruments such as foil activation, proton recoil counters, and ionization chambers to determine the total neutron dose.

**The Vacuum in the Chamber Containing the Target–Collector Assemblies during Irradiation.** This factor affects the deposition of the material ejected from the target surface on the collector surface, for example, by changes in the sticking probability (2). In many of the earlier experiments poor vacuum conditions prevailed, and only recently have experiments been performed under ultrahigh vacuum (*15, 16, 17, 18, 26*).

**The Handling of the Targets and Collectors between Irradiation and Analysis.** Often it is necessary to store the collectors before analysis because of high activation levels. The collectors should be stored under ultrahigh vacuum to avoid significant contamination of the deposits because some of the analytical techniques used, such as Auger spectroscopy, are very sensitive to contamination buildup. For example, if extensive oxidation and carbon buildup occurs on the sputtered deposits prior to Auger analysis, the signals typical for the sputtered deposits can be completely masked by such contamination. The requirement for keeping the deposits on the collector surfaces free of contamination does not imply that the target surfaces need to be free of contamination.

**Blank Experiments.** In order to ascertain that the collector surfaces have not been contaminated with the target material during the handling procedures it is essential to conduct blank experiments with unirradiated target–collector assemblies that have been handled similarly to the irradiated ones.

**Analysis of the Sputtered Deposits on the Collectors.** This can be done by various sensitive analytical techniques such as neutron activation analysis (NAA) (*17, 18, 21, 26*). Although this technique can give accurately the total amount of material deposited on the collector, it is difficult to determine if the deposits exist as atoms, atom clusters, or particulate matter (chunks). Recently, surface analytical techniques such as Rutherford backscattering (RBS), ion microprobe mass analysis

(IMMA), Auger electron spectroscopy (AES), and scanning electron microscopy (SEM) in conjunction with x-ray spectrometry have been used to analyze sputtered deposits (*15, 16, 27, 28*).

In considering the neutron sputtering results, it is necessary to distinguish between those experiments which have used neutrons with different energy spectra. Neutron sputtering results according to the type of neutron sources used are reviewed below.

### Neutrons from Fission Reactors

The random flux of neutrons present in fission reactors has been used by several investigators for neutron sputtering experiments (*21, 24, 25, 29–32*). The neutron energy distribution depends on the reactor design but generally ranges from thermal energies to several MeV.

"Fission," i.e., reactor spectrum neutron sputtering yields of gold have been measured by Norcross et al. (*21*), Verghese (*24*), and Kirk et al. (*25*). Norcross et al. (*21*) used a target of polycrystalline gold foil together with an aluminum collector foil. The assembly was irradiated in the Battelle research reactor in a vacuum of $\sim 10^{-5}$ torr to a fast ($> 0.1$ MeV) neutron fluence of $4 \times 10^{17}$ neutrons/cm$^2$. The amount of gold deposited on the aluminum collector foil was determined by NAA by measuring the 0.412-MeV gamma radiation from the decay of $^{198}$Au. A sputtering yield of $(1.0 \pm 0.3) \times 10^{-4}$ atom/neutron was reported (*21*).

Verghese (*24*) made neutron sputtering yield measurements of gold using three types of spectra—well thermalized, predominantly fast ($> 1$ MeV), and both thermalized and fast neutrons, using the Oak Ridge National Laboratory Bulk Shielding Reactor. The aluminum collector foils and the gold target foils were sealed in a quartz tube at $< 10^{-5}$ torr pressure. The gold deposits were analyzed by NAA and by autoradiography to determine the uniformity of deposits. The gold was found to be deposited uniformly over the collector area. The fast neutron ($> 1$ MeV) sputtering yield was $[1.83 \pm 0.56] \times 10^{-6}$ atom/neutron.

Recently Kirk et al. (*25*) irradiated monocrystalline gold samples which were placed in the cryogenic irradiation facility of the CP-5 reactor at Argonne National Laboratory (ANL). The collector was a cylindrical-shaped high purity aluminum foil. The encapsulated target–collector assembly was pumped to a seal-off pressure of $\sim 1 \times 10^{-5}$ torr. The capsule was cooled to liquid helium temperature during the irradiation. The total fast neutron fluence ($E > 0.1$ MeV) ranged from $2.1 \times 10^{17}$ to $5.5 \times 10^{17}$ neutrons/cm$^2$. The amount of sputtered gold deposits was determined by NAA. The detection sensitivity of gold by their technique was quoted as $10^{10}$ gold atoms/cm$^2$. The neutron sputtering yields originally reported (*25*) ranged from $1 \times 10^{-3}$ to $6 \times 10^{-3}$

atom/neutron, that is about three orders of magnitude higher than that reported by Verghese (24) and more than an order of magnitude higher than those by Norcross et al. (21). However, this sputtering yield has been revised to $1.8 \times 10^{-5}$ atom/neutron (32).

None of the experimenters cited above (21, 24, 25) have characterized their targets with respect to surface roughness and the degree of cold rolling in the case of the polycrystalline gold samples. Such target parameters, however, can affect neutron sputtering yields, as shown in experiments using 14.1-MeV neutrons (12) to be discussed later. Furthermore, there is no indication to what degree the collector surfaces were free of contaminants, especially from the target material, prior to irradiation. In addition, there is no indication of any blank experiments. For these reasons it is difficult to assess which of the above-listed neutron sputtering yield values for gold (polycrystalline, monocrystalline) with fission neutrons can be considered reliable.

Garber et al. (29) also studied fast neutron (> 1 MeV) sputtering of polycrystalline and monocrystalline targets of many elements (Be, Al, Si, Ti, Cr, Fe, Co, Cu, Zn, Ge, Zr, Nb, Mo, Ag, Cd, Sb, Ta, W, Au, Pb, and Bi) using the VVR-M reactor of the Institute of Physics, Academy of Sciences, Ukrainian USSR. They monitored a collector current (collector facing the target) which was caused only in part by the ions sputtered from the target. The collector currents were related to the intensity of the positive ions sputtered from the target. However, from their paper it is not clear how additional currents contributing to the total collector current [e.g., caused by secondary electrons and sputtered ions leaving the collector] have been taken into account. They found that the collector current was larger for monocrystalline targets than for polycrystalline targets by a factor of 1.5–2.5. Moreover, a periodic dependence of the collector current on the atomic number of the target element was observed. There were maxima in the collector currents for Cu, Zn, Ag, and Au. No sputtering yields for the elements studied were given.

Baer and Anno (30) studied neutron sputtering of iron using "fission" (reactor spectrum) neutrons from the Batelle Research Reactor. The target was a 1-mil thick iron foil of 1-cm² exposed area, and the collectors were quartz plates with a 500-Å aluminum coating. The target–collector assembly was placed in a quartz tube, sealed at $1 \times 10^{-6}$ torr, and irradiated to a total neutron fluence (for neutrons with $E > 0.1$ MeV) of $2 \times 10^{18}$ neutrons/cm². The aluminum coating together with the sputtered iron deposits were removed from the quartz collectors by an acid bath. The iron was then precipitated out and dried. Subsequently, the amount of the iron deposit was estimated from 1.29- and 1.10-MeV $\alpha$- decay of $^{59}$Fe after applying corrections for counter efficiency and the time between the end of the irradiation and the counting. A fast neutron sputtering yield for iron of $S_n = (5.7 \pm .8) \times 10^{-3}$ atom/neutron was reported (30). This value was later revised (31) to $S_n = (4.5 \pm 0.7) \times 10^{-3}$ atom/neutron after correcting for the effects of thermal sputtering ($\approx 1\%$ of $S_n$), for the contribution of direct recoils from the $^{58}Fe\ (n,\gamma)\ ^{59}Fe$ reaction ($\approx 20\%$ of $S_n$), and for the thermal desorption ($< 0.1\%$ of $S_n$). These authors also measured fast neutron sputtering yields of iron (average neutron energy $\sim 2\,MeV$) at $200°-340°C$ and obtained values ranging from $3 \times 10^{-3}$ to $8 \times 10^{-3}$ atom/neutron with no uniform variation with temperature (31). These sputtering yield values are larger than theoretical estimates would indicate, as discussed later. These experimenters do not report the characteristics of their targets (e.g., surface stresses, surface roughness), the cleanliness of the collector surfaces, or the conduction of blank experiments.

### Neutrons from the (d, t) Reaction

Since the neutron energy spectrum in a D-T fusion reactor will have a strong 14.1-MeV component, several sputtering measurements have been performed using the 14.1-MeV neutrons from the $(d, t)$ reaction (15–18, 23, 27, 28, 33–38). A D-T neutron source usually consists of a tritiated metal target that is bombarded with high energy deuterons from an accelerator. The angular distribution of the emitted neutrons is to a good approximation isotropic. At present, one of the most intense 14.1-MeV neutron sources available for neutron sputtering measurements is at Lawrence Livermore Laboratory, a source which has been described in detail elsewhere (19, 20, 39). In this source the neutrons are produced in a thin tritiated titanium target which is bombarded with 400-keV $D^+$ ions, and typically a total neutron emission rate (over the $4\pi$ solid angle) of $3 \times 10^{12}-5 \times 10^{12}$ neutrons/sec can be obtained. This D-T neutron source has been used by several investigators (13, 15–17, 34–38).

Keller (23) attempted to measure sputtering yields of Au, W, Cu, and In by 14.1-MeV neutrons. Plastic collectors were placed in front and in back of each target foil and were placed in a glass bell jar which was pumped continuously to $\sim 10^{-5}$ torr. The 14.1-MeV neutron dose at the target foils ranged from $1.46 \times 10^{13}$ neutrons/cm$^2$ to $2.4 \times 10^{13}$ neutrons/cm$^2$. The collectors were analyzed by NAA, and no sputtered atoms were detected. However, upper limits of sputtering yields were estimated, based on the minimum detectable amount of sputtered deposits. These upper limits ranged from $6 \times 10^{-4}$ atom/neutron for gold to $3.9 \times 10^{-2}$ atom/neutron for copper. For fast-neutron sputtering of gold (using $\sim 4$-MeV neutrons from a Pu–Be source) Keller and Lee (22) had reported earlier a high yield value of 0.5 atom/neutron. The authors later suggested (23) that there may have been large systematic errors in those observations.

Garber et al. (33) studied 14.1-MeV neutron sputtering of a 2500-Å monocrystalline gold foil irradiated to a total dose of $2.32 \times 10^{13}$ neutrons/cm². The gold foil and an aluminum-coated glass screen which served as a collector were placed in a glass ampule and sealed off at

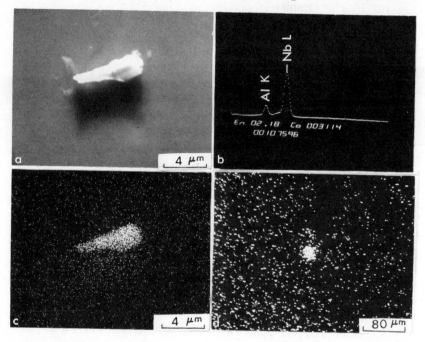

Surface Effects in Controlled Fusion

*Figure 1. Niobium deposited on an aluminum substrate when a niobium target at ambient temperature (estimated to be near room temperature) was irradiated with 14-MeV neutrons to a total dose of $4.6 \times 10^{15}$ neutrons/cm². (a), scanning electron micrograph (backscattered electron images) of a niobium chunk deposited on an aluminum surface; (b), a x-ray energy spectrum of the niobium chunk in (a); (c), Nb L x-ray image of the same areas as in (a); (d), secondary ion ($^{93}Nb^+$) micrograph of the same area as in (a). The secondary ion micrograph was taken at much lower magnification than (a) and (c) (16).*

$1 \times 10^{-6}$ torr. From neutron activation analysis of the sputtered deposits on the collector, a sputtering yield of $3 \times 10^{-3}$ atom/neutron was estimated, a value that is five times the upper limit given by Keller and Lee (22). In addition, the angular distribution of sputtered atoms was determined by autoradiography. Preferential sputtering along low index crystallographic directions was reported.

Extensive measurements have been conducted by Kaminsky et al. (13, 15, 16, 27, 28, 34, 35) on the erosion of surfaces of cold-rolled and annealed polycrystalline metals (e.g., Nb and V) and of SiC with various surface finishes under 14.1-MeV neutron bombardment. The samples were irradiated at ambient temperature with 14.1-MeV neutrons from the RTNS at LLL to total doses ranging from $1.7 \times 10^{15}$ to $1.5 \times 10^{16}$ neutrons/$cm^2$, values that were more than two orders of magnitude higher than those used in the experiments by Keller (23) and by Garber et al. (33). The samples were kept in an ultrahigh vacuum of $\sim 2 \times 10^{-9}$–$5 \times 10^{-10}$ torr during irradiation. The type and the amount of the material deposited on collector surfaces were determined with the aid of RBS ($\sim 0.0005$ ML), IMMA ($\sim 0.001$ ML), AES ($\sim 0.01$ ML), and SEM in conjunction with an energy dispersive x-ray spectrometer all of which had been calibrated with standards prepared by vapor deposition. The detection sensitivities of some of these techniques are listed in brackets as fractions of a monolayer (ML) of niobium deposited on a silicon substrate. Ten independent neutron irradiation runs were performed, and for each type of cold-rolled target at least two independent irradiations were made. The observations reveal several striking and important results.

**Forms of Deposits.** The deposits of such materials as Nb (13, 15, 16, 34), V (13), stainless steel (15), Si (15, 16), and SiC (13, 28) appear in two forms. One form covered the substrate surface in atomic form as a fractional atomic layer as is commonly observed in ion sputtering. The other form was discovered (13, 15, 16) as chunks of various sizes and irregular shapes. Figure 1 shows an example of a niobium chunk on a silicon (111) substrate. The scanning electron micrograph in Figure 1a shows a niobium chunk. A secondary ion ($^{93}Nb^+$) micrograph of the chunk (Figure 1d) confirms that it is niobium. Figure 1b shows an x-ray energy spectrum of the chunk, where only Nb L and Si K peaks can be identified and Figure 1c shows a NbL x-ray image, further confirming that the chunk is niobium.

**Distribution of Deposits.** Both types of deposits are not uniformly distributed over the collector area but sometimes are clustered along streaks or appear in patches. For example, as illustrated in the secondary ion micrograph in Figure 2, the deposits can appear as wide streaks (e.g., the width $> 40$ $\mu$m, and the lengths $> 400$ $\mu$m) which are not associated with the microstructure of the substrate structure. The direction of the streaks appears to be parallel to the direction of the grinding. Figure 3 shows the surface structure of a cold-rolled sample with a coarse surface finish of 5–10 $\mu$m. The finish of these surfaces was intentionally kept rough, since plasma–wall interactions in future fusion reactors can cause surface roughening by blistering, physical and chemical sputtering, and other surface erosion processes. The patches containing high concentrations of niobium sometimes contain chunks, but sometimes only a

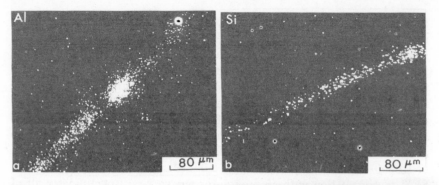

*Figure 2.   Secondary ion ($^{93}Nb^+$) micrographs of niobium deposits on (a) an aluminum substrate and (b) a silicon substrate.   Niobium released from cold-rolled niobium target under 14.1-MeV neutron impact (16).*

high concentration of atomic niobium can be found.  "Mini"-chunks of ≲ 100 Å diameter could not be resolved in the SEM studies.  The appearance of streaks and patches is much more pronounced for the cold-rolled niobium sample than for the vacuum-annealed samples.

**Concentration of Deposits.**  Measurements of the concentration profile of the fractional atomic layer deposit by RBS (beam diameter ∼ 300 μm) and IMMA (beam diameter ∼ 25 μm) reveal that the deposit concentration varies from < 5 × 10⁻⁴ ML to ∼ 5 × 10⁻² ML in patch and streak regions near chunks.  Even for the case with the maximum number

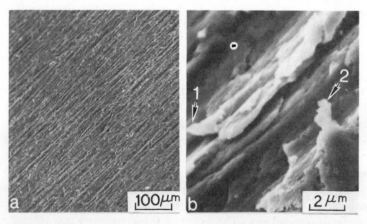

*Figure 3.   Scanning electron micrographs of surfaces of cold-rolled niobium (a) after optical polish and a light chemical etch (target type A) and (b) same surface at higher magnification, the arrow points to a protrusion (34)*

of deposited chunks, areas of 10–30 mm² can be found on the collector surface with an "average" deposit concentration of $< 5 \times 10^{-4}$ML.

**Size, Shape, and Density of Chunks.** As early as 1974 it was pointed out (*13*) that the number, the size, and the shape of the chunks depend very strongly on the tensile stresses in the surface regions (e.g., from cold rolling) and on the degree of microstructure of the irradiated surface (e.g., microprotrusions and microcracks). More recently, some improved values for the chunk deposit density (chunks/cm²) and the chunk size distribution have been reported (*34*), and these are listed in Table I for niobium targets with total neutron fluences of $1.69 \times 10^{15}$–$1.5 \times 10^{16}$ neutrons/cm². Since the chunk deposits are not uniformly dis-

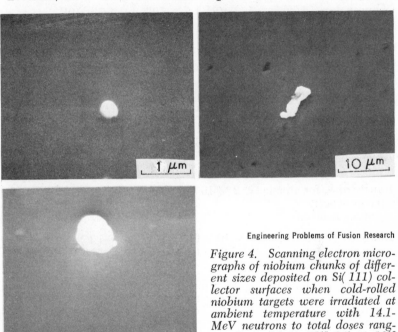

Engineering Problems of Fusion Research

*Figure 4. Scanning electron micrographs of niobium chunks of different sizes deposited on Si( 111) collector surfaces when cold-rolled niobium targets were irradiated at ambient temperature with 14.1-MeV neutrons to total doses ranging from 4.6 × 10¹⁵–1.5 × 10¹⁶ neutrons cm⁻² (35)*

tributed over the collector area but are sometimes clustered along streaks or in patches, only ranges for the chunk deposit density have been quoted. The ranges indicate the smallest and largest number of chunks which can be found more typically on a contiguous surface area of 1 cm² within the total collector area. The chunk deposit densities in the repeat runs fall within the ranges listed; some extreme values are listed in brackets. The listed values illustrate the fact that chunk ejection is a relatively rare event compared with the total number of primary knock-on events produced by 14.1-MeV neutrons in near-surface regions (*40*). No chunks could be observed for monocrystalline niobium targets.

Table I.    Distribution of Size and Density
Observed in 14.1-MeV Neutron

*Surface Conditions*

| | Cold-rolled, lightly chemically etched targets (Type A)[a] (~ 5–10μm micro-finish) | Cold-rolled, lightly electropolished targets (Type B)[a] (~ 5μm microfinish) |
|---|---|---|
| Total collector area investigated (cm²) | 12 | 8 |
| Neutron dose (n/cm²) | $5 \times 10^{15}$–$1.5 \times 10^{16}$ | $4.6 \times 10^{15}$ |
| Ranges for chunk deposit density (chunks/cm²)[a] | 5–25 (250)[b] | 0–30 (1000)[b] |
| Ranges for average chunk diameters (μm) | 0.3–1 (some cylindrical chunks up to 5μm long and 1.5μm dia.) | 0.3–5 (cylindrical chunks up to 15μm long and 3μm dia.) |
| Distribution of chunk size | 45%: 0.3–0.5μm<br>45%: 0.5–1.0μm<br>10%: > 1μm | 35%: 0.3–1μm<br>50%: 1–5μm<br>15%: > 5μm |
| Ranges of $S_a$ for atomic layer deposits (atom/neutron) | $< 2 \times 10^{-4}$–$6 \times 10^{-4}$ | $< 2 \times 10^{-4}$–$6 \times 10^{-4}$ (a few areas near streaks had $8 \times 10^{-3}$) |
| Ranges of $S_c$ from chunk deposits (atom/neutron) | $5 \times 10^{-5}$–$1.6 \times 10^{-4}$ $(1.6 \times 10^{-3})$[b] | 0–$1.1 \times 10^{-3}$ $(3.6 \times 10^{-2})$[b] |
| Total $S_n$ from chunk and background (atom/neutron) | $< 2 \times 10^{-4}$–$7.6 \times 10^{-4}$ $(2.2 \times 10^{-3})$[b] | $< 2 \times 10^{-4}$–$1.7 \times 10^{-3}$ $(3.7 \times 10^{-2})$[b] $[2.5 \times 10^{-1}]$[c] |

[a] Smallest and largest number of chunks that can be found on a contiguous surface of 1 cm² on the collector.
[b] Highest values observed in a run.
[c] $S_n$ values reported earlier (*15, 16*) for an estimated larger value for the average chunk size.

In order to estimate the contribution of the niobium chunks to the total deposit, it is also necessary to know the distribution of chunk sizes. Since only relatively few chunks are observed for the various types of niobium surfaces, the size distribution of the chunks is poorly known. Table I lists some ranges for the average chunk diameters which are typically observed. Some extreme values reported earlier (*13*) are indicated in the parenthesis. Since the chunks are irregular in shape,

**of Chunk Deposits and Sputtering Yields**
**Irradiation of Nb (34)**

Surface Conditions

| *Annealed for 2 hr at 1200°C in high vacuum, electropolished targets (Type C) (~ 2μm microfinish)* | *Annealed for 2 hr at 1200°C in high vacuum, electropolished targets (Type D) (~ 0.5μm microfinish)* | *Monocrystalline Nb (111) electropolished targets (Type E) (~ 0.5μm microfinish)* |
|---|---|---|
| 4 | 4 | 4 |
| $4.6 \times 10^{15}$ | $4.3 \times 10^{15}$ | $1.69 \times 10^{15}$ |
| 0–20 (100) [b] | 0–10 (40) [b] | 0 |
| 0.3–1 | 0.05–1 | — |
| 100%: 0.3–1μm | 100%: 0.05–1μm | — |
| $< 2 \times 10^{-4}$–$5 \times 10^{-4}$ | $< 2 \times 10^{-4}$–$5 \times 10^{-4}$ | $< 2 \times 10^{-4}$ |
| $0$–$2 \times 10^{-5}$ $(1 \times 10^{-4})$ [b] | $0$–$7 \times 10^{-6}$ $(2.8 \times 10^{-4})$ [b] | 0 |
| $< 2 \times 10^{-5}$–$5.2 \times 10^{-4}$ $(6 \times 10^{-4})$ [b] $[5 \times 10^{-2}]$ [c] | $< 2 \times 10^{-4}$–$5.1 \times 10^{-4}$ $(7.8 \times 10^{-4})$ [b] $[1 \times 10^{-2}]$ [c] | $< 2 \times 10^{-4}$ |

[a] The targets denoted here as types A and B were erroneously reversed in Table II in Reference *13*. For these two types of targets the inclusions (e.g., SiC, Al2O3 particles) on the surface left from optical polishing are not completely removed by light etching or electropolishing.

Journal of Nuclear Materials

the average diameter of either a roughly spherical or cylindrical chunk is defined as the diameter of a circle whose area is equal to that of the projection of the chunk onto the surface plane. The chunks vary from ~ 0.5 μm to 5 μm in average diameter (for example, *see* Figure 4), the larger size chunks (~ 1 μm) being observed for the cold-worked samples having the roughest surface finish. Even small changes in the average radius of such chunks will drastically influence the calculated volume

($\propto r^3$) and thereby the total number of atoms per chunk. Since both the chunk deposit density and the size distribution are poorly known, only crude estimates of the contribution by chunks to the total neutron sputtering yield $S_n$ have been made. Table I lists ranges of neutron sputtering yield $S_c$ from chunk deposits alone. These values were obtained from the upper and lower limit values for the chunk deposit density and the chunk size distribution. The yield values listed in parenthesis were obtained from the extreme values for the chunk deposit density listed in parenthesis and from the average chunk diameters.

Table I also lists ranges for the contributions of fractional atomic layer deposits to the neutron sputtering yield. For the niobium targets listed, the estimated neutron sputtering yield $S_a$ from atomic deposits ranges from $< 2 \times 10^{-4}$ to $6 \times 10^{-4}$ atom/neutron, but near streaks and patches, values as high as $8 \times 10^{-3}$ atoms/neutron have been observed (15, 16). To what extent this relatively high yield could have been caused by the deposition of atomic clusters and "mini" chunks with sizes below the detection ability ($\sim 100$ Å) of the SEM is difficult to assess.

Table I also lists ranges for the total yield $S_n$ from chunk deposits and atomic deposits. The upper and lower limits for the total yields were obtained by adding the upper and lower limit values for $S_c$ and $S_a$, respectively. The yields marked by an asterisk were estimated for the extreme value of the chunk deposit contribution. The values in the brackets were quoted recently by Kaminsky et al. (14, 15), in which a higher average chunk size value than the one used in Table I for the lower and upper limit values had been assumed.

**Estimate of Chunk Velocity.** The average velocity $\bar{v}$ of the chunks has been estimated (16) to be $\leq 10$ cm/sec. For chunks with a mass $\simeq 8 \times 10^{-11}$ g and $\bar{v} = 10$ cm/sec, the value of the kinetic energy $E$ is $4 \times 10^{-9}$ ergs, a value which would correspond to about 0.5% of the energy deposited by a 600-keV recoiling niobium atom or about 1% of the damage energy. As pointed out earlier (13, 15, 16), the stored energy per volume of chunk (e.g., from cold rolling) can be several orders of magnitude higher (the order of $10^{-4}$ ergs/chunk volume) than the energy deposited by the primary recoil.

**Discussion of Additional Results.** The results of Kaminsky et al. (13, 15, 16, 34) discussed above indicate that chunk emission from niobium under 14.1-MeV neutron irradiation is a rare event and that it appears to depend upon stresses and certain surface microstructures (e.g., microprotrusions) on the surface. A model for chunk emission that explains these observations qualitatively will be discussed in "Model for Chunk Emission," below. In order to ensure that the phenomenon of chunk emission is caused by D-T neutron irradiation and not by any contamination or mechanical handling, blank experiments were performed by Kaminsky et al. (*16, 34*). Blank samples of cold-worked niobium and cold-worked vanadium of the type used in these experiments, together with highly polished silicon (111) collectors, were mounted in the ultrahigh-vacuum transfer chamber (used for the transport from LLL to ANL) in the same way as the actual target–collector assembly. These blank samples were transported, together with the irradiated samples, under identical conditions. For several irradiation runs an accelerometer was attached to the support frame of the transfer chamber to register possible shocks and vibrations to which the assembly might be exposed during transport. For a typical transport from LLL to ANL only 39 shocks of 0.5–2.5 g acceleration (vertical direction) and 71 shocks of 0.5–2.5 g acceleration (horizontal direction) were measured. The sensitivity of the accelerometer was such that shocks of 0.25-g acceleration with frequencies up to 100 Hz in both the vertical and the horizontal direction could be detected. It was determined in independent experiments using cold-rolled niobium and vanadium samples and silicon (111) collectors mounted in a chamber which was held at ultrahigh vacuum that shocks of the number and magnitude listed above did not release any chunks.

Behrisch et al. (*18*) also measured D-T neutron sputtering yields for cold-rolled niobium and gold with an unspecified surface finish. Silicon collectors were placed on either side of the target foils and the target–collector assembly was irradiated with D-T neutrons at the University of Mainz, West Germany. The vacuum in the chamber containing the target–collector assembly was in the $10^{-9}$-torr range, and the total neutron doses were $1.9 \times 10^{14}$ neutrons/cm$^2$ for gold and $1.0 \times 10^{14}$ neutrons/cm$^2$ for niobium. These doses are two orders of magnitude lower than the maximum dose used by Kaminsky et al. (*34*). Neutron activation analysis of the sputtered gold deposits gave sputtering yields of $3.3 \times 10^{-4}$ atom/neutron and $2.6 \times 10^{-4}$ atom/neutron for forward and backward sputtering, respectively. For niobium, only the $^{92}$Nb isotope from the (*n, 2n*) reaction deposited on the collector was measured, and a yield for niobium from the (*n, 2n*) reaction, sometimes referred to as sputtering yield from radioactive recoil atoms (*38*), of $7.8 \times 10^{-8}$ atom/neutron was reported (*18*). The collectors were examined only in SEM (magnification and area scanned were not specified), and no chunks were detected. However, since the surface condition of the niobium target used in the experiments was not reported (e.g., the degree of surface finish and the extent of microprotrusions present on the surface), it is difficult to know whether one should expect chunk emission from such a surface. Moreover, since the neutron fluence was very low compared with the experiments conducted in Ref. *34*, the probability of chunk emission would be correspondingly low.

Harling et al. ($17, 36$) also investigated D-T neutron sputtering of niobium targets with different surface finish and of annealed gold targets. The targets and high purity silicon collectors were irradiated with the RTNS at LLL to total fluences ranging from $5 \times 10^{14}$ neutrons/cm$^2$ to $2.6 \times 10^{16}$ neutrons/cm$^2$, the vacuum in the target–collector chamber being in the range $10^{-6}$–$10^{-8}$ torr range. Sputtered deposits on most of the collectors were analyzed by NAA. For several collectors containing niobium deposits, SEM analysis was also carried out. A total area of 1 cm$^2$ was scanned at 300X on two separate collectors which faced the cold-worked niobium targets, and the authors failed to detect any niobium chunks larger than $\sim 0.5 \mu$m. The authors quote ranges for total sputtering yields for different types of niobium targets, for forward and backward directions. For example, for their niobium I target (Marz grade, cold-worked niobium, mechanically polished and electropolished, surface roughness 1–5 $\mu$m), they quoted ranges of sputtering yield of $1.1 \times 10^{-5}$–$5.9 \times 10^{-4}$ atom/neutron (forward direction) and of $1.5 \times 10^{-5}$–$1.3 \times 10^{-3}$ (backward direction). For a cold-rolled, chemically etched ($\sim 10 \mu$m removed per surface) niobium target (niobium II), the sputtering yields ranged from $6.4 \times 10^{-5}$ to $< 2.3 \times 10^{-4}$ atom/neutron for forward sputtering and from $< 4.4 \times 10^{-5}$ to $< 1.0 \times 10^{-3}$ atom/neutron for backward sputtering. A niobium target of type A (see Table I) used in the experiments by Kaminsky et al. ($13, 34$) was also included in their runs; no niobium chunks were observed by SEM analysis (1.2 cm$^2$ collector area scanned at 1000X), and a value of $< 6 \times 10^{-5}$ atom/neutron was reported for backward sputtering.

More recently, the emission of chunks from cold-rolled niobium surfaces (of type A in Table I used by Kaminsky et al. ($13, 34$)) under 14.1-MeV neutron irradiation has been confirmed in observations made by Cafasso et al. ($37$) and Meisenheimer ($38$). Meisenheimer later reported ($41$) the observation of 39 niobium chunks/cm$^2$ with an average diameter of 2 $\mu$m for niobium target which would correspond to a type A target in Table I. A total neutron sputtering yield of $1.45 \times 10^{-4}$ atom/neutron was given. Both the values for the chunk density and the total neutron sputtering yield fall within the ranges of values given in Table I for type A targets. Both authors used IMMA in addition to SEM analysis to detect chunks. Meisenheimer ($38$) also observed nonhomogeneous deposits of niobium on a silicon collector (i.e., patches containing a high concentration of niobium) similar to those observed by Kaminsky et al. ($15$). The failure to observe niobium chunks from type A target in Table I in the experiments by Harling et al. ($17$) may in part be caused by the detection techniques used. Neutron activation analysis without autoradiography allows one to determine the total amount of sputtered deposits but not to determine the spatial distribution and the type of such deposits (chunks or atomic). While it should be possible to detect $\geq 0.5$-$\mu$m diameter chunks in SEM at 1000X magnification, a few chunks of this diameter present on the entire collector surface may be readily overlooked during scanning because of their small size. At this magnification the chunk diameter will appear as 0.5 mm. On the other hand, in IMMA one can readily obtain secondary ion images of very small chunks ($\sim 0.2$ $\mu$m), even if one scans at a much lower magnification ($\sim 250$X), as shown earlier in Figure 1d.

The absence of chunk emission from heavily etched cold-rolled niobium targets used by Harling et al. (17, 36) is not very surprising in the light of the results of Kaminsky et al. (13, 34), since as the surfaces become smoother and less stressed the chunk emission disappears. From the results on D-T neutron sputtering of niobium described above, it appears that chunk emission is very likely limited to stressed surface regions having protrusions. As is discussed later, such surface conditions appear necessary for chunk emission according to a model of chunk emission.

### Neutrons from the (d,Be) Reaction

The $^9$Be ($d$, $n$) reaction has been used as a fast neutron source. A cyclotron is generally used to produce high energy deuterons which then strike a beryllium target. Unlike the D-T neutrons, the D-Be neutrons have a broad neutron energy distribution which depends on the incident deuteron energy. Furthermore, the D-Be neutrons have an anisotropic angular distribution which is strongly peaked parallel to the direction of the incident beam.

Jenkins et al. (26) used D-Be neutrons [using 40-MeV deuterons from the Oak Ridge cyclotron (ORIC)] for sputtering from monocrystalline niobium, from discharge-machined monocrystalline niobium and from two types of gold targets, one type in the as-rolled condition, the other electropolished. The neutron energy spectrum had an intensity peak at $\sim 15$ MeV, and the targets received an integrated fluence of $2.4 \times 10^{16}$–$3.5 \times 10^{16}$ neutrons/cm$^2$ for niobium and $2.3 \times 10^{16}$–$4.0 \times 10^{16}$ neutrons/cm$^2$ for gold. High purity silicon discs were used as collectors, and the vacuum in the target chamber was in the $10^{-9}$-torr range. The sputtering yields for both types of gold were measured by NAA, and only upper limits of the total sputtering yield values were given. The yield values for both types of gold targets were $< 7 \times 10^{-5}$ atom/neutron for forward sputtering and $< 3 \times 10^{-5}$ (for electropolished gold) and $< 1.3 \times 10^{-4}$ (for cold-rolled gold) for backward sputtering. The collectors facing the niobium targets were examined by mass spectroscopy and by backscattering using 36.7-MeV argon ions, and no niobium deposits were detected. The upper limit for sputtering yields has been estimated from the neutron fluence and from the detection sensitivities of the techniques used and is given as $< 3 \times 10^{-5}$ atom/neutron for both forward and backward sputtering. They examined the collectors for niobium chunks by optical and electron microscopy and did not observe any chunks. Neither the magnification nor the area scanned were reported. The absence of chunk emission from monocrystalline niobium surfaces irradiated with D-Be neutrons is similar to the results obtained by Kaminsky et al. (*13, 34*) on D-T neutrons. In the case of discharge-machined monocrystalline niobium it is difficult to assess whether the surface stresses and the surface roughness (e.g., existence of protrusions) are high enough for chunk emission or not.

More recently, Jenkins, et al. (*42*) reported (*d,* Be)-neutron sputtering yields for type A niobium targets (*see* Table I) of $\leq 5.6 \times 10^{-5}$ atom/neutron (forward direction) and $\leq 4.0 \times 10^{-5}$ atom/neutron (backward direction). Chunk deposits were not observed.

**Table II.   Fast Neutron Sputtering Yields for Different Metals**

| Neutron Source | | Target Metal[a] | Neutron Energy[b] |
|---|---|---|---|
| $^{235}$U Fission | Battelle Research Reactor | Au | spectrum |
| | Bulk Shielding Reactor, ORNL | Au | " |
| | CP-5, ANL | Au (monocrystal) | " |
| | Battelle Research Reactor | Fe | " |
| | | Fe (at 200°–340°C) | " (average ~ 2.0 MeV) |
| Pu–Be | | Au | mean ~ 4.2 |
| D–Be | (40-MeV D⁺, cyclotron) | Au (electropolished) | spectrum [0–35 MeV |
| | | Au (cold-rolled) | peak ~ 15 MeV] |
| | | Nb (monocrystalline) | |
| | | Nb (discharge machined) | |

[a] All targets are polycrystalline unless stated otherwise.
[b] Only neutrons with energies larger than 0.1 MeV were considered as contributing to sputtering and accounted for in the total dose value quoted. The maximum neutron energy can go up to 7 MeV, but the fluence at the high energy tail is 2–3 orders of magnitude smaller than at 0.1 MeV.
[c] Revised to $1.8 \times 10^{-5}$ atom/neutron (*32*).

*Summary of Experimental Neutron Sputtering Yields*

The sputtering yield values for metals irradiated with neutrons having a broad energy spectrum are summarized in Table II, whereas those for nearly monoenergetic D-T neutrons are listed in Table III. The scatter in the experimental sputtering yields $S_n$ for gold, one of the most widely studied metals, is very large. For example, for neutrons from fission reactors total sputtering yields for gold range from $1.8 \times 10^{-6}$ atom/neutron (*24*) to $6 \times 10^{-3}$ atom/neutron (*25*). Since many of the earlier experiments may have suffered from collector contamination with the target material (no blank experiments were reported), the more recent experiments are considered more reliable. The most probable forward sputtering yield for gold with 14.1-MeV neutrons is in the range $1.2 \times 10^{-5}$–$4.5 \times 10^{-4}$ atom/neutron (*17*). For 14.1-MeV neutron sputtering of niobium, there is a somewhat better agreement between the reported $S_n$ values for targets of comparable surface finish and microstructure. For example, for a cold-rolled, lightly etched, niobium surface

**Irradiated with Neutrons from Fission Processes and the ($d$,Be) Reaction**

| Neutron Dose (neutrons/$cm^2$) | Sputtering Yield $S_n$ (atom/neutron) | Reference |
|---|---|---|
| $4 \times 10^{17}$ | $(1.0 \pm 0.3) \times 10^{-4}$ | *21* |
| not stated | $(1.83 \pm .56) \times 10^{-6}$ | *24* |
| $2.1 \times 10^{17}$–$5.5 \times 10^{17}$ | $1 \times 10^{-3}$–$6 \times 10^{-3\,e}$ | *25* |
| $2 \times 10^{18}$ | $(5.7 \pm 0.8) \times 10^{-3}$ | *30* |
| | $4.5 \pm 0.7 \times 10^{-3\,d}$ | *31* |
| $7.8 \times 10^{18}$–$1.4 \times 10^{19}$ | $3 \times 10^{-3}$–$8 \times 10^{-3}$ | *31* |
| $3.6 \times 10^{11}$–$7.3 \times 10^{11}$ | $0.5^{\,e}$ | *22* |
| $4 \times 10^{16}$ | $< 7 \times 10^{-5\,f}, < 3 \times 10^{-5\,g}$ | |
| $2.5 \times 10^{16}$ | $< 7 \times 10^{-5\,f}, < 1.3 \times 10^{-4\,g}$ | *26* |
| $3.5 \times 10^{16}$ | $< 3 \times 10^{-5\,f}$ | |
| $3.3 \times 10^{16}$ | $< 3 \times 10^{-5\,f,\,g}$ | |

$^{d}$ Revised value of the one in Ref. *30*.
$^{e}$ The same authors later suggested (*23*) that there may have been large systematic errors in this value.
$^{f}$ Forward sputtering yield.
$^{g}$ Backward sputtering yield.

**Table III.  Summary of** $(d,t)$ **Neutron**

| Target Metal[b] | Neutron Dose (neutrons/cm²) |
|---|---|
| Au (monocrystalline) | $2.3 \times 10^{13}$ |
| Au | $2.4 \times 10^{13}$ |
| Au | $1.9 \times 10^{14}$ |
| Au (annealed) | $5 \times 10^{14}$–$2.6 \times 10^{16}$ |
| Nb (cold-rolled, lightly etched, 5–10 μm micro-finish) | $5 \times 10^{15}$–$1.5 \times 10^{16}$ |
| | $5 \times 10^{15}$ |
| Nb (cold-rolled, lightly electropolished, ~ 5 μm microfinish) | $4.6 \times 10^{15}$ |
| Nb (annealed, electropolished) | $4.3$–$4.6 \times 10^{15}$ |
| Nb (monocrystalline) | $1.7 \times 10^{15}$ |
| Nb (cold-rolled, electropolished, 1–5 μm micro-finish) | $5 \times 10^{14}$–$2.6 \times 10^{16}$ |
| Nb (cold-rolled, heavily etched, 1–4 μm micro-finish) | $5 \times 10^{14}$–$2.6 \times 10^{16}$ |
| Nb (cold-rolled, lightly etched, 1–4 μm micro-finish) | $5 \times 10^{14}$–$2.6 \times 10^{16}$ |

[a] The mean energy of neutrons is 14.1 MeV.
[b] All targets are polycrystalline metals unless stated otherwise.
[c] Backward sputtering yield.

( ~ 5–10 μm microfinish) values of $S_n$ ranging from $< 2 \times 10^{-4}$ to $7.6 \times 10^{-4}$ atom/neutron (34) and $< 6 \times 10^{-5}$ atom/neutron (17) have been reported. For the case of cold-rolled, electropolished niobium, values of $S_n$ ranging from $< 2 \times 10^{-4}$ to $1.7 \times 10^{-3}$ atom/neutron (34) and $1.1 \times 10^{-5}$ to $5.9 \times 10^{-4}$ atom/neutron (17) have been reported. In the following section the experimental sputtering yield values will be compared with some theoretical estimates.

## Theoretical Estimates of Fast Neutron Sputtering Yields

Most of the theoretical treatments of neutron sputtering are based on the concept of neutron-induced displacement collision cascades, a concept which has also been used for the treatment of ion sputtering. The existing theories (43, 44, 45) used to describe neutron sputtering do not take into account the surface topography, such as surface roughness or microprotrusions on the surface and the nature and amount of stresses

**Sputtering Yields for Different Metals**[a]

| Sputtering Yield $S_n$ (atom/neutron) | Reference |
|---|---|
| $3 \times 10^{-3}$ | *33* |
| $< 6 \times 10^{-4}$ | *23* |
| $3.3 \times 10^{-4e}, 2.6 \times 10^{-4e}$ | *18* |
| $2.5 \times 10^{-5}\text{–}4.5 \times 10^{-4e}, 1.2 \times 10^{-5}\text{–}2.0 \times 10^{-4e}$ | *17* |
| $< 2 \times 10^{-4}\text{–}7.6 \times 10^{-4}$ $(2.2 \times 10^{-3})$ [e] [includes contribution from chunk deposits of $5 \times 10^{-5}\text{–}1.6 \times 10^{-4}$ $(1.6 \times 10^{-3})$ [e]] | *34* |
| $< 6 \times 10^{-5}$ | *17* |
| $< 2 \times 10^{-4}\text{–}1.7 \times 10^{-3}$ $(3.7 \times 10^{-2})$ [d] [includes contribution from chunks of $0\text{–}1.1 \times 10^{-3}$ $(3.6 \times 10^{-2})$ [d]] | *15, 16, 34* |
| $< 2 \times 10^{-4}\text{–}5.2 \times 10^{-4}$ $(6\text{–}8 \times 10^{-4})$ [d] [contribution from chunks, $0\text{–}2 \times 10^{-5}$ $(1\text{–}3 \times 10^{-4})$ [d]] | *17* |
| $< 2 \times 10^{-4}$ | |
| $1.1 \times 10^{-5}\text{–}5.9 \times 10^{-4e}, 1.5 \times 10^{-5}\text{–}1.3 \times 10^{-3c}$ | *17* |
| $6.4 \times 10^{-5}\text{–} < 2.3 \times 10^{-4e}, < 4.4 \times 10^{-5}\text{–} < 1.0 \times 10^{-3c}$ | *17* |
| $< 3.6 \times 10^{-5e}, 8.5 \times 10^{-4e}$ | *17* |

[d] These are highest values observed in a run.
[e] Forward sputtering yield.

in near-surface regions, which can significantly influence the neutron sputtering process as shown experimentally (*34*). Only very recently has the surface topography been considered in ion sputtering theories (*46*). We will first consider some of the neutron sputtering theories which are based on the concept of displacement collision cascades and discuss some of the deficiencies of these theories.

Taimuty (*43*) provided one of the earliest theoretical estimates of fast (fission) neutron sputtering yields by adapting the theory of sputtering by high energy ions developed by Goldman and Simon (*47*). The neutron sputtering yield $S_n$ was given as:

$$S_n = \frac{0.17 \, \bar{\nu}}{\pi R^2 \cos \Psi} \, \bar{\sigma}_d \left( 1 - \frac{A E_d}{4 \, E_n} \right) \qquad (3)$$

where $\bar{\nu}$ is the average number of secondary displacements per primary knock-on atom, $\bar{\sigma}_d$ is the average fast neutron scattering cross section leading to displacements, $A$ is the atomic weight of the irradiated material, $E_d$ is the threshold energy for an atomic displacement, $R$ is the distance of closest approach in hard sphere collisions between displaced atoms and lattice atoms at rest, and $E_n$ is the average neutron energy. With certain assumptions about the value of the various parameters Taimuty (43) estimated a sputtering yield value of $1.4 \times 10^{-6}$ atom/neutron for copper irradiated with fission neutrons. Taimuty's estimate agrees within a factor of 10 with the one (45) based on Sigmund's sputtering theory, (44) as will be discussed in the following.

In a theory for ion sputtering Brandt and Laubert (48) suggested that the number of sputtered atoms is proportional to the number of secondary displaced atoms formed by primary knock-on atoms and is also proportional to the ratio of the mean range of the displaced atoms to the depth of penetration of the particle at which the primary knock-on event occurs. Subsequently, Sigmund (44) developed an ion sputtering theory which also used the concept that the release of sputtered particles is a result of the formation of atomic collision cascades. He applied transport theory to describe the formation of such cascades. His theory was extended to include sputtering of a target by energetic neutrons via sufficiently energetic recoil atoms in collision cascades. The primary recoil atoms are produced by elastic collisions between energetic neutrons and lattice atoms. The energy deposited into damage $\mu (E)$ in a thin region near the target surface is calculated. The energy lost by electronic excitation is excluded. Then the sputtering yield $S_n$ for fast neutrons is given by the formula (44):

$$S_n = \alpha \Lambda N \sigma(E_n) \langle \mu(E) \rangle \tag{4}$$

where $\alpha$ is a constant which takes into account the fact that part of the energy is not deposited inside the target surface, $\Lambda$ is a parameter dependent on the property of the target material and the state of the surface, $N$ is the atomic density (number of atoms/cm$^3$) of the target, $\sigma (E_n)$ is the total cross section for the production of displaced atoms by neutrons with energy $E_n$, and $\langle \mu(E) \rangle$ is the average of the energy deposited into damage over the spectrum of recoil energies. The parameter $\Lambda$ is given by:

$$\Lambda = \frac{3}{4 \pi^2} \frac{1}{N C_0 U_0} \tag{5}$$

where the constant $C_0$ (which depends on the interaction potential of the target atoms) is given by:

$$C_0 = \frac{\pi}{2} \lambda_0 a^2 \tag{6}$$

which for $\lambda_0 = 24$ and $a = 0.219$ Å yields:

$$C_o = 1.8 \, [\text{Å}]^2 \tag{7}$$

The parameter $U_o$ is the surface binding energy, which for metals can be taken to be equal to the cohesive energy or equal to the heat of sublimation. The terms $\sigma(E_n)$ and $\langle \mu(E) \rangle$ have often been combined into one parameter $\langle \sigma E_D \rangle$ called "specific damage energy" (8) or "damage energy cross section" (49, 50), where the "damage energy" $E_D$ is the energy available for displacements and has been estimated by several authors (8, 49, 50) for different materials and for different neutron energies using available neutron cross section data from ENDF/B files. (Evaluated Nuclear Data File, National Neutron Cross Section Center, Brookhaven National Laboratory, Upton, N.Y., available on magnetic tapes.) Thus, we can write:

$$S_n = \alpha \, \frac{3}{4\pi^2} \frac{\langle \sigma E_D \rangle}{C_o U_o} \tag{8}$$

The values of the factor $\alpha$ have been given by Sigmund (44) as follows; for forward sputtering $\alpha \approx 1$, for backward sputtering $\alpha \approx 0.06$, and for sputtering with random neutrons (e.g., a target immersed in an isotropic flux of neutrons in the core of a reactor) $\alpha \approx 2$. Table IV lists some theoretical sputtering yields for several metals for 14-MeV neutrons,

**Table IV.  Fast Neutron Sputtering Yields for Some Materials Calculated Using Sigmund's Theory**

| | | | *Sputtering Yields* $S_n$ *(atom/neutron)* | |
|---|---|---|---|---|
| *Target Metal* | $U_o$[a] *(ev/atom)* | $\langle \sigma E_D \rangle$[b] *for 14-MeV Neutrons (ev $-$ cm$^2$ $\times 10^{20}$)* | *Forward* $S_n$ *for 14-MeV Neutrons*[c] | *Random* $S_n$ *for* $^{235}U$ *Fission Neutron Spectrum*[d] |
| C | 7.4 | 4.26 (2.8) | $2.4 \times 10^{-6}$ ($1.6 \times 10^{-6}$) | — |
| Al | 3.36 | 16.4 (17.5) | $2.1 \times 10^{-5}$ ($2.2 \times 10^{-5}$) | $2.4 \times 10^{-5}$ |
| Ti | 4.89 | 19.8 | $1.7 \times 10^{-5}$ | — |
| V | 5.33 | 23.7 | $1.9 \times 10^{-5}$ | — |
| Cu | 3.52 | (25.3) | ($3.0 \times 10^{-5}$) | $1.9 \times 10^{-5}$ |
| Nb | 7.59 | 27.3 (24.5) | $1.5 \times 10^{-5}$ ($1.4 \times 10^{-5}$) | $8.9 \times 10^{-6}$ |
| Au | 3.80 | (18.2) | ($2.0 \times 10^{-5}$) | $1.1 \times 10^{-5}$ |

[a] From Ref. 51.

[b] Values from Parkin et al. (50); values in parentheses from Robinson (8).

[c] Values in parenthesis were calculated using the $\langle \sigma E_D \rangle$ values given by Robinson (8); others were calculated using $\langle \sigma E_D \rangle$ values given by Parkin et al. (50).

[d] Calculated by Robinson in Ref. 45.

using Equation 6 together with published data on specific damage energy (8, 50) and the heat of sublimation (51) at 298°K. The sputtering yields for fission neutrons listed in Table IV were calculated by Robinson (45) for a spherical target immersed in an isotropic flux of neutrons.

It can be seen from Table IV that theoretical estimates of neutron sputtering yields using Sigmund's theory for 14-MeV neutrons and $^{235}U$ fission neutrons are generally lower than the experimental values listed in Tables II and III. For example, for 14.1-MeV neutron sputtering of gold the forward $S_n$ value ($3.3 \times 10^{-4}$ atom/neutron) reported by Behrisch et al. (18) and the upper limit of the range in $S_n$ ($4.5 \times 10^{-4}$ atom/neutron) reported by Harling et al. (17) are more than an order of magnitude higher than the theoretically estimated value in Table IV ($2.0 \times 10^{-5}$ atom/neutron). The fission (reactor spectrum) neutron sputtering yield for gold of $1 \times 10^{-3}$–$6 \times 10^{-3}$ atom/neutron reported by Kirk et al. (25) is more than two orders of magnitude higher than the estimated theoretical sputtering yield of $1.1 \times 10^{-5}$ atom/neutron (Table IV). Kirk et al. suspected an error in their values and tried to improve the accuracy of their values. The more recent value reported by Kirk (32) et al. is $1.8 \times 10^{-5}$ atom/neutron. In the case of 14.1-MeV neutron sputtering of gold, the reported lower limit on $S_n$ values agrees somewhat better with the theoretical estimate. Similarly, for niobium the reported (16, 34) lower limit on sputtering yields agrees more closely with the theoretical estimate, whereas the upper limit values are more than one order of magnitude higher than estimated. Another theoretical estimate (11, 52) of $S_n$ for 14.1-MeV neutron sputtering of niobium, using a modified theory of Pease (53), gave a value of $S_n \approx 6 \times 10^{-5}$ atom/neutron, which is a factor of four higher than estimated in Table IV using Sigmund's theory but is lower than the experimental values quoted in Ref. 34 for both cold-rolled and annealed niobium samples.

Since the experimental values of $S_n$ are higher than the theoretical estimates based on the collision cascade concepts discussed above, the application of a "thermal spike" concept (see Ref. 54) to estimate neutron sputtering yields has been considered (18, 45). It has been shown (18, 45) that use of thermal spikes can yield values of $S_n$ that are greater than those obtained, for example, by Sigmund's theory. For fission-neutron sputtering of gold (assuming a mean neutron energy of 2 MeV (45)), the use of a thermal spike theory yields a value of $S_n$ as large as $3 \times 10^{-4}$ atom/neutron (cf. $1.1 \times 10^{-5}$ atom/neutron in Table IV).

Sigmund's theory predicts a backward sputtering yield that is ~ 1/17th of the forward sputtering yield, whereas the experimental yield values listed in Table III do not show any significant difference. In fact, in one case a higher yield value has been reported (17) for backward sputtering (see Table III for niobium) than for forward sputtering. The surface condition of the target (e.g., microprotrusions or surface stresses in annealed vs. cold-worked targets) is not considered in Sigmund's theory. Moreover, the inhomogeneity observed in the spatial distribution of sputtered deposits of niobium (e.g., patches, streaks) on the collector surfaces cannot be explained by the sputtering theories based on displacement collision cascade concepts. The phenomenon of chunk ejection from niobium surfaces during 14.1-MeV neutron irradiation also cannot be explained by these theories. Dynamic interference between cascades has been considered by Robinson (45), but his estimates show that one cannot account for the ejection of chunks by such interference.

### Model for Chunk Emission

Kaminsky et al. suggested (15) that the energy deposited by 14.1-MeV neutrons could lead to localized thermal spikes which in turn could cause the generation of shock waves. These shock waves in near-surface regions could set up stresses large enough to release stored energy by initiating submicroscopic cracks or by propagating already existing microcracks, and thus cause the emission of chunks. J. Beeler (55) and J. E. Robinson (56) suggested independently that the existence of shock waves in near-surface regions of rough and stressed ("technical") surfaces could lead to chunk ejection. Guinan (40) investigated theoretically to what extent shock waves and transient thermal stresses generated by 14.1-MeV neutron impact-induced collision cascades can account for the phenomenon of chunk emission. He found that substantial transient thermal and shock stresses can be generated by collision cascades initiated by primary knock-on atoms (PKA) with maximum energies of 600 keV and mean energies of 180 keV for the case of 14.1-MeV neutron impact on niobium. However, the energy available in transient thermal stresses and the duration ($\approx 10^{-12}$ sec) of the radiating shock pulse are insufficient to create micron-size cracks and the release of large size chunks. In a later publication Weertman (57) also assessed the possibility of creating microcracks by the energy supplied by the collision cascades. His calculations showed that the growth of an existing crack in a stress free surface would be very limited if one assumed that the energy needed for crack growth would be supplied by the collision cascades alone.

Guinan (40), however, pointed out that a collision cascade resulting from a single 14.1-MeV neutron collision event can produce large enough local stresses and nucleate a penny-shaped crack of the size of the cascade by fulfilling the Griffith criterion:

$$\sigma^2 L = C \qquad (9)$$

where $\sigma$ is the tensile stress, $L$ is the crack length, and $C$ is a constant. The cascade size will depend on the PKA energy, and in niobium for PKA energies above 150 keV, the predicted cascade volume is $\sim 60\%$ of a sphere with diameter $R_p$ (i.e., projected range $R_p$ of PKA). Guinan (40) showed that if such a penny-crack is nucleated near (within 100–200 Å) the head of a dislocation group having a long range stress field, (e.g., unstressed single-ended dislocation pile-ups or a dislocation climb configuration at a discontinuous subboundary) the crack can grow almost as long as the size of the dislocation group, provided that the Griffith criterion is initially satisfied. In other words, the collision cascade provides a trigger for the release of stored internal energy, as was speculated earlier (15). Figure 5 illustrates schematically how crack growth can occur in the presence of an unstressed single dislocation pileup and lead to chunk emission, for example, from a protrusion. In cold-rolled samples the presence of macroscopic tensile stresses can help to extend the crack considerable distances towards the surface. Thus, the size of the chunk will vary depending on where the crack nucleated in a protrusion on the surface (Figure 5b). More recently, Robinson, et al. (58), have extended Guinan's model to provide arguments for the dependence of chunk emission on protrusions or steps on the surface. They also suggested oxide intrusions as an alternative source for large internal stresses.

The development of shock waves and thermal stresses will depend strongly on the damage energy density (i.e., the damage energy/unit volume) of the cascade. Guinan (40) estimated the magnitude of radial and tangential stresses as a function of cascade radius for different initial damage energy densities. From linear elastic analysis it was shown that the combination of developing thermal stresses and outgoing shock waves could produce a penny-crack with a diameter slightly larger than the cascade for energy densities $> 0.5$ eV/atom. In the presence of externally aided stresses the required energy density reduces to $\sim 0.25$ eV/atom. Guinan (40) showed that such damage energy densities can indeed be obtained readily in the case of 14.1-MeV neutron irradiation of niobium (e.g., for a PKA energy of 100 keV, the energy density is 0.33 eV/atom for a random collision cascade).

In order to account for the more typical number of chunks, for example 30/cm² observed for cold-rolled niobium in Table I, a minimum of 30 dislocation groups/cm² would be required within approximately 3 $\mu$m of the surface (taking a 3-$\mu$m diameter chunk). A density of $1 \times 10^5$ groups/cm³ containing a total dislocation density of $4 \times 10^2$ lines/cm² would be adequate. Assuming 40 dislocations/group with a length of each dislocation of 1 $\mu$m, a dislocation spacing of 250 Å gives a length of 1 $\mu$m for the dislocation group. Such dislocation density appears highly plausible, since it represents only a small fraction ($4 \times 10^{-9}$) of the total dislocation density of $\sim 10^{11}$ lines/cm$^2$ in cold-rolled niobium and a fraction of ($\sim 4 \times 10^{-6}$) of the dislocation density of $\sim 10^8$ lines/cm$^2$ for annealed niobium.

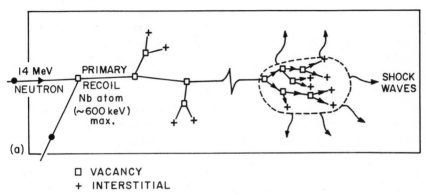

□ VACANCY
+ INTERSTITIAL

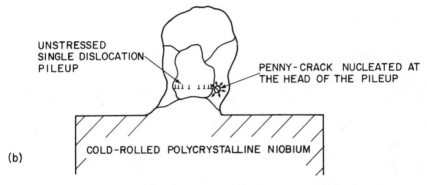

*Figure 5. Schematic diagrams showing (a) generation of shock waves in a collision cascade formed near the end of the range of PKA and (b) nucleation of a penny-crack at the head of an unstressed single dislocation pileup present in a protrusion on the niobium surface. The penny-crack can grow, and the protrusion can be ejected.*

Another important consideration deals with the momentum required for a chunk to reach the collector surface. Guinan (40) pointed out that the energy stored in a dislocation group could be released as surface free energy during crack propagation, and as much as one fourth of the stored energy could be converted directly to kinetic energy normal to the direction of the crack. For a crack covering an area of $10^{-8}$ cm$^2$, this could amount to a momentum transfer of $10^{-8}$ g cm/sec. For a niobium chunk of 1-$\mu$m diameter this would give a velocity of $2.2 \times 10^3$ cm/sec—a velocity which is several orders of magnitude higher than the one estimated from experimental observations ($\lesssim 10$ cm/sec) (16). In other words, there is more than enough momentum available for the transfer of the chunk from the target to the collector.

For the same type of cold-rolled and rough niobium surfaces (e.g., target types A and B in Table I), chunk emission should also be observable under ion irradiation which simulates the PKA spectrum of 14.1-MeV neutrons for the higher PKA energy region. The PKA spectrum induced by 16.4-MeV protons in niobium is very similar to that for 14.1-MeV neutrons for the high energy region ($E \lesssim 80$ keV) of the PKA spectrum (60).

Attempts have been made by Logan et al. (61) and Robinson et al. (62) to observe chunk emission from niobium under 16-MeV protons. These authors failed to observe chunk emission, but the proton sputtering yields for niobium reported by them were $\leq 10^{-4}$ atom/proton, which are close to the lower limit values of the ranges of 14.1-MeV neutron sputtering yields given in Table III. To what extent differences in irradiation conditions and detection techniques in the proton sputtering experiments and in the 14.1-MeV neutron sputtering experiments in which chunk emission was observed are responsible for the difference in the results in the two types of experiments is difficult to assess at this time.

In this connection the emission of atom clusters and chunks from uranium and uranium dioxide, respectively, during passage of fission fragments which can form during irradiation with thermal neutrons should be mentioned. Rogers and Adam (63) and Rogers (64, 65, 66) observed atom clusters containing up to a few thousand atoms ejected by the fission fragments from uranium metal and thin films of $UO_2$. Subsequently, Nilsson (67, 68) reported ejection of uranium atoms by the fission fragments from sintered $UO_2$ and electropolished uranium metal, but he did not observe clusters. Later on Verghese and Piascik (69) provided evidence for ejection of uranium atom clusters by fission fragments from uranium metal. More recently Biersack et al. (70) reported ejection of chunks of $\sim 0.5$ $\mu$m in diameter from a $UO_2$ film by fission fragments. Blewitt et al. (71) investigated sputtering by fission fragments of niobium doped with 0.1 at % $^{235}U$ and irradiated with thermal neutrons. These authors did not observe ejection of particles with sizes larger than 0.1 $\mu$m. They feel that with their detection techniques, they should have been able to detect particles of such sizes. It has been argued by Blewitt et al. (71) that since the energy of the fission fragments (167 MeV divided between two fission fragments) and the number of fission events is higher in their experiments than the corresponding maximum PKA energy and the number of primary knock-on events during 14.1-MeV neutron irradiation of niobium, one should be able to observe emission of niobium chunks in their experiments. However, Guinan pointed out (40) that it is not the initial energy of the primary recoil, but the damage energy density (energy deposited into damage/unit volume) that is important for nucleation of cracks and subsequent chunk emission. For the case of fission fragments from uranium moving in niobium, preliminary estimates by Guinan (72) indicate that the energy deposited into damage is ~ 7 MeV, and that only towards the end of the range of the fission fragment the niobium recoil atom spectrum is similar to that induced by 14.1-MeV neutron irradiation of niobium. In the initial portion of the path of the fission fragment most of the energy loss will result from electronic excitation and low angle collisions without causing any displacements. Some recent calculations by Marwick (73) for 33-MeV niobium (its mass is comparable with that of one of the fission fragments) irradiation of niobium, the PKA spectrum near the end of the range (at depth of 4 $\mu$m, the range of 33-MeV niobium ion in niobium being ~ 4.4 $\mu$m) is very similar to that expected from 14.1-MeV neutron irradiation of niobium for high recoil energies (> 10 keV). Because these recoil spectra are similar only near the end of the range of such heavy projectiles in niobium, one can expect damage energy densities comparable only in this range with those formed by 14.1-MeV neutrons.

### Conclusions

The upper limit values of 14.1-MeV neutron sputtering yields for gold and niobium surfaces of different microstructures quoted by several authors are about one to two orders of magnitude ( ~ $10^{-4}$–$10^{-3}$ atom/neutron) larger than calculated values ( ~ $10^{-5}$ atom/neutron) based on theories using the concept of displacement collision cascades. In turn, several of the lower limit values are of the same order of magnitude as such theoretical estimates. These findings indicate the existing uncertainty in the experimentally determined yields and the difficulty in assessing the validity of the theoretical estimates. The yield values reported for fission neutron sputtering of gold and iron surfaces with different microstructures are also one to two orders of magnitude higher than the theoretically estimated ones. Some of the earlier reported yields with values of >> $10^{-3}$ atom/neutron for both fission and (d, t)-neutron sputtering may have suffered from contamination of the collectors with the target material.

In recent studies of 14.1-MeV neutron sputtering of cold-rolled and annealed niobium surfaces with coarse surface finishes (technical surfaces), two types of deposits were discovered on collector surfaces facing the irradiated targets. One type appeared as a fractional atom layer covering the surface, the other in the form of chunks. The observation that the number of chunks/unit collector area and the average chunk size was larger for cold-rolled samples than for annealed ones and that no chunks were observed for a monocrystalline sample suggested that stresses in the surface regions affected chunk emission. In addition, a comparison of the sputtering results for samples with different surface roughness revealed that those with a coarser surface finish emitted more chunks. The chunk emission appears to be a relatively rare phenomena compared with the total number of neutron events occurring in near-surface regions. The contribution of chunks to the total neutron sputtering yield can range from being insignificant to being larger than the contribution from atomic layer deposits.

Chunk emission has been observed by several different authors for 14.1-MeV neutron irradiation of rough and stressed niobium surfaces. However, it has not been observed by other authors using either 14.1-MeV neutrons, neutrons from the $(d, \text{Be})$-reaction, or 16-MeV protons for niobium surfaces with similar or different microstructures. A model for the emission of chunks has been offered by Guinan. He showed, that stressed surfaces which contain sufficient concentrations of certain types of defects (e.g., unstressed single-ended dislocation pile-ups) can emit chunks by triggering the release of stored internal energy by propagation of penny-cracks nucleated by collision cascades at the head of a dilocation group. From the fluence levels used in the experiments in which chunk emission was observed and from an estimate of such defect densities in both cold-rolled and annealed niobium, the observed number of chunks/unit area appears plausible according to Guinan. The observation of no chunk emission from monocrystalline niobium targets appears plausible on the basis of this model. The contribution of neutron sputtering to the surface erosion of fusion reactor components is not considered to be significant, but the effect of chunk emission on plasma contamination needs to be assessed.

*Acknowledgments*

We are grateful to F. Cafasso and R. Meisenheimer for their permission to quote their results prior to publication. We would also like to thank North-Holland Publishing Co., Amsterdam and IEEE Nuclear and Plasma Science Society, New York for their permission to use certain illustrations in this manuscript.

*Literature Cited*

1. Behrisch, R., *Ergeb. Exakten. Naturwiss.* (1964) **35**, 295.
2. Kaminsky, M., "Atomic and Ionic Impact Phenomena on Metal Surfaces," Springer-Verlag, New York, 1965.

3. Carter, G., Colligon, J., "Ion Bombardment of Solids," American Elsevier, New York, 1968.
4. Nelson, R. S., "The Observation of Atomic Collisions in Crystalline Solids," North-Holland, Amsterdam, 1968.
5. Sigmund, P., *Rev. Roum. Phys.* (1972) **17**, 1079.
6. Dearnaley, G., Freeman, J. H., Nelson, R. S., Stephen, J., "Ion Implantation," Chap. 3, North-Holland, Amsterdam, 1973.
7. McCracken, G. M., *Rep. Prog. Phys.* (1975) **38**, 241.
8. Robinson, M. T., *Proc. Br. Nucl. Energy Soc. Eur. Conf. Nucl. Fusion Reactors* (1969) 364.
9. Lindhard, J., Nielsen, V., Scharff, M., *Mat. Fys. Medd. Dan. Vid. Selsk.* (1968) **36** (10).
10. Kinchin, G., Pease, R., *Rep. Prog. Phys.* (1955) **18**, 1.
11. Kaminsky, M., *Proc. Int. Work. Sess. Fusion Reactor Technol.* (1971) **CONF-710624**, 86.
12. Kaminsky, M., *IEEE Trans. Nucl. Sci.* (1971) **18**, 208.
13. Kaminsky, M., *Plasma Phys. Controlled Nucl. Fusion Res., Proc. Inf. Conf.* (1975) **2**, 287.
14. Behrisch, R., Kadomstev, B. B., *Plasma Phys. Controlled Nucl. Fusion Res. Proc. Int. Conf.* (1975) **2**, 229.
15. Kaminsky, M., Peavey, J. H., Das, S. K., *Phys. Rev. Lett.* (1974) **32**, 599.
16. Kaminsky, M., Das, S. K., "Surface Effects in Controlled Fusion," H. Wiedersich, M. Kaminsky, and K. M. Zwilsky, Eds., North-Holland (1974) 162; also *J. Nucl. Mater.* (1974) **53**, 162.
17. Harling, O., Thomas, M. T., Brodzinski, R. C., Ranticelli, L. A., *Phys. Rev. Lett.* (1975) **34**, 1340.
18. Behrisch, R., Gähler, R., Kalus, J., *J. Nucl. Mater.* (1974) **53**, 183.
19. Goldberg, E., Griffith, R., Logan, C., *Lawrence Livermore Lab. Rep.* (1972) **UCRL-51317**.
20. VanKonynenburg, R. A., *Lawrence Livermore Lab. Rep.* (1973) **UCRL-51393**.
21. Norcross, D. W., Fairand, B. P., Anno, J. N., *J. Appl. Phys.* (1966) **37**, 621.
22. Keller, K., Lee, Jr., R. V., *J. Appl. Phys.* (1966) **37**, 1890.
23. Keller, K., *Plasma Phys.* (1968) **10**, 195.
24. Verghese, K., *Trans. Am. Nucl. Soc.* (1969) **12**, 544.
25. Kirk, M. A., Blewitt, T. H., Klank, A. C., Scott, T. L., Malewicki, R., *J. Nucl. Mater.* (1974) **53**, 179.
26. Jenkins, L. H., Noggle, T. S., Reed, R. E., Saltmarsh, M. J., Smith, G. J., *Appl. Phys. Lett.* (1975) **26**, 426.
27. Kaminsky, M., Das. S. K., Proceedings of the Fifth Symposium on Engineering Problems of Fusion Research, *IEEE Pub.* (1974) No. **73 CHO 843-3-NPS**, 37.
28. Kaminsky, M., Das, S. K., *Proc. Top. Meet. Technol. Controlled Nucl. Fusion 1st* (1974) USAEC Conference Series **CONF-74002**, 508.
29. Garber, R. I., Karasev, V. S., Kolyada, V. M., Fedorenko, A. I., *Sov. At. Energy* (1970) **28**, 510 [English translation of *At. Energ.* (1970 **28**, 400].
30. Baer, T. S., Anno, J. N., *J. Appl. Phys.* (1972) **43**, 2453.
31. Baer, T. S., Anno, J. N., *J. Nucl. Mater.* (1974) **54**, 79.
32. Kirk, M., Conner, Jr., R. A., Proceedings of International Conference on Fundamental Aspects of Radiation Damage in Metals (1976) **CONF-751006-PI**, National Technical Information Service, Springfield, Va., 171.
33. Garber, R. I., Dolya, G. P., Kolyada, V. M., Modlin, A. A., Fedorenko, A. I., *JETP Lett.* (1968) **7**, 296 [English translation of *Pis'ma Zh. Eksp. Teor. Fiz.* (1968) **7**, 375].
34. Kaminsky, M., Das, S. K., *J. Nucl. Mater.* (1976) **60**, 111.

35. Kaminsky, M., Das, S. K., Proceedings of the Sixth Symposium on Engineering Problems of Fusion Research, *IEEE Pb.* No. **75CH1097-5-NPS** (1976) 1141.
36. Harling, O. K., Thomas, M. T., Brodzinski, R., Rancitelli, L., Proceedings of the Third Conference on Application of Small Accelerators, USERDA Conference Series, Vol. 1 (1975) **CONF-741040-Pl**, 298.
37. Cafasso, F., Dudey, N., Johnson, C. E., Argonne National Laboratory, private communication.
38. Meisenheimer, R., private communication and *Lawrence Livermore Lab. Rep.* (1976) **UCRL-78322**.
39. Booth, R., Barschall, H. H., *Nucl. Instrum. Methods* (1972) **99**, 1.
40. Guinan, M., *J. Nucl. Mater.* (1974) **53**, 171.
41. Meisenheimer, R., *Lawrence Livermore Lab. Rep.* (1976) **UCRL-78396**.
42. Jenkins, L. H. et al., *J. Nucl. Mater.* (1976) **63**, accepted for publication.
43. Taimuty, S. I., *Nucl. Sci. Eng.* (1961) **10**, 403.
44. Sigmund, P., *Phys. Rev.* (1969) **184**, 383.
45. Robinson, M. T., *J. Nucl. Mater.* (1974) **53**, 201.
46. Sigmund, P., *J. Mater. Sci.* (1973) **8**, 1545.
47. Goldman, D. T., Simon, A., *Phys. Rev.* (1959) **111**, 383.
48. Brandt, W., Laubert, R., *Nucl. Instrum. Methods* (1967) **47**, 201.
49. Parkin, D. M., Goland, A. N., *Brookhaven Nat. Lab. Rep.* (1974) **BNL-50434**.
50. Parkin, D. M., Goland, A. N., Berry, H. C., *Proc. Top. Meet. Technol. Controlled Fus.*, *1st* (1974) Vol. 1, USAEC Conference Series, **CONF-740402-Pl**, 339.
51. Gschneider, K., *Solid State Phys.* (1964) **16**, 344.
52. Kaminsky, M., *Proceed. Symp. Fusion Technol.*, *7th* (1972) Report No. **EUR 4938 e**, 41.
53. Pease, R., Rendiconti, S. I. F., *Corso* (1960) **13**, 158.
54. Thompson, M. W., "Defects and Radiation Damage in Metals," p. 245, Cambridge University, London, 1969.
55. Beeler, J., North Carolina State University, private communication, 1973.
56. Robinson, J. E., Argonne National Laboratory, private communication, 1973.
57. Weertman, J., *J. Nucl. Mater.* (1975) **55**, 253.
58. Robinson, J. E., et al., *J. Nucl. Mater.*, in press.
59. Das, S. K., Kaminsky, M., *J. Appl. Phys.* (1973) **44**, 25.
60. Logan, C. M., Anderson, J. D., Mukherjee, A. K., *J. Nucl. Mater.* (1973) **48**, 223.
61. Logan, C. M., *J. Vac. Sci. Technol.* (1975) **12**, 536.
62. Robinson, J. E., Thompson, D. A., *Phys. Rev. Lett.* (1974) **33**, 1569.
63. Rogers, M. D., Adam, J., *J. Nucl. Mater.* (1962) **6**, 182.
64. Rogers, M. D., *J. Nucl. Mater.* (1964) **12**, 332.
65. *Ibid.* (1965) **15**, 65.
66. *Ibid.* (1965) **16**, 298.
67. Nilsson, G., *J. Nucl. Mater.* (1966) **20**, 215.
68. *Ibid.*, **20**, 231.
69. Verghese, K., Piascik, R. S., *J. Appl. Phys.* (1969) **40**, 1976.
70. Biersack, J. P., Fink, D., Mertens, P., *J. Nucl. Mater.* (1974) **53**, 194.
71. Blewitt, T. H., Kirk, M. A., Busch, D. E., Klank, A. C., Scott, T. L., *J. Nucl. Mater.* (1974) **53**, 189.
72. Guinan, M., private communication (1976).
73. Marwick, A. D., *J. Nucl. Mater.* (1974) **56**, 355.

RECEIVED June 16, 1976. Work performed under the auspices of the U.S. Energy Research and Development Administration.

# 4

# Helium and Hydrogen Implantation of Vitreous Silica and Graphite

G. J. THOMAS, W. BAUER, and P. L. MATTERN

Sandia Laboratories, Livermore, Calif. 94550

B. GRANOFF

Sandia Laboratories, Albuquerque, N.M. 87115

*The effects of 150–300-keV helium and hydrogen implantation have been studied by a number of techniques—optical and scanning electron microscopy, gas reemission measurements, and microtopography—on two materials which have potential CTR applications, graphite and vitreous silica. In graphite, changes in surface structure were observed over the entire implant temperature range of −160° to 1200°C. At higher temperatures, there was evidence for hydrogen–carbon chemical reaction. Vitreous silica was stable under helium and hydrogen implantation, with no surface deformation at any implant temperature from ~ 100° to ~ 800°C. Volume compaction measurements were made at high dose levels, and evidence was found for chemical bonding during hydrogen implantation.*

Controlled thermonuclear reactor (CTR) components, including the first wall assembly, limiters, and divertors, will be subjected to bombardment by helium and hydrogen isotope atoms. Because of the limited range of these particles, only the surface and near-surface regions of exposed materials will be affected. In metals(1) gas implantation causes blistering and other surface deformation which can lead to rapid surface erosion and plasma contamination. This phenomenon is related to the insolubility of helium or low solubility of hydrogen and the radiation-induced production of point defects.

Nonmetallic materials such as glasses, ceramics, and graphite will also be used in several CTR applications. For example, diagnostic feedthrough assemblies and optical ports will require insulating or transparent substances which can withstand the radiation environment. As an extreme example, the theta-pinch reactor will use an insulating first wall which must survive direct exposure to the fusion plasma (2). Low atomic number materials such as graphite (3) and SiC (4) have recently been proposed for use in Tokamak reactors to isolate metallic first wall structures from the plasma.

In this paper, recent experimental results on helium and hydrogen implantation of graphite and vitreous silica are discussed. These materials were studied because they represent two distinct types of nonmetals which may be used as conducting (graphite) and insulating (glass) liners. Gas reemission measurements, microtopography, and scanning electron microscopy (SEM) were used to examine implanted samples of bulk, composite, and oriented pyrolytic graphites and several types of commercial grade and high purity synthetic vitreous silica. In contrast to metals, the effects of particle bombardment appear to be complex in these materials. In graphite the formation of blisters is inhibited by the reemission of implanted gas by processes which may be determined by porosity and other structural features. Surface erosion, however, may be enhanced by beam-induced chemical reactions which produce volatile hydrocarbons.

Blistering has been observed in a number of ceramic materials including MgO, spinel, and sapphire following $H^+$ and $He^+$ irradiation at room temperature (5). In these studies interference microscopy was used to deduce blister cross sections and exfoliation depths. More recently (6), scanning electron microscopy techniques have shown blistering in thin coatings of barium aluminosilicate glass following 100-keV and 250-keV $He^+$ implantation at room temperature. However, vitreous silica has shown no indication of beam-induced blistering (7), presumably because of the relatively high values of solubility and diffusivity of hydrogen and helium. Although surface deformation was not observed, near-surface properties were affected by an increase in density (radiation-induced compaction) (8, 9, 10, 11) and the production of high concentrations of hydroxyl (7).

In the next section, the experimental results of helium and hydrogen implantation of graphite are presented, and in the following section the silica data are described.

*Graphite*

**Low Temperature Implantations.** Graphite can be fabricated in many forms. Some physical and crystallographic properties which were measured on the three different graphite materials used in the present study are summarized in Table I. Briefly, ATJ is a molded, fine grain

**Table I.  Properties of Graphite Samples**

| Material | Density (g/cm³) | Porosity (%) | Permeability (Darcy) | Basal Interplanar Spacing (Å) | Crystal Length (Å) |
|----------|-----------------|--------------|----------------------|-------------------------------|--------------------|
| ATJ | 1.74 | 14 | 0.015 | 3.371 | 200 |
| STC-10 | 1.73 | 16 | 0.05–0.1 | 3.381 | 335 |
| Pyroid | 2.25 | ∼ 0.5 | ⩽ 10⁻⁴ | 3.358 | 250 |

material fabricated by Union Carbide, the STC-10 material is a composite of carbon-felt and pyrolytic carbon matrix, and the pyroid is a highly oriented bulk pyrolytic graphite. Both basal-oriented and edge-oriented (basal planes perpendicular to the surface) samples of pyroid were implanted. From Table I it is seen that the physical parameters of the ATJ and STC-10 materials are similar in most respects. The higher permeability of STC-10 indicates somewhat greater interconnection of voids. The pyrolytic material closely approaches single crystal graphite in density (2.266 maximum density) and interplanar spacing (3.354 Å minimum $d_{002}$).

Gas reemission and surface deformation during 300-keV helium implantation of these materials were examined at implantation tempera-

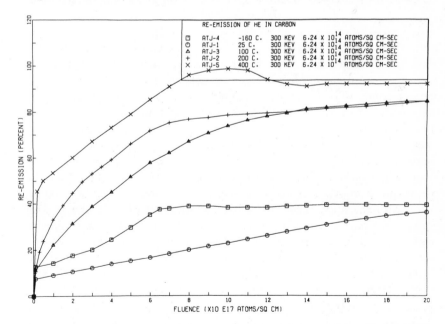

*Figure 1. Helium reemission as a function of dose for various implantation temperatures in ATJ graphite. Beam energy was 300 keV, and a flux of 6 × 10¹⁴ He⁺/cm² sec was used in all cases.*

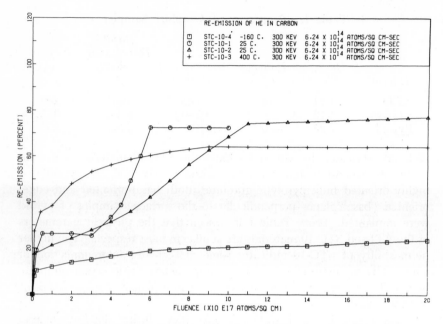

*Figure 2. Helium reemission as a function of dose for several implantation temperatures in STC-10 graphite. The same beam parameters were used as for ATJ, Figure 1.*

tures ranging from $-160°$ to $1200°C$. The low temperature ($< 400°C$) data are discussed first. The experimental arrangement for helium reemission measurements has been described previously (1). Reemission results from ATJ at sample temperatures from $-160°$ to $400°C$ are shown in Figure 1. These plots represent the fraction of the incoming beam which is not retained during implantation. The behavior of this material is in sharp contrast to the characteristic helium reemission found in most metals in this temperature range. Generally, the initial reemission in metals is quite low until a critical dose is reached, at which time the reemission abruptly increases. This onset of gas release has been correlated with the appearance of surface deformation (1). In the present case, however, the reemission smoothly increases with dose, with greater reemission at higher implantation temperatures. By $400°C$, the steady state reemission is approximately 100%, indicating that the flux of gas leaving the sample is equal to the implanted flux. A similar smoothly varying behavior with increasing dose was found in STC-10, shown in Figure 2. However, the reemission at $25°C$ is greater in STC-10 than ATJ and does not increase significantly at $400°C$. The low temperature behavior ($- 160°C$) in the ATJ appears to be anomalously high and is not understood at this time.

The surface structural changes following implantation in both ATJ and STC-10 samples were small and nearly identical. Scanning electron micrographs of as-polished STC-10 and samples implanted at $-160°$, $25°$, and $400°C$ are shown in Figure 3. No gross structural changes could be discerned at any implant temperature after doses to $2 \times 10^{18}$ He/cm², although a slight surface roughening may have occurred. Measurements of the surface topography by a Gould microtopographer, a stylus instrument used in the compaction measurements described in the next section,

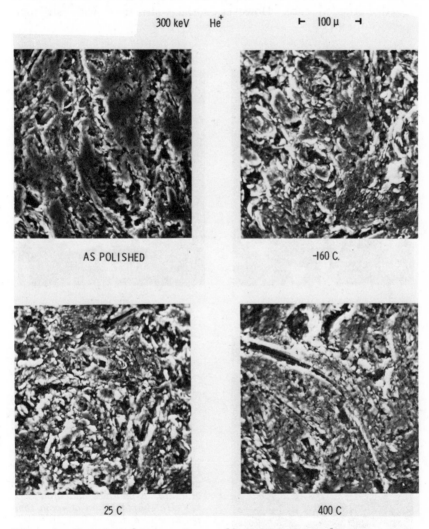

*Figure 3. Scanning electron micrographs of STC-10 graphite prior to implantation and after implantation at $-160°$, $25°$, and $400°C$ to a dose of $2 \times 10^{18}$ He⁺/cm². All micrographs are at the indicated scale.*

showed that the implanted region was raised slightly ($\sim 0.5\ \mu m$) relative
to the unimplanted surface. Thus any surface erosion or sputtering
which may have occurred was more than offset by a volume expansion.
The lack of blisters or flaking in these materials is consistent with the
smooth reemission characteristics.

The pyroid graphite exhibited similar smooth reemission behavior
at room temperature, with a high dose ($2 \times 10^{18}$ He/cm$^2$) value of
$\sim 10\%$ reemission for the basal orientation and $\sim 40\%$ reemission in
the edge-oriented case. However, surface structures were strikingly

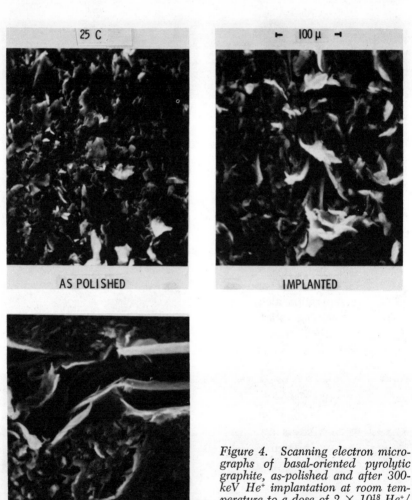

*Figure 4. Scanning electron micro-
graphs of basal-oriented pyrolytic
graphite, as-polished and after 300-
keV He$^+$ implantation at room temperature to a dose of $2 \times 10^{18}$ He$^+$/
cm$^2$. All micrographs are at the
same scale. Two regions of the
implanted area indicate the observed
variation in peeled layer size.*

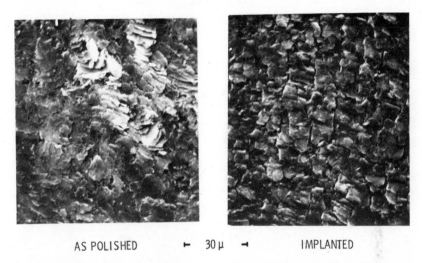

AS POLISHED     ⊢ 30 μ ⊣     IMPLANTED

*Figure 5. Scanning electron micrographs of edge-oriented pyrolytic graph-ite, as-polished and after 300-keV $He^+$ implantation at room temperature to a dose of $2 \times 10^{18} He^+/cm^2$*

different, as seen in Figures 4 and 5. Different magnifications were used in both sets of micrographs. The edge-oriented material contained a series of parallel cracks running perpendicular to the surface, whereas flaking was produced in the basal-oriented materials. The thicknesses of some of the peeled layers were as much as an order of magnitude greater than the expected penetration range of 300-keV $He^+$ ions. It appears from the SEM and reemission results that deformation occurs mainly by separation along basal planes and is not directly associated with helium agglomeration into bubbles.

SEM surface observations were also made during implantation using an in-situ irradiation facility (*12*). In all cases, changes occurred slowly with increasing dose, rather than abruptly as in the case of metals. No evidence was found for emission of particles (e.g., blister lids) from ATJ or STC-10, as might be expected from results on brittle materials (*12*). Similarly, the peeling in basal-oriented graphite occurred gradually, suggesting that stress accumulation or a volume change, rather than accumulated gas pressure, was responsible for the deformation. Several woven graphite fibers also were examined in the in-situ facility, but these broke readily from heating effects upon exposure to the ion beam.

**High Temperature Implantations.** Chemical effects appear to play an increasingly important role during helium and hydrogen implantation at 800°C and above. Before describing these results, however, certain negative aspects of the presence of graphite in high vacuum systems,

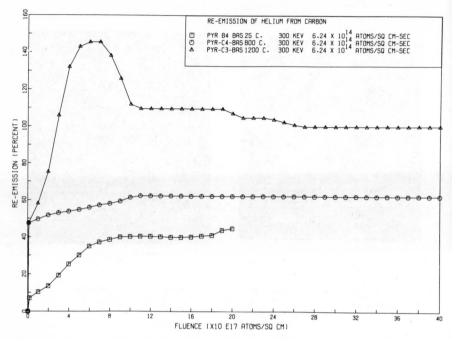

*Figure 6. Helium reemission as a function of dose for three implantation temperatures in basal-oriented pyrolytic graphite. The same beam parameters were used as described for Figure 1.*

particularly at high temperatures, should be mentioned. It was found that carbonaceous materials were released into the vacuum during high temperature irradiations causing a large increase in residual gas pressure and impairing the utility of the system. In fact, subsequent implantation of glass in the same chamber used in these experiments resulted in a thick ($\sim$ 10 nm) layer of carbon contamination on the surface. Clearly, the possibility of contamination must be considered if graphite materials are used in CTR systems.

The helium reemission from basal-oriented pyrolytic graphite for three different implant temperatures is plotted in Figure 6. At 800°C, the reemission rises immediately and remains essentially constant to a dose of $4 \times 10^{18}$ He/cm². The 1200°C data, however, exhibit a well defined peak which exceeds 100% reemission prior to attaining a 100% steady state value. This indicates that some of the helium retained at the beginning of the implantation is released at a dose level centered at about $5 \times 10^{17}$ He/cm².

SEM examination of the surface after helium implantation at 1200°C reveals the structure shown in Figure 7. The white features could be interpreted as evidence for blistering. However, they appear to emanate predominantly from existing voids or cracks and may be growth features or redeposition of volatilized material.

In addition to determining helium reemission, the residual gas spectrum was monitored up to mass 60 during helium implantation of pyroid at 1200°C. A portion of a spectrum in Figure 8 shows the complexity of the ambient vacuum near irradiated graphite. Analysis of the peaks indicates that in addition to gases normally found in vacuum systems, significant amounts of methane and its derivatives as well as propane were present. Similar measurements were made during $H^+$ implantation at 470°, 640°, and 850°C. Qualitatively, the spectra were similar, except for the helium peak, to that shown in Figure 8. Absolute determinations of the hydrogen reemission under these conditions are extremely difficult. However, crude estimates of the relative amounts of some components were made by comparing residual gas spectra obtained with the $H^+$ beam on the graphite and on a heated Mo flag. At low temperatures ($\sim$ 450°C), some hydrogen was retained, while at intermediate temperatures ($\sim$ 650°C), significant amounts of methane and its derivatives were produced in addition to molecular hydrogen release. At higher temperatures ($> 800°C$), much of the hydrogen was released in the form of propane and its derivatives. It appears, therefore, that radiation-enhanced chemical reactivity was important at temperatures between 800° and 1200°C with the implantation conditions used during these experiments.

*Figure 7. Scanning electron micrograph of the surface of pyrolytic graphite after 300-keV $He^+$ implantation at 1200°C to a dose of $4 \times 10^{18}$ $He^+/cm^2$*

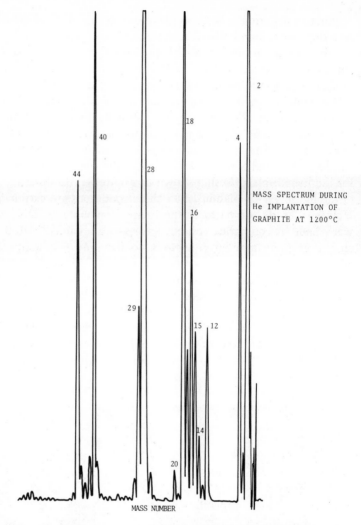

Figure 8.  Residual gas spectrum obtained during helium implantation of pyrolytic graphite. The mass number of selected peaks are indicated. The vertical scale is not calibrated.

## Implantations in Vitreous Silica

The effects of energetic (150–300-keV) He⁺ and H⁺ implantation on vitreous silica have been examined as a function of fluence ($\leq 2 \times 10^{19}/cm^2$) on several varieties of natural and synthetic vitreous silica. A hot filament electron source flooded the sample surface during implantation to enhance conductivity. Irradiated samples were studied using optical and scanning microscopy, gas reemission, and microtopography.

Sample temperatures at the implanted surface were monitored by an infrared thermometer designed to operate near 5 $\mu$m. At this wavelength vitreous silica is sufficiently opaque and the sapphire window in the sample chamber is sufficiently transparent for accurate measurements. Implanted region temperatures ranged from $\sim 100°$ to $\sim 1000°$C.

One of the most striking features found in all types of vitreous silica is the complete lack of surface deformation attributable to gas agglomeration. No indication of surface deformation attributable to implantation was found after doses as much as two orders of magnitude greater than necessary to cause blistering and exfoliation in a variety of structural and refractory metals. These results are in direct contrast to the helium blistering found by Kaminsky et al. (6) on barium aluminosilicate glass and may be accounted for by the relatively large values of helium diffusivity in vitreous silica. Shelby (13) has measured helium permeation in both materials in the temperature range $\sim 200–300°$C. Extrapolation of his data to lower temperatures indicates that the permeation rate may be as much as five orders of magnitude greater in the silica than in the barium aluminosilicate glass at the lowest implant temperature. In addition, his data indicate that the difference in permeation rate is greatly reduced at $300°$C, lending support to the interpretation by Kaminsky et al. (6) concerning reduced blister densities at higher temperature.

Prior to implantation, samples $\sim 1–2$ mm thick were mechanically polished to an optically flat finish. This preparation enabled the topography of the implanted surface to be measured with a Gould microtopographer, a stylus instrument which can detect surface variations as small as $\sim 10$ nm. By means of an automated drive system, repeated scans can be made to produce a complete topographical map of a sample surface. An example is shown in Figure 9, obtained from a sample implanted with $1 \times 10^{18}$ 300-keV $H^+/cm^2$. The slight circular indentation corresponds to the implanted region while the deep ridges in the upper right hand corner indicate a fiducial mark scribed onto the sample surface. As indicated on the figure, the implanted region is depressed by about 40 nm. A number of samples were implanted with 150-keV and 300-keV $H^+$ and $He^+$ to doses ranging from $1 \times 10^{16}/cm^2$ to $1 \times 10^{18}/cm^2$ at a temperature of $\sim 150°$C.

The observed depressions can be caused by two processes—radiation compaction and physical sputtering. The magnitude of the depressions as a function of dose and particle energy cannot be accounted for satisfactorily with present models for physical sputtering. Primarily, the sputtering ratios necessary to describe the data are approximately two or three orders of magnitude larger than are reasonable. It is believed, instead, that the depressions result from a volume compaction of the

HORIZONTAL:  X20        VERTICAL:  0.1 inch = 130 nm

**40 nm**

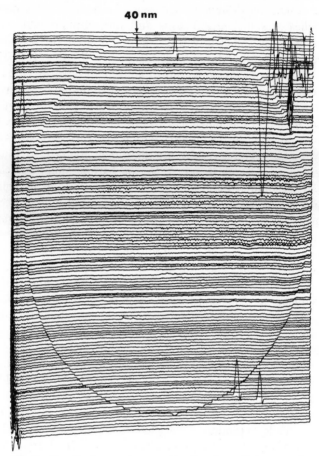

*Figure 9. Surface microtopograph of an $SiO_2$ sample implanted with $1 \times 10^{18}$ 300-keV $H^+/cm^2$. The implanted area corresponds to the circular indentation, and the feature in the upper right corner is a fiducial mark. The depth of the indentation is indicated.*

silica throughout the implanted layer. Primak (*10*) and co-workers have studied volume compaction in silica for several years, beginning as early as 1953. Recently, EerNisse and Norris (*11*) have measured compaction in silica at low doses using ions and at high doses using electron irradiation. In these studies, volume change was calculated from the magnitude of induced lateral stress, measured by a cantilever beam technique.

In Figure 10, our high dose data points determined by microtopography are plotted with results published by Primak (*10, 14*) and EerNisse (*11*). The present values of $\Delta V/V$ and deposited energy density were calculated using LSS range values. $\Delta V/V$ was set equal to $\Delta l/l$, where $\Delta l$ is the measured indentation, and $l$ is the calculated ion range. For consistency with previous analyses (*10, 11*), the $H^+$ data are plotted vs. ionization dose, whereas the $He^+$ results are given as a function of dose into atomic processes. Our measured compaction, plotted in Figure 10, agrees well with the saturation value of isotropic compaction observed in vitreous silica after an extensive neutron irradiation (*12*). The compaction after $H^+$ implantation is consistently less than that produced after $He^+$ implantation for comparable particle fluences. This effect may result from alteration of the silica network by the formation of a large concentration of hydroxyl, a point which will be discussed below. It should be mentioned, however, that somewhat anomalous results have been reported (*8, 9*) on $H^+$-implanted samples and that the low fluence hydrogen implant data of Figure 10 (at $10^{22}$ keV/cm$^3$) appear to differ from the electron irradiation curve by more than reasonable experimental error. Additional measurements are necessary to resolve these discrepancies.

Gas reemission measurements were also made during $He^+$ and $H^+$ implantations. At temperatures $< 200°C$, the $He^+$ reemission rose to

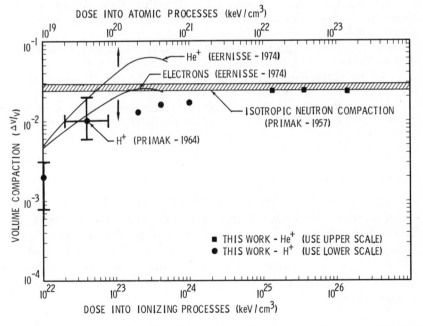

*Figure 10. Plot of volume compaction $\Delta V/V$ as a function of deposited energy density in vitreous silica. The solid curves are from EerNisse and Norris, while the present helium and hydrogen results are plotted as data points. The vertical bar on the hydrogen implant at $10^{22}$ keV/cm$^3$ indicates the estimated uncertainty for that point only.*

essentially 100% release within the response time of the system ($\sim$ 1 sec). These data indicate a lower limit to helium diffusivity in this material of $D \sim 10^{-8}$ cm$^2$/sec, consistent with permeation data. Estimates of hydrogen diffusivity from the reemission are also in substantial agreement with hydrogen permeability values. In the case of H$^+$ implantation to a fluence of $2 \times 10^{18}$ H$^+$/cm$^2$ at temperatures $\sim$ 100°C, a small but experimentally significant amount of gas retention ($\sim$ 3%) was observed. The quantity of retained hydrogen agrees with the observation of OH formation in the implanted layer by optical absorption measurements (7, 8, 9). Moreover, in a series of irradiations, the growth of the OH peak near 2700 nm could be correlated with H$^+$ dose (7, 8, 9).

## Summary

The main points of this chapter may be summarized as follows.

(1) Helium release characteristics and surface erosion and deformation in graphite implanted at low temperatures ($<$ 400°C) are believed to result from cracking and other stress effects, rather than from helium bubble formation. Deformation in highly oriented graphite appears to be more severe than in fine-grained material.

(2) Evidence for radiation-induced hydrocarbon formation was observed in residual gas spectra during helium and hydrogen implantation in graphite at 800°–1200°C.

(3) Vacuum system contamination by graphite at high temperature may be a serious problem with respect to CTR applications.

(4) From the standpoint of stability under He$^+$ and H$^+$ implantation, vitreous silica appears to be an ideal material for CTR applications.

(5) A saturation in the amount of volume compaction from high doses of He$^+$ and H$^+$ irradiation was determined. H$^+$ irradiation produced approximately one half as much compaction compared with He$^+$ or electrons at high fluences.

## Acknowledgments

The authors gratefully acknowledge the aid of F. W. Clinard in citing previous work on ceramic materials and wish to thank M. Kaminsky for the use of his manuscript prior to publication. The experimental assistance of D. H. Morse, L. A. Brown, and G. K. Kawamoto is greatly appreciated. Microtopograph measurements were performed by C.. E. Shinneman. Discussions with J. E. Shelby are gratefully acknowledged.

## Literature Cited

1. *See,* for example, Bauer, W., Thomas, G. J., Proc. Conf. on Defects and Defect Clusters in B.C.C. Metals and Their Alloys, Nucl. Mat., R. J. Arsenault, Ed., p. 225, Vol. 18, Gaithersburg, Md., 1973.

2. Bunch, J. M., Clinard, F. W., *Proc. Top. Meet. Technol. Controlled Nucl. Fusion, 1st* (1974) **II**, 448.
3. Kulcinski, G. L., Conn, R. W., Lang, G., *Nucl. Fusion*, in press.
4. Hopkins, G. R., *Proc. Top. Meet. Technol. Controlled Nucl. Fusion, 1st* (1974) **II**, 437.
5. Primak, W., Luthra, J., *J. Appl. Phys.* (1966) **37**, 2287.
6. Kaminsky, M., Das, S. K., Ekern, R., *Proc. Am. Nucl. Soc. Winter Meet.* (1975), in press.
7. Mattern, P. L., Bauer, W., Thomas, G. J., *Bull. Am. Phys. Soc.* (1974) **19**, 1117.
8. Mattern, P. L., Bauer, W., Thomas, G. J., *Bull. Am. Ceram. Soc.* (1975) **54**, 440.
9. Mattern, P. L., Bauer, W., Thomas, G. J., *J. Vac. Sci. Technol.* (1976) **13**, 430.
10. Primak, W., *J. Appl. Phys.* (1972) **43**, 2745.
11. EerNisse, E. P., Norris, C. B., *J. Appl. Phys.* (1974) **45**, 5196.
12. Thomas, G. J., Bauer, W., *Proc. Ann. Meet. Electron Microsc. Soc. Am. 33rd* (1975) 262.
13. Shelby, J. E., private communication.
14. Primak, W., *Phys. Rev.* (1957) **110**, 1240.

RECEIVED January 5, 1976. This work supported by the U.S. Energy Research and Development Administration.

# 5

# Radiation Blistering in Metals and Alloys

S. K. DAS and M. KAMINSKY

Argonne National Laboratory, Argonne, Ill. 60439

*Radiation blistering in solids leads to damage and erosion of irradiated surfaces. Major parameters governing the blistering process in metals and some metallic alloys include the type of projectile and its energy, total dose, dose rate, target temperature, channeling condition of the projectile, orientation of the irradiated surface plane, and target material and microstructure. Both experimental results and models proposed for blister formation and rupture are reviewed. The blistering phenomenon is important as an erosion process in applications such as fusion reactor technology (plasma–wall interactions) and accelerator technology (erosion of components and targets). There are several methods for reducing surface erosion caused by blistering.*

The irradiation of metal surfaces with energetic particles causes a variety of surface phenomena such as physical and chemical sputtering, secondary electron emission, x-ray emission, optical photon emission, release of absorbed and adsorbed gases, backscattering of particles, trapping and reemission of trapped particles, and radiation damage. For recent reviews, *see* Refs. *1, 2, 3, 4* and related articles in this volume. If such energetic particles penetrate a metal lattice, they may displace lattice atoms from their sites and create vacancies and interstitials. When the incident particles have slowed down sufficiently, they may be trapped in the lattice either interstitially or substitutionally.

Depending on the type of both implanted particle and the surrounding lattice atoms and on the impact parameters, the interaction between them may be either physical, or chemical, or both. For example, during the irradiation of titanium with hydrogen isotopes, metal hydrides were formed (*5, 6, 7*) leading to partial trapping of the incident particles. Such chemical trapping processes are discussed in Chapter 2 (*8*). In other cases the interaction between the implanted atoms and the lattice atoms may not lead to strong chemical bonds and compound formation, but is more physical in nature. For example, when the implanted particles are inert gas atoms, they can combine with the vacancies created by lattice displacements and nucleate as gas bubbles.

The formation of gas bubbles has been observed by Barnes et al. (9–11) after annealing of copper and aluminum which had been irradiated with 38-MeV $\alpha$ particles to total doses of $7 \times 10^{16}$–$1.7 \times 10^{17}$ $\alpha$ particles cm$^{-2}$. These gas bubbles were observed in the bulk material. If the gas bubbles form in near-surface regions and the gas pressure is high enough, bubbles may plastically deform the surface layers above them, and when the deformation is extreme, the surface layers may rupture. This phenomenon of surface deformation associated wtih gas bubbles formed because of irradiation has been called radiation blistering (12, 13, 14). It is of historical interest to note that an early indication of this phenomenon was obtained by Stark and Wendt (15) in 1912, when irradiating insulating materials such as calcite and calcium fluoride with $\sim$ 10-keV hydrogen ions. However this work went unnoticed, and in the early 1960's Primak (14, 16) and Kaminsky (12) first reported experimental evidence for blister formation in insulators irradiated with 100–140-keV protons and helium ions and in metals irradiated with 125-keV deuterons, respectively. Primak et al. (17) first observed flat-bottomed pits in silicon irradiated with 100-keV protons, using optical interferometry for surface examination. The pits were caused by ruptured blisters. Kaminsky (12, 13) first reported mass spectrometric observations of gas bursts from ruptured blisters during irradiation of copper with 125-keV deuterons. He also observed pitting of surface regions where blisters had exploded, using surface replica techniques in conjunction with transmission electron microscopy. The number of gas bursts correlated well with the number of pits observed on the surface, indicating that the pits were indeed caused by the rupture of gas bubbles.

After the early work by Kaminsky and Primak et al. the effect of space radiation on materials was studied because a major component of space radiation consists of energetic protons from cosmic rays and solar winds (18). During the last five years the interest in the radiation blistering phenomenon has increased greatly because of its importance in the operation of controlled thermonuclear fusion devices and reactors (19, 20, 21). In a fusion reactor having D–T plasma, energetic D, T, and He particles (formed by the D–T fusion reaction) can leak out of the confining magnetic field either as charged particles or as neutrals (formed, for example, by charge exchange near the plasma edge) and strike the surfaces of reactor components and form blisters. More recently, interest in radiation blistering has also developed in connection with other applications such as possible erosion of containers of short-lived transuranic nuclides (e.g., $^{242}Cm$ or $^{252}Cf$) caused by exposure to high alpha radiation fluxes over extended periods of time (23).

The radiation blistering phenomenon should be distinguished from other types of blistering phenomena observed in metals which are caused by entirely different processes. For example, blistering is quite common in aluminum castings (24), and it arises from exposing molten aluminum to a   gaseous environment containing hydrogen. This is caused by the relatively high solubility of hydrogen in molten aluminum and the low solubility in the solid aluminum resulting in the precipitation of excess hydrogen from the metal lattice. The ratio of solubility in the liquid phase to that in the solid phase at the freezing point is approximately 20 to one (25). Blister formation has also been observed in silver annealed first in an environment containing a high partial pressure of oxygen and then in an environment containing high partial pressures of hydrogen (26). In hydrogen embrittlement studies, blisters have been observed in iron electrolytically charged with hydrogen (27, 28).

We discuss first some of the general aspects of the experimental techniques used in radiation blistering studies. Experimental results obtained for various target–projectile systems under different irradiation conditions are described. However, the description will be limited to metals and alloys. Radiation blistering in nonmetals is discussed in Chapter 4 (29). Data reported after October 1975 are not included.

### General Aspects of Experimental Techniques

To detect even small changes in the surface topography caused by radiation blistering, it is often desirable to start with a high degree of optical finish on the surface by mechanical polishing (30, 31), electro-polishing (32), chemical polishing (33), or by some combinations of these three processes. Irradiations of the targets are normally done under high or ultrahigh vacuum conditions to avoid serious surface contamination (30, 32, 33). Various types of ion accelerators have been used to irradiate the targets. Duoplasmatron sources have been used for the low energy range (1–30 keV) (30), Cockroft–Walton generators for the medium energy range (25–500 keV) (14, 16, 17), and Van deGraaff accelerators for high energy (100–2000 keV) (11, 12, 16, 18) irradiations. The facilities used to produce and manipulate ion beams for implantation studies have been the subject of many recent reviews (2, 3) and will not be discussed here.

In certain blistering studies the gas released from irradiated targets has been measured by mass spectrometry by Kaminsky (12), Daniels (34), Bauer et al. (31, 35, 36), and Erents and McCracken (37). There are also other studies, not connected with radiation blistering studies, on gas release from surfaces during ion irradiation and during high temperature annealing after ion implantation. Reviews of these studies can be found elsewhere (2, 4).

The surface topography after irradiation can be examined by various techniques; however no attempt is made here to review all the available ones. In the early work on radiation blistering the irradiated surfaces were examined by optical microscopy (14, 17), interferometric techniques (14, 17), and by using surface replicas in transmission electron microscopes (12, 18). More recently, and in most of the studies to be reviewed here, scanning electron microscopy has been used. Transmission electron microscopy (TEM) has been used (38, 39, 40, 41) to study gas bubble and blister formation in near-surface regions. Most examinations of the blistering of irradiated surfaces have been done after irradiation. Blewer and Maurin (42) used a hot stage in a scanning electron microscope to make in situ observations of blister formation during heating of thin films of rare earth metals after helium ion implantation. Thomas and Bauer (43) have recently constructed a scanning electron microscope facility to observe surfaces undergoing ion bombardment. However, in such in situ observations the secondary electron emission during ion bombardment constrains one to use backscattered electrons for imaging, and this limits the resolution.

Roth et al. (33, 44) have used Rutherford backscattering techniques to study blister formation in monocrystalline niobium surfaces. They derived information about the thickness of the misaligned region, which they equate with the thickness of the blister skin, from dechanneling measurements.

To understand the basic mechanism of blister formation, it is important to know the depth profile of the implanted gas. Backscattering techniques have been used (45–50) to depth profile low-Z implanted gases in metals, and this technique is reviewed in Chapter 11 (50). Nuclear reaction techniques have been used (51, 52, 53, 54) to measure the depth profiles of implanted gases in metals, and discussions on this topic can be found in the chapters by Overley et al. (55) and Terreault et al. (56). More recently, a very good depth resolution in the 20–30-Å range has been claimed for $^3$He implanted in niobium using the $d$ ($^3$He, $p$) $^4$He reaction (57).

### Blistering Parameters

For a description of the various features of radiation blistering the following terms are commonly used:

(1) The "blister diameter" and the "distribution of blister diameters," in cases where the blisters are irregular in shape an "average blister

diameter" is given (32) which is defined as the diameter of a circle having approximately the same area as the blister.

(2) The "blister height" and the "distribution of blister heights."

(3) The number of blisters which are visible per unit irradiated area, often called "blister density." This number depends strongly on the resolution of the instrument used for surface examination, e.g., whether a scanning electron microscope or an optical microscope is used.

(4) The fraction of the total irradiated area occupied by blisters, often called "degree of blistering."

(5) The "blister shape."

(6) The "blister skin thickness."

(7) The amount of blister skin material lost from exfoliation of blister skin. This quantity has been often expressed as "erosion yield" which is equal to the number of target atoms lost from blister skin exfoliation per incident projectile ion.

### Table I.   Major Parameters That Can Affect Radiation Blistering

*Target-Related Parameters*
    type of target metal or alloy
    target temperature
    target microstructure
        grain size
        initial defect density (e.g. cold-worked vs. annealed structures)
        effects of precipitates—size distribution, volume fraction, and
            type of precipitates
    yield strength and rupture strength of target material
    crystallographic orientation of irradiated surface
    target surface finish

*Projectile-Related Parameters*
    type of projectile
    projectile energy
    total dose (fluence)
    dose rate (flux)
    channeling condition of projectile
    angle of incidence of projectile

*Parameters Affected by Target-Projectile Combinations*
    diffusivity and solubility of projectile in metals and alloys
    critical dose for blister appearance

Some authors (58, 59) distinguish between "surface bubbles," which refer to more circular surface features, and "blisters," which refer to irregular surface features. In this discussion no such distinction will be made, and all types of surface features, irrespective of their shape or size, resulting from surface deformation from gas bubbles will be referred to as blisters.

Many studies have been conducted to determine how one or several of the blistering features listed above depend on various parameters which are related to the target material and to the irradiation conditions chosen. Table I lists some of the major parameters which appear to affect radiation blistering. Some of the parameters are more target related, some are more projectile related, and some depend on the target–projectile combinations chosen. Many of the parameters in Table I are interdependent with the other parameters. For example, the yield strength of the target strongly depends on the type of target, target temperature, and the microstructure. At this time, the relative importance of many of the parameters listed in Table I on the blistering of metals and alloys has not been fully established.

*Projectile–Target Systems*

Most of the blistering studies on metals and alloys have been conducted with light projectiles such as H, D, and He ions, mainly because of the recent strong interest in the blistering effect in connection with the controlled thermonuclear fusion program (*19, 20, 21, 22*). Many of the target materials that have been investigated in recent years are materials proposed for use in controlled thermonuclear fusion devices. Very little data are available for materials irradiated with heavier ions such as $Ar^+$, $Xe^+$, etc. A list of radiation blistering studies for various target metals and alloys which have been irradiated with hydrogen isotope ions (e.g., $H^+$, $H_2^+$, $D^+$) is given in Table II. A similar list is given in Table III for various target materials which have been irradiated with helium ions. In both tables some of the irradiation conditions are also listed. More metals and alloys have been investigated for helium ion irradiation than for hydrogen-isotope ion irradiation. One notices that for irradiations with hydrogen isotope ions, the ion energies range from 10 keV for certain metals (Al and Nb) to 3.8 MeV for monocrystalline Sn (100). For helium ion irradiations the ion energies range from 1 keV (for Nb and Nb (100)) to 5.8 MeV (for Pd). Two additional studies are listed in Table II where type 304 stainless steel and platinum were irradiated with $\alpha$ particles from a 50% pure $^{242}$curium oxide source with an energy of 6 MeV (*23*). The total dose values reported for both hydrogen-isotope ion and helium ion irradiations range from $1.4 \times 10^{16}$ ions cm$^{-2}$ (for $He^+$ on Nb (100)) to $7 \times 10^{20}$ ions cm$^{-2}$ (for $He^+$ on polycrystalline nobium). The target temperatures range from $-196°C$ (for $H^+$ on Al) to $1327°C$ (for $He^+$ on Mo).

Solubility and diffusivity of the implanted gas in metals are two of the important parameters affecting the blistering process (Table I). In general, hydrogen isotopes have higher solubility and diffusivity in many metals than inert gases such as helium, and thus differences in blistering behavior for irradiation with hydrogen isotope ions and helium ions are

## Table II.   Targets for Which Radiation Blistering Has
## Irradiation

| | | Irradiation Conditions |
|---|---|---|
| Projectile Type | Target Metal or Alloy [a] | Projectile Energy or Energy Range (keV) |
| $H_3^+$ | Be | 100 |
| $H^+$ | Al | 10–200 |
| | Ti | 20 |
| | V | 150 |
| | type 302 st. steel | 20 |
| | type 316 st. steel | 20, 150 |
| | type 304 st. steel | 5 |
| | Cu | 70–140 |
| $H_2^+$ | Nb (100) | 10 |
| $H^+$ | Mo | 15, 150 |
| | Sn (100) | 2200, 3800 |
| | Au | 50, 100 |
| $D^+$ | Be | 15 |
| | Cu (100) | 125 |
| | Cu (110) | 125, 800 |
| | Ni | 150–400 |
| | type 4301 st. steel | 15 |
| | Nb | |
| | | 250, 300 |
| | Mo | 15 |
| | Mo (100) | 150 |

[a] All targets are polycrystalline materials unless otherwise indicated.

to be expected. An example (60) of the difference in the blistering behavior of niobium for helium ion and deuteron irradiation is shown in Figures 1a and b. There are two sizes of blisters formed during irradiation at 700°C with 250-keV $^4He^+$ ions (1a) after a total dose of $6.2 \times 10^{18}$ ions $cm^{-2}$. The larger size blisters have an average diameter of 5–8 $\mu$m while the smaller size blisters have an average diameter of ~ 0.5$\mu$m. The blisters formed during 250-keV $D^+$ ion irradiation at the same temperature after an even higher dose of $1.3 \times 10^{19}$ ions $cm^{-2}$ are also of two types but are much smaller (1b) than the helium blisters. The larger size deuterium blisters are more elongated, and their average length is 1–3 $\mu$m while the smaller size blisters have an average diameter of ~ 0.15$\mu$m. The observation of different blister shapes in Fig. 1a and b can be related to the orientation of the grains with respect to the ion beam, as discussed in the section on channeling conditions, below. Most of the deuterium blisters are unruptured (1b), whereas some of the large helium blisters have ruptured. The observation that blister size for deuteron irradiation is smaller than for helium ion irradiation under nearly

**Been Investigated for Hydrogen Isotope Irradiation with the Conditions**

*Irradiation Conditions*

| Dose or Dose Range (ions cm$^{-2}$) | Irradiation Temperature (°C) | References |
|---|---|---|
| $2 \times 10^{19}$ | room temp. | *154* |
| $5 \times 10^{16}$–$3.5 \times 10^{17}$ | −196 to 200 | *18, 119* |
| $5 \times 10^{18}$–$5 \times 10^{19}$ | 45 to 150 | *155* |
| $1.5 \times 10^{19}$–$2.5 \times 10^{19}$ | −105 to 115 | *65, 66* |
| $1 \times 10^{19}$–$2 \times 10^{20}$ | 75 to 150 | *155* |
| $1 \times 10^{19}$–$2 \times 10^{19}$ | −93 to 350 | *65, 66, 155* |
| $4.9 \times 10^{18}$–$9.9 \times 10^{19}$ | room temp. | *22* |
| $1.5 \times 10^{18}$–$8 \times 10^{19}$ | room temp. | *14, 154, 156* |
| $2 \times 10^{19}$ | room temp. | *33* |
| $2 \times 10^{18}$–$6 \times 10^{18}$ | −115 to 100 | *64, 65, 66* |
| $6.3$–$6.8 \times 10^{17}$ | room temp. | *157* |
| not known | room temp. | *119* |
| $1 \times 10^{18}$ | room temp. | *64* |
| $1.8 \times 10^{19}$ | room temp. | *12* |
| — | room temp. | *158* |
| $6 \times 10^{18}$–$1.3 \times 10^{19}$ | −153 | *71* |
| $5 \times 10^{18}$–$6 \times 10^{18}$ | 27 and 343 | *64* |
| | 550 to 700 | *60, 61, 107, 159, 160, 161* |
| $1 \times 10^{19}$–$1.9 \times 10^{19}$ | room temp. | *64* |
| $8 \times 10^{18}$ | room temp. | *64* |

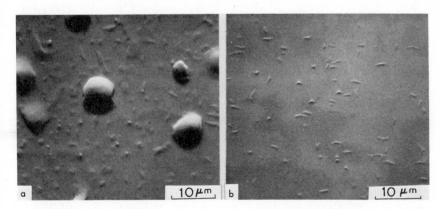

Journal of Nuclear Materials

*Figure 1.    Scanning electron micrographs (SEMs) of annealed polycrystalline niobium surfaces irradiated at 700°C (a) with 500-keV He$^+$ ions to a total dose of $6.2 \times 10^{18}$ ions cm$^{-2}$, (b) with 250-keV D$^+$ ions to a total dose of $1.25 \times 10^{19}$ ions cm$^{-2}$ (60)*

**Table III.   List of Targets for Which Radiation Blistering Has Irradiation**

| Target Metal or Alloy [a] | Irradiation Conditions Projectile Energy or Energy Range (keV) |
|---|---|
| Be and sintered Be powder | 100 |
| Al and sintered Al powder | 100–1000 |
| V | 100–1000 |
| | |
| V (111) | 500 |
| V-20% Ti | 500 |
| Type 4301 stainless steel | 15 |
| Type 304 stainless steel | 100–1500 |
| | 6000 |
| | ($\alpha$-particles from $^{242}$Cm oxide) |
| Type 316 stainless steel | 300 |
| Ni | 36 |
| | 40–140 |
| Ni (110) | 500 |
| Cu | 140 |
| Zr | — |
| Nb | 1–15 |
| | 100–1500 |
| | |
| Nb (100) | 1–15 |
| | |
| Nb (100), (110), (111) | 500–1500 |
| Mo | 7–80 |
| | 15 |
| | 300 |
| Mo (100) | 150 |
| Pd | 300 |
| | 4800–5700 |
| Sn (100) | 1500–3800 |
| Er | 160 |
| Er D$_{1.4}$, Er D$_{1.8}$ | 160 |
| W | — |
| Re | 20 |
| Pt | 36 |
| | 6000 |
| | ($\alpha$-particles from $^{242}$Cm oxide) |
| Au | 100–1500 |

[a] All target metals and alloys are polycrystalline unless otherwise indicated.

comparable irradiation conditions (the dose for deuteron irradiation was higher) has been related (*60, 61*) to the fact that the gas buildup is greatly reduced for deuterium in niobium, since the deuterium permeability (determined by the solubility and diffusivity) is many orders of magnitude larger than that of helium. For example, the diffusion coeffi-

**Been Investigated for Helium Ion Irradation Together with Conditions**

*Irradiation Conditions*

| Dose or Dose Range (ions cm$^{-2}$) | Irradiation Temperature (°C) | References |
|---|---|---|
| 3.1–6.2 × 10$^{18}$ | room temp. to 600 | 153 |
| 3.1–6.2 × 10$^{18}$ | room temp. to 400 | 67, 150, 151 |
| ~ 1 × 10$^{17}$–6.2 × 10$^{18}$ | room temp. to 1200 | 38, 39, 60, 65, 69, 107, 111, 139 |
| 6.2 × 10$^{17}$–6.2 × 10$^{18}$ | 650 to 750 | 39 |
| 6.2 × 10$^{18}$ | room temp. to 900 | 111 |
| 3 × 10$^{18}$ | 27 | 64 |
| 6.2 × 10$^{17}$–6.2 × 10$^{18}$ | room temp. to 550 | 21, 60, 70, 112 |
| 3 × 10$^{18}$ | room temp. | 23 |
| 4 × 10$^{18}$ | −170 to 700 | 65, 66, 113 |
| — | room temp. | 37 |
| up to 4 × 10$^{18}$ | — | 14 |
| 6.2 × 10$^{18}$ | 500 to 950 | 72 |
| up to 4 × 10$^{18}$ | — | 14, 37 |
| — | — | 37 |
| 3.1 × 10$^{17}$–7 × 10$^{20}$ | room temp. to 700 | 22, 30, 108, 109, 137 |
| 6.2 × 10$^{16}$–6.2 × 10$^{18}$ | −170 to 1200 | 33, 35, 38, 39, 60, 61, 69, 75, 107, 110, 118 |
| 1.4 × 10$^{16}$–4 × 10$^{17}$ | −110 to 1000 | 33, 44, 52, 76, 108, 137, 162 |
| 6.2 × 10$^{17}$–6.2 × 10$^{18}$ | room temp. to 900 | 61, 73, 118, 120 |
| 1 × 10$^{16}$–1 × 10$^{18}$ | room temp. to 1327 | 37, 40, 41, 100 |
| 2.5 × 10$^{18}$ | room temp. | 64 |
| ~ 1 × 10$^{17}$–4 × 10$^{18}$ | 400 to 1200 | 38, 65, 66 |
| — | room temp. | 64 |
| 7.5 × 10$^{17}$–2 × 10$^{18}$ | −180 to 200 | 31, 163 |
| ~ 1 × 10$^{17}$ | room temp. | 36, 164 |
| 6.2 × 10$^{17}$–1.8 × 10$^{18}$ | room temp. | 157 |
| 5 × 10$^{16}$–1.5 × 10$^{18}$ | room temp. | 42, 58, 59 |
| 3 × 10$^{17}$ | room temp. | 58 |
| — | — | 37 |
| 9 × 10$^{16}$–3 × 10$^{18}$ | 27 | 165 |
| — | room temp. | 37 |
| — | — | |
| 5 × 10$^{17}$ | room temp. | 23 |
| 3.1 × 10$^{18}$– 9.4 × 10$^{18}$ | room temp. | 166 |

cient of deuterium in niobium (*62*) is $D_D = 1.3 \times 10^{-4}$ cm$^2$ sec$^{-1}$ at 800°C while that for helium in nobium (*63*) is $10^{-19}$–$10^{-14}$ cm$^2$ sec$^{-1}$ between 600° and 1200°C.

A difference in blistering behavior has also been observed by Verbeek and Eckstein (*64*) in molybdenum for D$^+$ and He$^+$ ion irradiations.

Figures 2a and b show some of their results on blistering in molybdenum after irradiation at room temperature with 15-keV $D^+$ ions to a dose of $2 \times 10^{18}$ ions $cm^{-2}$ and with 15-keV $He^+$ ions to a dose of $2.5 \times 10^{18}$ ions $cm^{-2}$. The number of blisters per unit area (blister density) is higher for helium ion irradiation (2a) than for deuteron irradiation (2b). They also observed more exfoliation of the blisters for helium ion irradiation (inset, 2a) than for deuteron irradiation (inset, 2b). However, the total dose for helium ion irradiation was slightly higher than for deuteron irradiation.

Thomas and Bauer (65, 66) studied proton and helium blistering in type 316 stainless steel. For irradiation with 150-keV protons to a dose of $1 \times 10^{19}$ ions $cm^{-2}$ at $-93°C$, they observed blisters with an average diameter of 5 $\mu$m. For proton irradiation at temperatures of $-78°$, $70°$, $270°$, and $350°C$ no blisters were observed even after irradiation to total doses of $1.3 \times 10^{19}$–$2 \times 10^{19}$ ions $cm^{-2}$. For 300-keV helium ion irradiation to a lower dose of $4 \times 10^{18}$ ions $cm^{-2}$, large blisters with severe exfoliation of the blister skin were observed for the range $-170°C$ to $500°C$.

No blisters were observed (65) for vanadium irradiated at $115°C$ with 150-keV protons to a dose of $1.5 \times 10^{19}$ ions $cm^{-2}$ and at $-105°C$ to a dose of $2.5 \times 10^{19}$ ions $cm^{-2}$. In contrast, blisters have been observed in vanadium during helium ion irradiation for much lower dose ranges ($2 \times 10^{18}$–$6.2 \times 10^{18}$ ions $cm^{-2}$) over a wide energy range of 100–1000 keV (67, 68) and a wide temperature range (38, 39, 60, 65, 69, 70).

Primak and Luthra (14) did not observe any blisters on nickel which had been irradiated with protons for total doses up to approximately $4 \times 10^{18}$ ions $cm^{-2}$, but they observed blisters after irradiation with $He^+$ ions up to the same total dose values in the energy range 40–140 keV. More

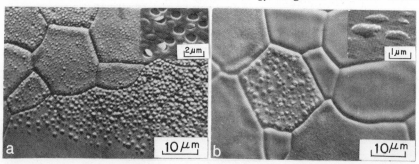

Application of Ion Beams to Metals

*Figure 2.   Optical micrographs of annealed polycrystalline molybdenum surfaces irradiated at room temperature (a) with 15-keV $He^+$ ions to a dose of $2.5 \times 10^{18}$ ions $cm^{-2}$, (b) with 15-keV $D^+$ ions to a dose of $2 \times 10^{18}$ ions $cm^{-2}$. The insets in (a) and (b) are SEMs showing enlarged views of some of the blisters (64).*

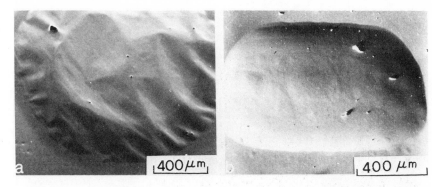

Ion Beam Surface Layer Analysis

*Figure 3.    SEMs of (a) monocrystalline Ni (110) surface irradiated at room temperature with 500 keV $^4He^+$ ions to a dose of 6.2 $\times 10^{18}$ ions $cm^{-2}$, (b) polycrystalline nickel surface irradiated at $-153°C$ with 400-keV $D^+$ ions to a dose of 1.25 $\times 10^{19}$ ions $cm^{-2}$ (71). The blisters in (b) were observed at room temperature after irradiation at $-153°C$.*

recently, blisters have been observed in nickel irradiated with 400-keV $D^+$ ions to total doses above $8 \times 10^{18}$ ions $cm^{-2}$ at a low temperature of $-153°C$ (*71*). For this low temperature deuteron irradiation the average blister diameter was the same order of magnitude as the blister diameter observed for 500-keV $^4He^+$ ion irradiation of a Ni (110) surface to a dose of 6.2 $\times 10^{18}$ ions $cm^{-2}$ at room temperatures (*72*) as can be seen in Figures 3a and b. In these irradiations one large blister appears to cover most of the irradiated area.

Several of the examples cited above illustrate the difference in the blistering behavior of a given target material to the implantation of hydrogen isotope ions and helium ions under otherwise nearly identical irradiation conditions. Some of the major parameters affecting the radiation blistering process are described below.

### Projectile Energy

The depth at which the ions are implanted in a solid depends on the projectile energy. In fact both the depth profile of the implanted ions and the energy deposited into damage are functions of the projectile energy and affect the gas bubble and subsequent blister formation significantly.

Several of the features which describe the blistering phenomena such as blister diameter, blister density, blister skin thickness, and critical dose for blister appearance increase with increasing projectile energy in a number of target projectile systems. Figures 4a–d illustrate this effect for vanadium irradiated at room temperature with helium ions in the energy

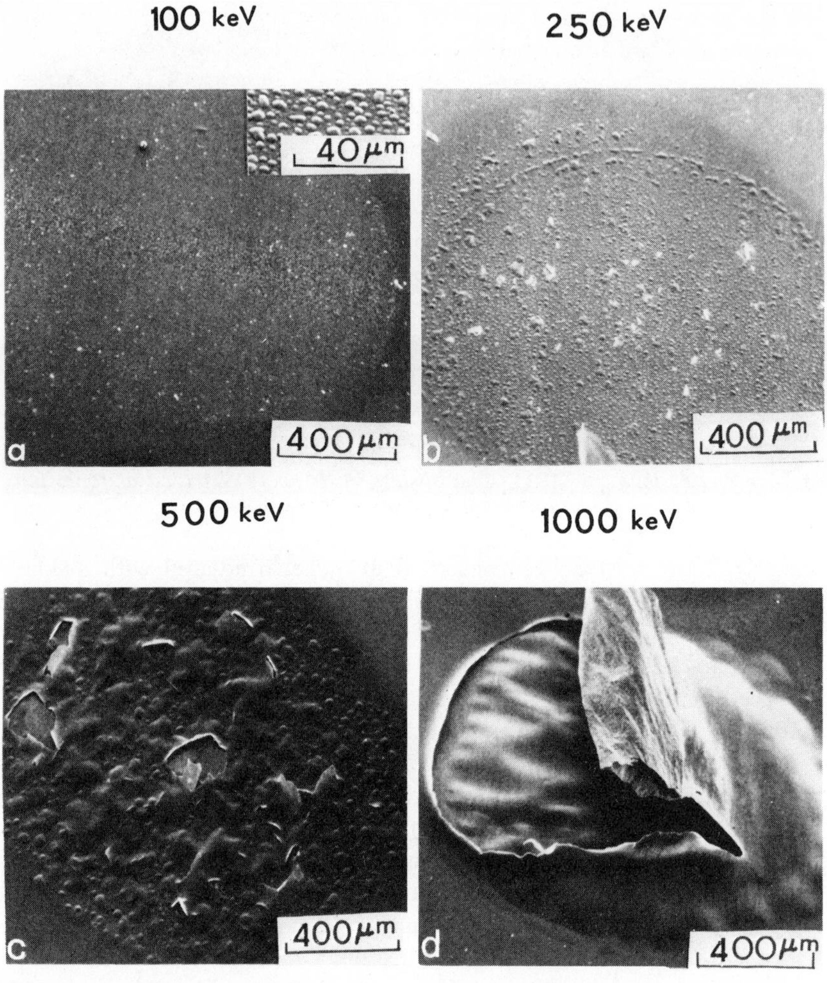

100 keV                    250 keV

500 keV                    1000 keV

*Figure 4.   SEMs of annealed polycrystalline vanadium surfaces irradiated at room temperature (a) with 100-keV $^4He^+$ ions to a dose of 3.1 ×10$^{18}$ ions cm$^{-2}$ (the inset shows an enlarged view of some of the blisters), (b) with 250-keV $^4He^+$ ions to a dose of 3.1 × 10$^{18}$ ions cm$^{-2}$, (c) with 500-keV $^4He^+$ ions to a dose of 6.2 × 10$^{18}$ ions cm$^{-2}$, (d) with 1000-keV to a dose of 6.2 ×10$^{18}$ ions cm$^{-2}$*

range 100–1000 keV. For irradiation with 100-keV $^4He^+$ ions to a dose of 3.1 × 10$^{18}$ ions cm$^{-2}$, the blisters are barely resolvable in the micrograph in 4a, but at higher magnifications (*see* inset) blisters with average diameters of 1–8 $\mu$m were observed. Irradiation with 250-keV $^4He^+$ ions to the same dose showed slightly larger blisters with diameters of 3–40 $\mu$m (4b). For 500-keV irradiation to a larger dose of 6.2 × 10$^{18}$ ions cm$^{-2}$ the average blister diameters range from 15 to 350 $\mu$m (4c). On further increasing the projectile energy to 1000 keV, only one large blister occu-

pied the entire bombarded area (4d), and this blister had already ruptured, and a second blister had formed.

Similar results have been obtained on niobium irradiated at room
temperature with helium ions for an even larger energy range of 2 keV–
1.5 MeV. Figure 5 is a plot of the available data (*30, 32, 33, 37, 60, 73*)
on average blister diameters as a function of helium ion energy. The bars
indicate the smallest and the largest average blister diameter observed at
a given energy. In general a range of blister diameters is observed in a
given irradiated area, and Figure 6 shows an example (*32*) of the distribution of blister diameters for polycrystalline niobium irradiated at room
temperature with 500-keV $^4He^+$ ions to a dose of $6.2 \times 10^{18}$ ions $cm^{-2}$.
In this particular example, the average blister diameters range from 9 $\mu$m
to as large as 500 $\mu$m, and the largest fraction of blisters has an average
diameter of $\sim$ 12$\mu$m. For the cases where such blister diameter distribu-

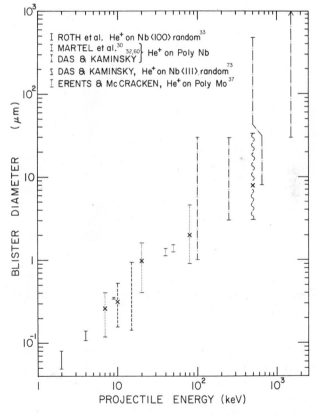

*Figure 5.   Blister diameter as a function of projectile
energy for niobium and molybdenum irradiated at room
temperature with helium ions*

tions are known, the average blister diameter corresponding to the peak in the distribution has been indicated in Figure 5. Even though there is a large range of average blister diameters observed at a particular projectile energy, Figure 5 shows that the average blister diameters increase with increasing projectile energy for niobium irradiated at room temperature with helium ions. At energies above 500 keV very large blisters covering nearly the entire bombarded area have been observed as indicated by the arrows. An increase in blister diameter in molybdenum

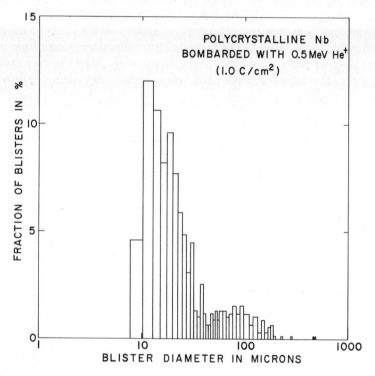

Journal of Applied Physics

*Figure 6. Histogram of the blister diameter distribution in cold-worked polycrystalline niobium irradiated at room temperature with 500-keV He⁺ ions to a dose of 6.2 × 10¹⁸ ions cm⁻² (32)*

irradiated with helium ions was observed by Erents and McCracken (37) with increasing ion energy from 7 keV to 80 keV (Figure 5). For nickel irradiated at − 153°C with deuterons in the energy range 150 keV–400 keV, a similar dependence was observed by Möller et al. (71).

The increase in the average blister diameter with increase in projectile energy may be understood (60, 73) qualitatively as follows. One can express the gas pressure p inside a blister as (32):

$$p = 4\sigma_y ht / (r^2 + h^2) \tag{1}$$

where $\sigma_y$ is the yield strength of the material, $t$ is the blister skin thickness, $r$ is the radius of the blister, and $h$ is the height of the blister. This equation was originally derived by Hill (74) for spherical bulging by plastic deformation caused by the gas pressure applied to one side of a circular metal diaphragm firmly clamped around its perimeter. It has been observed that the blister skin thickness $t$ is a strong function of ion energy $E$ and increases with increasing projectile energy as discussed below. If one assumes that the blister height $h$ and pressure $p$ are only weak functions of $E$, and $\sigma_y$ does not depend on $E$, a higher projectile energy (and correspondingly higher skin thickness $t$) will result in a larger blister radius $r$ (see Equation 1). The value of $\sigma_y$ in an irradiated sample may differ from the value for an unirradiated material because of radiation damage and may be a weak function of the projectile energy.

The blister skin thickness in all metals studied increases with increasing projectile energy over a wide range of energies. Figure 7 shows an example for vanadium irradiated at room temperature with helium ions of energies ranging from 100 keV to 1000 keV (68). It illustrates the increase of blister skin thickness for increasing helium ion energies of 100 keV, 250 keV, 500 keV, and 1000 keV. For helium ion irradiation of niobium the blister skin thicknesses have been measured (32, 33, 60, 75–77) for a much larger energy range (1 keV–1500 keV), and the measured skin thickness for different projectile energies have been compared with corresponding calculated projected ranges of $^4He^+$ ions incident on niobium. Figure 8 is a plot of projected ranges of helium ions in niobium for different energies calculated according to the method formulated by Schiøtt (78). The dotted curve represents calculations using Thomas–Fermi nuclear stopping cross sections together with the electronic stopping cross sections of Lindhard et al. (79), which are proportional to projectile velocities. This latter assumption of velocity proportional electronic stopping is strictly true for velocity $v$ smaller than $v_0 Z_p^{2/3}$ (here $v_0 = e^2/\hbar$, $Z_p =$ atomic number of projectile), which corresponds to helium ion energies less than 250 keV. The solid curve was calculated using Thomas–Fermi nuclear stopping cross sections together with Brice's (80) expression for electronic stopping which uses the semi-empirical values of stopping power from Ziegler and Chu (81). The available data on measured blister skin thickness values for niobium irradiated with helium ions of different energies are plotted in Figure 8. The measured skin thickness values for the energy range of 250 keV–1500 keV (75) agree within 15% with the calculated projected range values according to Brice. For the energy range 60–80 keV (77) the agreement is better with the calculated projected range values according to Schiøtt. For He$^+$

ion energies below 40 keV the blister skin thickness values (33, 76, 77) are higher than the projected range values, the difference increasing with decreasing ion energy. For 15-keV He⁺ ion value obtained by Behrisch et al. (76) agrees closely with the one obtained by St.-Jacques et al. (77). Blister skin thickness values of Roth et al. (33) for the low energy range were indirectly obtained from Rutherford backscattering measurements, whereas values of Kaminsky et al. (75) and of St.-Jacques et al. (77)

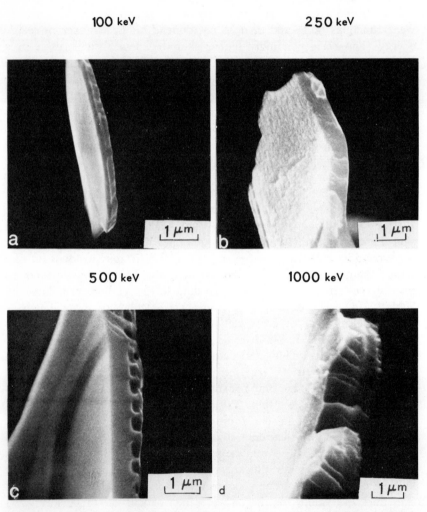

**100 keV**    **250 keV**

**500 keV**    **1000 keV**

*Figure 7.   SEMs of blister skins in polycrystalline vanadium irradiated at room temperature with ⁴He⁺ ions with energies of (a) 100 keV, (b) 250 keV, (c) 500 keV, and (d) 1000 keV. The total dose in (a) and (b) was 3.1 × 10¹⁸ ions cm⁻² and in (c) and (d) was 6.2 × 10¹⁸ ions cm⁻².*

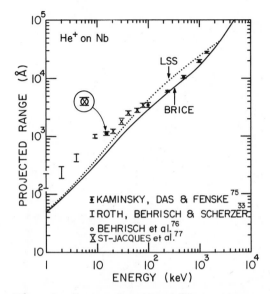

*Figure 8.   Projected ranges of ⁴He⁺ ions in nio-
bium as a function of projectile energy.   The
dotted curve marked "LSS" was calculated using
Thomas–Fermi nuclear stopping and the velocity
proportional electronic stopping cross sections
of Lindhard, et al. solid curve marked "Brice"
was calculated using Thomas–Fermi nuclear
stopping and Brice's semiempirical expression
for electronic stopping.   The data points with
error bars are measured blister skin thicknesses
for niobium irradiated at room temperature with
He⁺ ions, as measured by various authors.*

were directly measured from the scanning electron micrographs of blister
skins.

In the unfolding of Rutherford backscattering data for a determination of blister skin thickness, however, uncertainties in the values for
differential energy losses are reflected in corresponding uncertainties in
the thickness values. Figure 9 shows a typical example of energy distributions of backscattered protons from which the estimates of skin thickness were made by Roth et al. (33, 44) for blisters resulting from 4-keV
helium ion irradiation of Nb (100) surface in random direction. The
backscattered spectra were taken in $<100>/<111>$ double alignment
with 150-keV protons before and after bombardment at room temperature
with different doses of 4-keV He⁺ ions. A random backscattered spectrum
is also shown in Figure 9. The backscattered energy distributions have
been transformed into a depth distribution as shown by assuming a
differential energy loss. One can see that for a dose of $1.4 \times 10^{17}$ ions
cm⁻², a surface layer about 500 Å thick becomes misaligned. This thick-

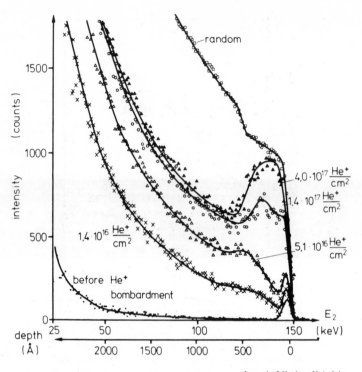

Journal of Nuclear Materials

*Figure 9.* $<100>/<111>$ *double aligned backscattered energy distribution of 150 keV protons from a niobium single crystal after irradiation at room temperature with 5-keV He⁺ ions in a random direction (incident at 12° off the surface normal) for different doses as indicated (33)*

ness of the misaligned region has been interpreted by Roth et al. (*33*) to equal the thickness of the blister skin.

The fact that the observed blister skin thickness for niobium irradiated with low energy helium ions (1–15 keV) was higher than the theoretical projected range was initially attributed (*33, 44*) to channeling of a large fraction of incident helium ions along close-packed lattice directions or planes. Subsequently the same authors (*76*) explained this difference between the blister skin thickness and the calculated projected range as being caused by the blister skin separation at the end of the range. An alternative explanation suggested by other authors (*77*) is that the observed difference may in part be caused by the more uniform distribution of helium bubbles for a large portion of the implant depth leading to volumetric changes (swelling from helium bubbles) of that region for sufficiently high doses, as is discussed under "Models for Blister Formation," below.

The measurements of blister skin thicknesses on vanadium (Figure 7) irradiated at room temperature with helium ions in the energy range 100–1000 keV also show a good correlation with the calculated projected ranges (*68*). In contrast, for aluminum irradiated at room temperature with helium ions for the same energy range, the same authors obtained blister skin thickness values that were slightly lower than the calculated projected ranges (*67*).

It is interesting to determine whether the blister skins separate from the bulk closer to the region of maximum lattice damage or the region where the concentration of the implanted ions is at a maximum. Recent studies by Kaminsky et al. (*68, 75*) and Das et al. (*67*) have provided some insights here. Figure 10 compares blister skin thickness values for aluminum, vanadium, and niobium with the corresponding damage energy distributions and projected range distributions for the energy range above 100 keV (*67*). The data points marked a–d in Figures 10a, b, and c are blister skin thickness values for aluminum, vanadium, and niobium, respectively. The solid curves in Figures 10a–c are calculated damage energy distributions, and the dashed curves are projected range distributions for the helium ion energies indicated. The distributions were obtained using the tabulated projected range, straggling, and damage energy values for helium ions according to Brice (*82*). These tabulated values for damage energy distributions were calculated using an intermediate method developed by Brice (*83*) which includes the spatial distribution of the energy transferred by primary recoil atoms. The total error in the calculation of the projected range distribution and the damage energy distribution was estimated to be 12% (*67, 68, 75*). The results in Figure 10a show that the average skin thickness values for aluminum correlate more closely with the maxima in the damage energy distributions than with the maxima in the projected range distributions. On the other hand, for vanadium (10b) and niobium (10c) the average blister skin thickness values correlate more closely with the maxima in the projected range distributions than with the maxima in the damage energy distributions. In this connection, one should consider the possibility that for a ductile metal like aluminum, the blister skin thickness could be lower because of the uniform elongation of the skin during plastic deformation and possible necking during rupture. However, measurements of blister skin thickness in aluminum made near the base of a blister and near a blister top by Das et al. (*67*) showed no significant differences in the thickness values within the experimental error.

More recently, blister skin thicknesses have been measured in nickel irradiated at − 153°C with D$^+$ ions in the energy range 200–400 keV by Möller et al. (*71*). They compared the blister skin thickness values with experimental projected ranges, which were obtained from depth profile

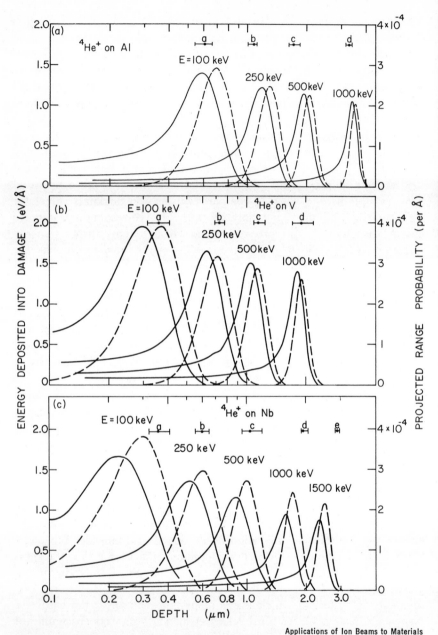

*Figure 10.  Plots of projected range probability distributions (– – –) as a func-tion of depth into (a) Al target, (b) V target, and (c) Nb target for different energies of the incident ⁴He⁺-ions. The solid curves are plots of energy de-posited into damage as a function of depth into the targets for ⁴He⁺ ions with various projectile energies. The data points with error bars are measured blister skin thicknesses for helium ion energies of (a) 100 keV, (b) 250 keV, (c) 500 keV, (d) 1000 keV, and (e) 1500 keV (67).*

measurements (*54*) and calculated maxima in the damage energy distributions, which were computed with a Monte-Carlo program, taking into account primary events only. The blister skin thickness values were closer to the maxima in the damage energy distribution than the maxima in the projected range distribution, a result similar to those described above for aluminum (*67*).

### Critical Dose for Blister Formation

In considering the effect of dose on blister formation a distinction should be made between the critical dose $C_b$ needed for the formation of gas bubbles in the bulk of the material during irradiation and the critical dose $C_{bl}$ for the appearance of blisters on the surface of an irradiated material. The determination of both of these parameters depends strongly on the sensitivity and the resolution of the detection technique used. For example, the more commonly used transmission and scanning electron microscopic techniques used to detect gas bubbles and blisters on the surface, respectively, often use commercially available instruments with resolutions ranging from 5–10 Å and 50–100 Å, respectively. Other techniques such as Mössbauer effect (*see* Ref. *84*) and positron annihilation techniques (*see* Ref. *85, 86*) have been used to study vacancy clusters in metals and may be used to detect gas bubbles. It is to be expected that the critical dose for gas bubble formation will depend on several parameters such as the concentration of the gas in the implant region, the initial and the radiation-induced defect concentrations, the type of lattice impurities and their concentration, target temperature, and the permeability of the implanted gas. While systematic studies of the critical dose for bubble formation are lacking, some information about the existence of gas bubbles in different metals after irradiation with noble gas ions is available. For example, helium bubbles have been observed in many metals such as aluminum (*87, 88, 89*), vanadium (*90, 91, 92, 93*), stainless steel (*94, 95, 96*), copper (*9–11, 97, 98*), niobium (*38, 39, 69, 99*), molybdenum (*40, 41, 100*), silver (*101*), and gold (*98, 102, 103*) after helium ion irradiation for total doses ranging from $\sim 1 \times 10^{14}$ ions cm$^{-2}$ to $\sim 6 \times 10^{18}$ ions cm$^{-2}$. The helium concentrations range from $\sim 10$ ppm (*96*) to several atomic percent (*39*). In a few studies (*40, 100*) the bubbles were observed after room temperature irradiation while in other studies (*8–11, 38, 39, 69, 87–103*) either the irradiations were done at higher temperatures, or the targets were annealed at high temperatures to observe bubbles. In vanadium irradiated with 240-keV He$^+$ ions, $C_b$ for bubbles with diameters $\geq 40$ Å was observed using TEM to be $1 \times 10^{17}$ ions cm$^{-2}$ for 500°C irradiation, and it decreased markedly to $7 \times 10^{14}$ ions cm$^{-2}$ for a higher target temperature of 600°C (*93*). Evans et al. (*100*) could not resolve gas bubbles (the resolution in their observation was $\sim 15$–$20$

Å) in molybdenum implanted at room temperature with 36-keV He⁺ ions
to a dose of $1 \times 10^{16}$ ions cm⁻². However, at a dose of $3 \times 10^{17}$ ions cm⁻²
bubbles became readily visible in TEM which aligned themselves on a
bcc superlattice exactly analogous to the void lattice (104). For irradiation with low energy (5-keV) Ar⁺ ions, bubbles of ~ 20 Å in diameter
have been observed in gold after irradiation to a dose as low as $1.5 \times 10^{15}$
ions cm⁻² (105). Gas bubbles can contribute to volume swelling, and for
niobium irradiated at 800°C with 300-keV He⁺ ions to a dose of $5 \times 10^{17}$
ions cm⁻², a fractional volume increase $\Delta V/V$ of ~ 1% has been measured (69).

The critical dose for blister appearance $C_{bl}$ depends also on several
parameters, such as the concentration of the implanted gas in the implant
region, the projectile energy, permeability of the implanted gas in the
solid, target temperature, initial defect concentration, and yield strength
of the material. Systematic studies of the dependence of $C_{bl}$ is very
limited.

Even though gas bubbles have been observed (93) in metals at doses
as low as $7 \times 10^{14}$ ions cm⁻² for 240-keV He⁺ on V at 600°C as mentioned
above, the value for the critical dose for blister appearance has been
observed to be approximately two orders of magnitude higher. Only a
few data are available on the dependence of $C_{bl}$ on the projectile energy.
For example for helium ion irradiation of niobium and molybdenum,
Figure 11 shows a plot of available data (32, 33, 37, 52, 106) on critical
dose for blister appearance as a function of helium ion energy. One
notices a general trend—with increasing projectile energy the $C_{bl}$ values
increase for the type of metals and the energy ranges studied. The range
of $C_{bl}$ values shown in Figure 11 ($C_{bl} > 1 \times 10^{17}$ ions cm⁻²) is generally
larger than the dose ranges mentioned above for which gas bubbles have
been observed. A similar trend of increasing $C_{bl}$ values with increasing
projectile energy has also been observed in nickel irradiated at − 153°C
with D⁺ ions for the energy range 200–400 keV (71). This trend appears
plausible since with increasing projectile energy the blister skin thickness
increases. The critical pressure $p_{cr}$ in a gas bubble needed to deform the
surface can be written as (18, 31, 32):

$$p_{cr} = \frac{4\sigma_y t^2}{3r^2} \qquad (2)$$

where the symbols have the same meaning as in Equation 1. This expression is slightly different from the one given in Equation 1 and will be
derived later under "Models for Blister Formation," below. For a bubble
with a given radius $r$ and for a given yield strength of the material $\sigma_y$,
a larger skin thickness $t$ resulting from an increase in projectile energy
requires a larger $p_{cr}$, and this can be obtained at larger doses. However,

there are other complicating microstructural effects that may invalidate this simple relationship between critical dose and projectile energy. For example, the critical dose for blister appearance in niobium at room temperature has been found to depend on the initial dislocation density (32). In Figure 11 the $C_{bl}$ value for 500-keV He$^+$ ion irradiation of annealed niobium at room temperature is $\sim 2 \times 10^{18}$ ions cm$^{-2}$ while for cold-worked niobium, only an upper limit value of $6.2 \times 10^{17}$ ions cm$^{-2}$ can be given. This lowering of $C_{bl}$ may (32) be caused by easier nucleation of helium bubbles on the more numerous dislocations in cold-worked samples.

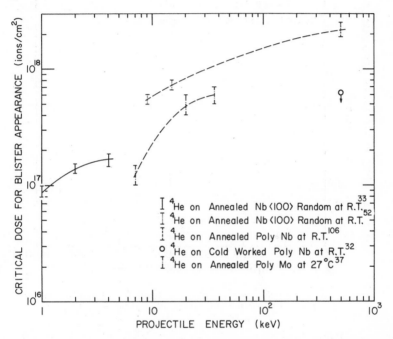

*Figure 11.   Critical dose for blister appearance $C_{bl}$ as a function of projectile energy for niobium and molybdenum irradiated at room temperature with $^4He^+$ ions*

For 20-keV ion irradiation of molybdenum at room temperature, Erents and McCracken observed $C_{bl} \sim 5 \times 10^{17}$ ions cm$^{-2}$, a value which is close to the interpolated one for annealed niobium for comparable irradiation conditions. For gases like hydrogen which have high permeability in metals, $C_{bl}$ is generally higher than for inert gases like helium. For example, for cold-worked polycrystalline niobium irradiated at room temperature with 500-keV D$^+$, no blisters were observed (61) after a dose of $6.2 \times 10^{18}$ ions cm$^{-2}$, whereas for helium ion irradiation under identical conditions, $C_{bl}$ was less than $6.2 \times 10^{17}$ ions cm$^{-2}$ (Figure 11).

Dependence of $C_{bl}$ on irradiation temperature has been observed. For example, for 500-keV He⁺ ion irradiation of niobium at room temperature, $C_{bl}$ is $\sim 2 \times 10^{18}$ ions cm⁻², and it decreases to $< 6.3 \times 10^{16}$ ions cm⁻² for irradiation at 900°C (61). However, for 4-keV He⁺ ion irradiation, the decrease in $C_{bl}$ with increasing irradiation temperature is not so large (see Figure 12). This trend can be understood since with an increase in temperature the yield strength $\sigma_y$ of most metals decreases, and thus for constant values of $r$ and $t$ in Equation 1, the value for $p_{cr}$ will decrease correspondingly.

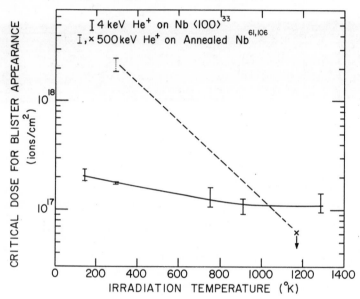

*Figure 12. Critical dose for blister appearance as a function of target temperature for niobium irradiated by helium ions*

### Effect of Total Dose

In this section we consider the effect of the total dose values which are above the critical dose $C_{bl}$ on the blistering phenomenon. Once the $C_{bl}$ values have been exceeded, the dose values will affect such features as blister diameter, blister density, and the exfoliation of the blister skin.

The ranges of average blister diameters depends strongly on the projectile energy (as described above), but less strongly on the total dose. For example, for cold-worked niobium irradiated at room temperature with 500-keV He⁺ ions, the average diameters of most of the blisters range from 10 to 30 μm for both total doses of $6.2 \times 10^{17}$ ions cm⁻² ($\sim 64\%$ of total number of blisters) and $6.2 \times 10^{18}$ ions cm⁻² ($\sim 75\%$ of total number of blisters (32). For annealed polycrystalline vanadium irradiated at 900°C with 500-keV He⁺ ions, as the dose is increased (*107*) by one order of magnitude from $6.2 \times 10^{17}$ ions cm⁻² to $6.2 \times 10^{18}$ ions cm⁻², the range of the values of average blister diameters does not change appreciably, but the mean value of the average blister diameters shows a small increase from 6.2 to 7.4 μm. Similar results have been obtained (*107*) for niobium irradiated at 900°C with 500-keV He⁺ ions, where the range of average blister diameters does not change appreciably for doses of $6.2 \times 10^{17}$ ions cm⁻² and $6.2 \times 10^{18}$ ions cm⁻². For 125-keV D⁺ irradiation of Cu (100) surface, the pits formed by blister rupture do not increase in size appreciably as the total dose is increased to above $1.9 \times 10^{19}$ ions cm⁻² (*19, 20*).

The blister density, i.e., the number of blisters per unit irradiated area, has been observed to increase with total dose. For palladium implanted at with 300-keV He⁺ ions Thomas and Bauer (*31*) observed an increase in blister density by a factor of five by increasing the total dose from $1 \times 10^{18}$ ions cm⁻² to $2 \times 10^{18}$ ions cm⁻², but the values for the blister diameters did not change. For molybdenum irradiated at room temperature with 36-keV He⁺ ions, Erents and McCracken (*37*) observed an increase in blister density by about a factor of four when the total dose was increased by a factor of two from $6 \times 10^{17}$ ions cm⁻² to $1.2 \times 10^{18}$ ions cm⁻². For higher temperature irradiation (at 900°C) of niobium with 500-keV He⁺ ions, Das and Kaminsky (*107*) observed an increase in the blister density from $\sim 6 \times 10^4$ blisters cm⁻² to $\sim 1.5 \times 10^5$ blisters cm⁻² with an increase in total dose from $6.2 \times 10^{17}$ ions cm⁻² to $6.2 \times 10^{18}$ ions cm⁻².

The increase in blister density with increasing total dose is observed only when there are numerous small blisters over the irradiated area. However, when there are only a few large blisters covering most of the irradiated area (as observed for high projectile energies, for example *see* Figures 4c, d), there is an increase in blister skin exfoliation with increase in total dose. Figures 13a–c illustrate this effect for type 304 stainless steel irradiated at 450°C with 500-keV He⁺ ions to a total doses of $6.2 \times 10^{17}$, $3.1 \times 10^{18}$, and $6.2 \times 10^{18}$ ions cm⁻², respectively (*60*). For a dose of $6.2 \times 10^{17}$ ions cm⁻² a large portion of the irradiated area is occupied by a single blister with an average diameter of $\sim 700$ μm (13a) which has ruptured. At a higher dose of $3.1 \times 10^{18}$ ions cm⁻² three exfoliated skin layers are observed (13b) as compared with one for a dose of $6.3 \times 10^{17}$ ions cm⁻² (13a). For an even higher dose of $6.2 \times 10^{18}$ ions cm⁻² the number of exfoliated skin layers increases to five in some regions (13c). One can estimate the erosion yields (*see* "Blistering Parameters," above) for these cases by measuring the area from which the blister skin has fallen off and the blister skin thickness. The estimated erosion yields for the three doses $6.2 \times 10^{17}$, $3.1 \times$

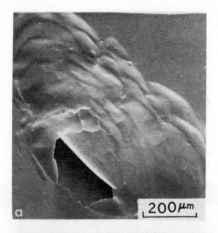

Journal of Nuclear Materials

*Figure 13. SEMs of surfaces of annealed type 304 stainless steel after irradiation at 450°C with 500-keV $^4He^+$ ions for total doses of (a) 6.2 × $10^{17}$ ions cm$^{-2}$, (b) 3.1 × $10^{18}$ ions cm$^{-2}$, and (c) 6.2 × $10^{18}$ ions cm$^{-2}$ (60)*

$10^{18}$, and $6.2 \times 10^{18}$ ions cm$^{-2}$ were $0.1 \pm 0.05$, $0.45 \pm 0.1$, and $0.8 \pm 0.2$ atoms per incident helium ion—an almost linear increase in erosion yield with total dose (*60*).

For low energy ($< 15$ keV) irradiations the successive exfoliation of blister skin with increasing total dose is not observed. Martel et al. (*30*) observed blisters appearing in niobium irradiated at room temperature with helium ions for energies of 5 keV, 10 keV, and 15 keV for doses or dose ranges of $3.1 \times 10^{18}$ ions cm$^{-2}$, $6.2 \times 10^{17} - 3.1 \times 10^{18}$ ions cm$^{-2}$, and $6.2 \times 10^{17} - 9.4 \times 10^{18}$ ions cm$^{-2}$ for the respective energies. However, when the dose values were increased to $\sim 6.2 \times 10^{18}$ ions cm$^{-2}$, $\sim 6.2 \times 10^{18}$ ions cm$^{-2}$, and $1.9 \times 10^{19}$ ions cm$^{-2}$ for 5-, 10-, and 15-keV helium ion irradiation, respectively, no blisters were observed. Figures 14a and b show surfaces of niobium irradiated with 15-keV He$^+$ ions to total doses of $3.1 \times 10^{18}$ ions cm$^{-2}$ and $6.2 \times 10^{19}$ ions $^{-2}$, respectively. Blisters are seen for the lower dose of $3.1 \times 10^{18}$ ions cm$^{-2}$ (14a) but are not observed at the higher dose of $6.2 \times 10^{18}$ ions cm$^{-2}$ (14b). The upper dose limit beyond which no blisters are observed has been termed "cut-off dose" and was found to be of the same order of magnitude as that was necessary to sputter off a thickness equivalent to the thickness of the blister skin. This behavior has been recently confirmed by Behrisch et al. (*22, 108*) for 9-keV He$^+$ ion bombardment of niobium at room temperature. They bombarded niobium surfaces with total doses up to $7 \times 10^{20}$ ions cm$^{-2}$, and observed formation of ridges and grooves on the surface at this highest dose after the blisters, observed initially at lower doses, had disappeared. Similar results have also been obtained (*22*) for type 304 stainless steel irradiated with 5-keV hydrogen ions, where blisters were initially observed for a dose of $4.9 \times 10^{18}$ ions cm$^{-2}$ but no blisters were observed at a dose of $4.9 \times 10^{19}$ ions cm$^{-2}$. The cut-off dose for blister disappearance has recently been observed to decrease with increasing irradiation temperature for helium ion irradiation of niobium (*109*). It has been suggested (*30*) that after a surface layer equivalent to the blister skin thickness has been sputtered off (e.g., for 10-keV He$^+$ irradiation of niobium a dose of $5.1 \times 10^{18}$ ions cm$^{-2}$ is needed, assuming a sputtering yield of 0.085 niobium atoms per helium ion and blister skin thickness of 800 Å) the resulting surface roughness and the damaged surface layers may increase the effective sputtering yield, and thus physical sputtering may prevent the formation of blisters. In addition, it has been suggested that the diffusion of the implanted gas through the highly damaged surface layers may contribute (*108*) to the observation of no blisters at high doses. Recently Evans (*109a*) suggested that the blister formation at these high doses of low energy ($< 15$ keV) helium ion irradiation may be prevented because of the increased surface roughness which hinders coalescence of small helium bubbles.

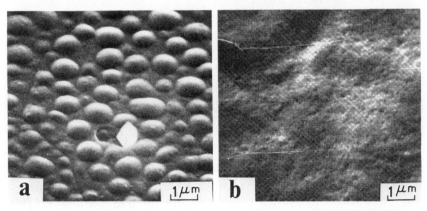

Journal of Nuclear Materials

*Figure 14.   SEMs of polycrystalline niobium irradiated at room temperature with 15-keV $^4$He$^+$ ions to total doses of (a) $3.1 \times 10^{18}$ ions cm$^{-2}$, (b) $6.2 \times 10^{19}$ ions cm$^{-2}$ (30). The target surface in (a) was annealed and electropolished, whereas in (b) it was cold-worked and mechanically polished prior to irradiation.*

## Dose Rate

The rate of gas buildup near the implant depth and the subsequent blister formation for a given irradiation temperature depend on the dose rate, i.e., the incident ion flux, and the rate of gas release from the surface. Depending on the balance of gas trapping and gas release, there may or may not be an effect of dose rate on blister formation. The blister density and the critical dose for blister appearance is affected by dose rate in some cases (*107*).

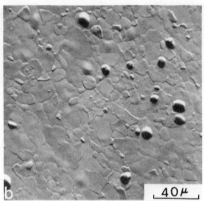

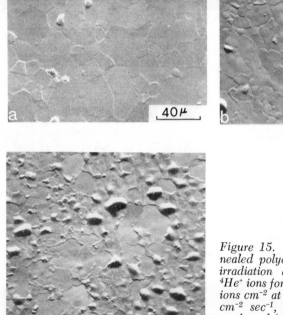

Applications of Ion Beams to Metals

*Figure 15.   SEMs of surfaces of annealed polycrystalline vanadium after irradiation at 900°C with 500-keV $^4He^+$ ions for a total dose of 6.2 × 10$^{17}$ ions cm$^{-2}$ at fluxes of (a) 1 × 10$^{13}$ ions cm$^{-2}$ sec$^{-1}$, (b) 1 × 10$^{14}$ ions cm$^{-2}$ sec$^{-1}$, and (c) 1 × 10$^{15}$ ions cm$^{-2}$ sec$^{-1}$ (107)*

For vanadium irradiated at 900°C with 500-keV helium ions, Das and Kaminsky (*107*) observed an increase in blister density with increasing dose rate. Figures 15a–c illustrate this for dose rates of 1 × 10$^{13}$, 1 × 10$^{14}$, and 1 × 10$^{15}$ ions cm$^{-2}$ sec$^{-1}$, respectively. The total dose in all three cases was the same, 6.2 × 10$^{17}$ ions cm$^{-2}$. There are two types of blisters observed in these micrographs, one type is very small in size, and the other is larger. The average diameters of blisters of both types do not seem to depend significantly on the beam flux. For the small size blisters the value for the average blister diameter ranges from $\sim 0.1$ $\mu$m to $\sim 0.3$ $\mu$m, and for the large size blisters the average diameter ranges from $\sim 3$ $\mu$m to $\sim 15$ $\mu$m for all three dose rates. The blister density for the small size blisters is about the same $[(5 \pm 2) \times 10^6$ blisters cm$^{-2}]$ for the three fluxes. For the larger size blisters, however, the blister density depends on the flux. The values increase from $(1.0 \pm 0.5) \times 10^4$, to $(3 \pm 1) \times 10^4$, and to $(7 \pm 2) \times 10^4$ blisters cm$^{-2}$ as the flux increases from $1 \times 10^{13}$, to $1 \times 10^{14}$, and to $1 \times 10^{15}$ ions cm$^{-2}$ sec$^{-1}$, respectively. For niobium irradiated at 900°C with 500-keV He$^+$ ions to a dose of $6.2 \times 10^{17}$ ions cm$^{-2}$, a similar increase in blister density has been observed (*107*) for the larger size blisters. Again, two types of blisters were observed, one type having smaller diameters (0.3–1.0 $\mu$m) and the other type having larger diameters (3–15 $\mu$m). The blister density for the large blisters increased from $(3 \pm 2) \times 10^4$ blisters cm$^{-2}$ to $(6 \pm 3) \times 10^4$ blisters cm$^{-2}$ as the flux was increased from $1 \times 10^{13}$ to $1 \times 10^{14}$ ions cm$^{-2}$ sec$^{-1}$, but a further increase in flux to $1 \times 10^{15}$ ions cm$^{-2}$ sec$^{-1}$ did not seem to change this value within the quoted error limits. However, for niobium irradiated at 700°C with 250-keV D$^+$ ions, Das and Kaminsky (*107*) observed an increase in blister size with increasing dose rate. Some of the blisters for these irradiations had "crow-foot" shape (as discussed under Channeling Condition, below), and the length of each of the three prongs of a particular blister increased by approximately a factor of three as the dose rate was increased from $1 \times 10^{14}$ ions cm$^{-2}$ sec$^{-1}$ to $1 \times 10^{15}$ ions cm$^{-2}$ sec$^{-1}$ at constant total dose.

The dependence of the critical dose for blister appearance $C_{bl}$ on dose rate has been investigated for 15-keV H$^+$ irradiation of molybdenum at room temperature by Verbeek and Eckstein (*64*). Figure 16a shows a plot of $C_{bl}$ as a function of the dose rate. $C_{bl}$ decreases with increasing dose rate, ranging from $2.9 \times 10^{14}$ ions cm$^{-2}$ sec$^{-1}$ (this is equal to the current density of $0.46 \times 10^{-4}$ A. cm$^{-2}$ plotted in Figure 16a) to $2.4 \times 10^{15}$ ions cm$^{-2}$ sec$^{-1}$. This can be understood if one considers that during the irradiation, the loss of hydrogen from the implant depth by diffusion competes with the incoming flux. Thus, $C_{bl}$ may be reached at a lower dose for a higher incident flux than for a lower incident flux. Recently, Möller et al. (*71*) studied the effect of dose rate for 300-keV and 350-keV deuteron irradiation of nickel at $-153$°C. Their results show [*see* Figure 16b] that $C_{bl}$ increases with increasing dose rate ranging from $3.7 \times 10^{15}$ ions cm$^{-2}$ sec$^{-1}$ ($\simeq 6 \times 10^{-4}$ A. cm$^{-2}$) to $1 \times 10^{16}$ ions cm$^{-2}$ sec$^{-1}$ ($\approx 16.9 \times 10^{-4}$ A. cm$^{-2}$). In Figure 16b, at low dose rates, the critical dose increases linearly and tends to level off at higher dose rates. These opposite trends, observed for the low energy proton irradiation of molybdenum (16a) and for high energy deuteron irradiation of nickel (16b),

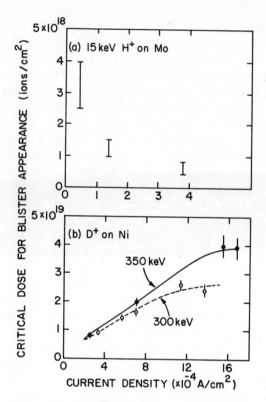

*Figure 16. The dependence of critical dose for blister appearance $C_{bl}$ on the incident ion flux for (a) 15-keV $H^+$ irradiation of molybdenum at room temperature (64), (b) 300- and 350-keV $D^+$ irradiation of nickel at $-153°C$ (71)*

cannot be readily understood at this time. According to Möller et al. (71) if one assumes that an increase in dose rate will increase the bubble density, and the volume of each bubble increases with increasing total dose, then for a given total dose the trapped gas will be distributed into a fewer number of gas bubbles in the case of low dose rate than in the case of high dose rate.

## Target Temperature

The target temperature is one of the most important parameters affecting the radiation blistering process. For example, the average blister diameter, blister density, and the exfoliation of blister skin (which determines the erosion yield) depend on target temperature. An example (39, 60, 70) of the effect of irradiation temperature on these parameters

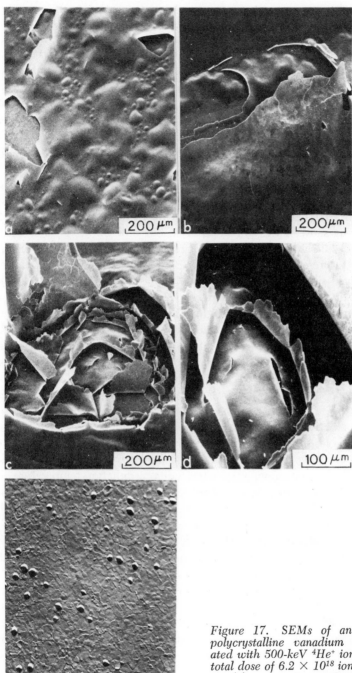

Figure 17. SEMs of annealed polycrystalline vanadium irradiated with 500-keV $^4He^+$ ions to a total dose of $6.2 \times 10^{18}$ ions $cm^{-2}$ at (a) room temperature, (b) 300°C, (c) 500°C, (d) 600°C, and (e) 900°C (60, 70)

is illustrated in Figure 17 for vanadium irradiated with 500-keV He$^+$ ions to a total dose of $6.3 \times 10^{18}$ ions cm$^{-2}$. For irradiation at room temperature, blisters with average diameters ranging from 15 $\mu$m to 350 $\mu$m can be seen in 17a. There is exfoliation of blister skins in some areas. As the irradiation temperature is increased from room temperature to 300°C (17b), the average blister diameter increases, and most of the irradiated area is occupied initially by one large blister, which subsequently exfoliates, and a second blister is formed and starts to exfoliate. A third blister has also been formed in the area which was exposed by the exfoliation of the first two blister skin layers (17b). Irradiation at 500°C (17c) shows an even stronger increase in blister exfoliation since six exfoliated skin layers are observed in certain regions. On further increasing the irradiation temperature to 600°C, there is still a large blister covering most of the irradiated area, but now the number of exfoliated blister skin layers decreases to three (17d). Increasing the irradiation temperature even higher to 800°C causes the average blister diameters to decrease significantly to only 10–20 $\mu$m, and no large scale exfoliation is observed (17e). Similar results have been obtained by Bauer and Thomas (38, 65, 69) for a slightly lower energy (300 keV) helium ion irradiation of vanadium to a lower dose of $2$–$4 \times 10^{18}$ ions cm$^{-2}$. For irradiation at 400°C to a dose of $4 \times 10^{18}$ ions cm$^{-2}$, they observed severe exfoliation of blister skins, but only a few small blisters with no severe exfoliation were observed for irradiation at 800°C to a dose of $2 \times 10^{18}$ ions cm$^{-2}$. They increased the irradiation temperature to 1200°C and observed no blisters after a dose of $2 \times 10^{18}$ ions cm$^{-2}$, but the surface had small holes with crystallographically oriented edges (see Figure 18). These holes may be caused by the intersection of large bubbles with the surface. Thus, the blister diameter and exfoliation of blister skin in vanadium is maximum at an intermediate irradiation temperature (300°–600°C). A similar temperature dependence of blister diameter and blister exfoliation has been observed for a number of other metals and alloys irradiated with helium ions as will be discussed below.

For niobium irradiated with 300-keV helium ions to a total dose of $4 \times 10^{18}$ ions cm$^{-2}$, severe exfoliation has been observed (38, 69) at 400° and 600°C. On increasing the irradiation temperature to 1000° and 1200°C, holes were observed on the surface (69, 110). For annealed polycrystalline niobium irradiated with 500-keV He$^+$ ions to a dose of $6.2 \times 10^{18}$ ions cm$^{-2}$, blister skin exfoliation was observed (111) at room temperature and 600°C, but at a temperature of 900°C only small blisters were observed (average diameters of $\sim 15$ $\mu$m) with no exfoliation of blister skin.

Irradiation of stainless steel type 304 (70, 112) and type 316 (65, 113) with 500-keV and 300-keV helium ions, respectively, showed a similar temperature dependence of blister skin exfoliation. Similarly, for molybdenum (38, 65) irradiated with 300 keV He$^+$ ions, there was severe exfoliation of blister skin at 400° and 600°C after a dose of $4 \times 10^{18}$ ions cm$^{-2}$, but at 800°C there were only small blisters, with average diameters of 20–30 $\mu$m and with no severe exfoliation after a dose of $2 \times 10^{18}$ ions cm$^{-2}$. At even a higher irradiation temperature of 1200°C some blisters and some holes were observed on the surface after a total dose of $2 \times 10^{18}$ ions cm$^{-2}$. A similar behavior has also been observed (37, 41) in molybdenum irradiated with low energy (36 keV) helium ions, where severe exfoliation was observed at an intermediate irradiation temperature of 527°C, but exfoliation was less severe at room temperature and at 827°C. At high irradiation temperature of 1327°C the irradiated surface was covered with "pin holes." These pin holes were similar to those shown in Figure 18 but did not have crystallographically oriented edges.

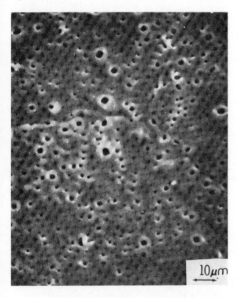

10μm

Journal of Nuclear Materials

*Figure 18. SEM of annealed polycrystalline vanadium irradiated at 1200°C with 300-keV $^4$He$^+$ ions to a dose of $2 \times 10^{18}$ ions cm$^{-2}$ (65)*

Since the exfoliation of blister skin strongly depends on target temperature, it is apparent that the erosion yields will also show a strong dependence on target temperature if the other irradiation conditions remain the same. The erosion yields have been estimated for a number of metals and alloys irradiated with helium ions as a function of temperature, and Figure 19 shows a plot of some of the available data (60, 65, 70, 111, 114). The temperature-dependence of erosion yield shows a maximum for type 304 stainless steel and vanadium. For type 304 stainless steel irradiated with 500-keV He$^+$ ions, the erosion yield goes through

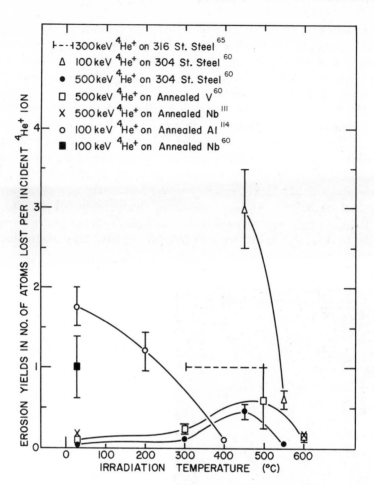

*Figure 19. Erosion rates for different metals and alloys as a function of irradiation temperature for different projectile energies and total doses. — — —, type 316 stainless steel with 300-keV $^4He^+$ for $4 \times 10^{18}$ ions $cm^{-2}$; $\triangle$, type 304 stainless steel with 100-keV $^4He^+$ for $3.1 \times 10^{18}$ ions $cm^{-2}$; $\bullet$, type 304 stainless steel with 500-keV $^4He^+$ for $3.1 \times 10^{18}$ ions $cm^{-2}$, $\square$, annealed vanadium with 500-keV $^4He^+$ for $6.2 \times 10^{18}$ ions $cm^{-2}$, X, annealed niobium with 500-keV $^4He^+$ for $6.2 \times 10^{18}$ ions $cm^{-2}$, $\bigcirc$, annealed aluminum with 100-keV $^4He^+$ for $6.2 \times 10^{18}$ ions $cm^{-2}$, $\blacksquare$, annealed niobium with 100-keV $^4He^+$ for $3.1 \times 10^{18}$ ions $cm^{-2}$.*

a maximum at about 350–550°C. For polycrystalline vanadium irradiated with 500-keV He$^+$ ions, the maximum erosion yield appears to be somewhere between 300° and 500°C. The temperature at which the erosion yield is a maximum depends on the type of target material. For annealed aluminum irradiated with 100-keV He$^+$ ions, it appears that the maximum in the erosion yield is at a much lower temperature compared with stainless steel and vanadium, possibly below room temperature.

The erosion yield depends on projectile energy for some materials. Figure 19 also shows the few available data for the erosion yields for niobium irradiated with 100-keV and 500-keV He$^+$ ions. The erosion yield for niobium at room temperature is about five times higher for 100-keV helium ion irradiation than for 500-keV irradiation. Similarly, for type 304 stainless steel one observes that as the helium projectile energy is reduced from 500 to 100 keV, the erosion yield increases by approximately a factor of seven for the same dose of $3.1 \times 10^{18}$ ions cm$^{-2}$ at $\sim 450°$C. The observation in niobium and stainless steel that the erosion yield is higher for the low ion energy than for the high one for the temperature and dose range studied does not seem to hold for all target materials. Some recent results (119) on annealed polycrystalline aluminum irradiated at room temperature with 100-keV, 250-keV, and 500-keV helium ions show erosion yields of $\sim 1.75 \pm 0.25$, $1.44 \pm 0.5$, and $1.67 \pm 0.8$, respectively. The reason for this difference in behavior in aluminum as compared with niobium or stainless steel is not clearly understood at present.

The observed change in erosion yield with temperature has been related (39, 60) to the strong temperature dependence of the yield strength of the material. For example, the yield strength of annealed type 304 stainless steel at 450°C is half of its value at room temperature (115). Similarly the tensile strength of vanadium at 900°C is less than a third of its value at room temperature (116). Thus, for a particular dose, the collected gas in a bubble can deform the surface skin more readily at high temperature, e.g., at 450°C in stainless steel, than at room temperature. In addition, the kinetic pressure of the gas in the bubble will increase with temperature and may also enhance this process. However, at very high temperatures, e.g., at 900°C in vanadium, helium may be released through the surface either by atomic diffusion or by migration of small bubbles. Therefore, the amount of helium trapped in the lattice will be affected by the helium release rate and the bubble nucleation rate and will, in turn, determine the size and density of helium bubbles and the subsequent blister size. Thus, the degree of blistering and exfoliation of blister skin is maximized if the temperature is high enough so that the surface can be deformed easily but low enough that the helium release from the surface is still very small.

## Temperature Dependence of Gas Reemission

The temperature dependence of blister rupture and exfoliation is closely related to the temperature dependence of the reemission of

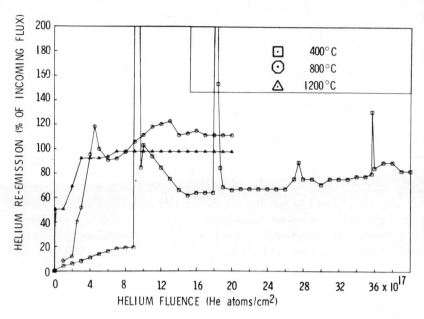

*Figure 20. Helium reemission from vanadium during 300-keV He⁺ implantation as a function of dose at three different implantation temperatures (66)*

trapped gas. Gas reemission measurements, associated with blister rupture, have been made by several authors for different target–projectile systems. For monocrystalline copper irradiated at room temperature with 125-keV $D^+$ ions, gas bursts were observed by Kaminsky (12) from blister rupture, as described at the beginning of this chapter. These early observations showed that the gas bursts were indeed caused by blister rupture, since the number of gas bursts correlated well with the number of pits observed on the surface. Extensive measurements of helium gas reemission from Nb, V, Mo, stainless steel, and Pd during irradiation have been made by Bauer et al. (31, 35, 38, 66) for different target temperatures and different total doses. Figure 20 shows an example of helium reemission from vanadium as a function of helium fluence for three different irradiation temperatures (66). For 400°C irradiation it can be seen that an abrupt change in reemission occurs after a certain dose is attained. The sudden onset of reemission, with a peak value well in excess of 100% of the incoming particle flux, occurs at nearly the same dose as that at which blister rupture is observed optically. The gas buildup prior to the sharp increase in the gas release explains the observed reemission rates in excess of 100% of the incoming flux. After the burst the reemission rate returns to a relatively low value.

With continued irradiation helium again accumulates resulting in the periodic reemission of "bursts", as seen in Figure 20 for 400°C. The number of reemission bursts observed have a one-to-one correspondence with the number of the large size exfoliated blister skin layers on the surface. At higher irradiation temperature (e.g., at 800°C) no large periodic bursts are observed, consistent with the observed changes in the surface topography. At a high temperature of 1200°C the reemission rate increases to $\sim 100\%$ almost immediately for small dose values and appears to remain constant thereafter. At this high irradiation temperature small pores were observed on the surface as shown earlier in Figure 18.

The first burst of gas in the reemission measurements is related to the exfoliation or rupture of the first blister skin. Figure 21 is a plot of the dose for the onset of first reemission burst for Nb, V, Mo, and type 316 stainless steel irradiated with 300-keV He$^+$ ions [taken from the helium reemission curves of Bauer et al. (38, 66)] as a function of irradiation temperature. One notices that the onset dose for the first reemission burst decreases with increasing target temperature. This can be expected since the critical dose for blister appearance also shows a decrease with increasing target temperature.

Hydrogen reemission from some metals during irradiation with H$^+$ ions shows a different behavior from helium reemission for comparable

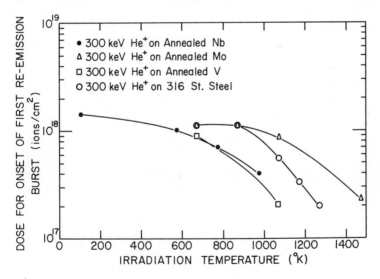

*Figure 21.   Dose for the onset of first reemission burst for Nb, Mo, V, and type 316 stainless steel as a function of irradiation temperature during implantation with 300-keV He$^+$ ions. These doses were taken from the helium reemission measurements by Bauer et al.* (38, 66).

irradiation conditions. The hydrogen reemission rises smoothly to an equilibrium value in many cases. For example, during 150-keV $H^+$ implantation in type 316 stainless steel at $-93°C$, $-75°C$, and $60°C$, Bauer and Thomas (66) observed that the hydrogen reemission rose smoothly as the dose was increased and attained an equilibrium value after a certain dose. The dose at which the equilibrium value was attained was lower at the higher temperature of $60°C$ than at $-93°C$. Only for irradiation at $-93°C$ were blisters observed on the surface (65). For 150-keV $H^+$ implantation in molybdenum a similar smooth rise in the reemission was observed for implantation at $0°C$ and in the range $25–100°C$, but an abrupt rise in reemission was observed for $-115°C$ irradiation, somewhat similar to the helium reemission behavior. No periodic reemission bursts, as seen for $He^+$ irradiation of many metals, have been reported for hydrogen irradiation of metals. This is consistent with the observation that no severe exfoliation of multiple blister skin layers occurs in most metals because of their high solubility and diffusivity for hydrogen.

## Channeling Condition of the Projectile and Crystallographic Orientation of the Irradiated Surface

In a monocrystalline solid a penetrating projectile can be guided by the regular lattice arrangement, e.g., through the spaces between the planes (planar channeling) or along channels formed by parallel rows of atoms (axial channeling) if the impact parameters are suitably chosen.

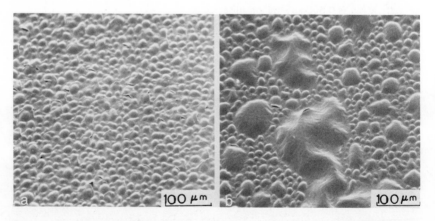

Journal of Applied Physics

Figure 22. SEMs of (111) surface planes of niobium monocrystal after room temperature irradiation to total dose of $6.2 \times 10^{18}$ ions $cm^{-2}$ of 500-keV $^4He^+$ ions (a) incident at $\sim 5°$ with respect to surface normal (unchanneled ions), (b) channeled axially in the [111] direction (73)

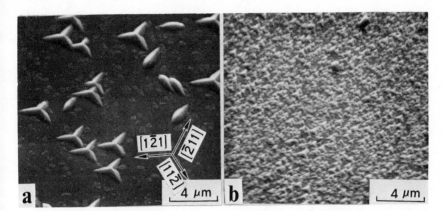

Applied Physics Letters

*Figure 23. SEMs of (111) monocrystalline niobium surfaces after irradiation at 900°C to total dose of 6.2 × 10^18 ions cm^{-2} of 500 keV He^+ ions (a) well channeled along the [111] axis, (b) incident at ~ 5° off the surface normal (un-channeled) (118)*

The channeling of projectiles affects the implant depth of the projectiles and the radiation-induced defect concentration. For a recent review of the various aspects of the channeling phenomenon *see* Ref. *117.* The blister formation on the surfaces of monocrystalline targets is expected to depend on the channeling condition of the projectile. For monocrys-talline Nb (111) surfaces irradiated at room temperature with 500-keV He^+ ions, Das and Kaminsky (73) observed blisters with larger average diameters for axially channeled ions than for unchanneled ions as shown in Figures 22a and b, respectively. This effect is somewhat similar to the effect of projectile energy on blister diameter discussed above, because the channeled ions are implanted at a greater depth from the surface than the unchanneled ions. Similar observations were also made by Verbeek and Eckstein (64) in Mo (100) surface irradiated with 150-keV He^+ ions. The average blister diameter was 6.0 $\mu$m for the channeled case as compared with 4.8 $\mu$m for the unchanneled case. They also found the blister skin thickness to be larger for the channeled case ( ~ 0.4 $\mu$m) than for the unchanneled case ( ~ 0.3 $\mu$m).

Channeling of the projectiles affects the blister density significantly (118). Figure 23 shows an example of blisters formed on Nb (111) surface irradiated at 900°C with 500-keV He^+ ions for ions channeled along [111] axis (a) and for non-channeled ions (b) for the same total dose of 6.2 × 10^18 ions cm^{-2}. These blisters have a threefold symmetry and have been called "crow-foot" blisters. Each of the three prongs of a blister has two distinct facets, which intersect along the lines whose projections on the (111) plane are in the [$\bar{1}2\bar{1}$], [$\bar{1}\bar{1}2$], and [$2\bar{1}\bar{1}$] directions.

No crow-foot blisters are observed whose prongs are pointed along the other equivalent [1$\bar{2}$1], [11$\bar{2}$], and [$\bar{2}$11] directions. The blister density is lower by approximately two orders of magnitude for the channeled helium projectiles (23a) as compared with the unchanneled ones (23b). Such a behavior is not unexpected since the channeling of projectiles helps to reduce the radiation damage in near-surface regions (corresponding to the initial part of the trajectory of the channeled projectiles), and thereby reduces the number of radiation-induced nucleation sites for bubble formation. The lower blister density for the channeled case helps to retain the well defined crystallographic character of the blisters (23a).

The effect of the crystallographic orientation of the irradiated surface on blister formation has not been investigated in as much detail as some of the other parameters described above. The density, diameter, and shape of the blisters depend on the crystallographic orientation of the surface with respect to the incident ion beam. Milacek and Daniels (119) studied orientation dependence of blister density in polycrystalline aluminum crystals with large grains (1–1.5 mm) bombarded with protons in the 10–100 keV range. The orientations of the individual grains were determined by the standard Laue backreflection technique but using a microfocus x-ray beam. They observed the lowest blister density in grains with orientation near {111}, and this was followed next by grains near {100} orientation. The grains near {110} and higher indices orientations had the maximum density of blisters. They attributed this to the orientation dependence of the range of ions; the greater the range of ions, the greater is the amount of retained gas and consequent blistering. Figure 24 shows an example of orientation dependence of blister diameter for polycrystalline niobium irradiated at room temperature with 15-keV He$^+$ ions. The variation of the blister diameter among the different grains can be readily seen. This may result from the fact that even in polycrystalline material certain grains may be favorably oriented so that the ions may be channeled to some extent.

For irradiation of monocrystalline copper with 125-keV D$^+$, Kaminsky (19) observed the pits from ruptured blisters to be more square, more rectangular, and more triangular for (100), (110), and (111) surfaces, respectively. Some recent results by Kaminsky and Das (120) on monocrystalline niobium surfaces irradiated at 900°C with 500-keV $^4$He$^+$ ions show different blister shapes for different orientation of the irradiated surface. The blisters formed on the Nb (111) surface had a "crow-foot" shape with three prongs as described in Figure 23. The blisters formed on the (110) plane had two major prongs, each with two facets. The two facets of one prong intersect along a line whose projection on the (110) plane is close to the [1$\bar{1}\bar{1}$] direction, and the two facets of the other prong intersect along a line whose projection is close to the [1$\bar{1}$3] direction. The blister alignment with respect to each other had a twofold symmetry on the (110) plane. The blisters formed on the (001) surface plane had one major prong with two facets intersecting along a line whose projection on the (001) surface was in the [110] direction. In addition, there were two lobes at one end of the blister. The orientation of the blisters with respect to each other had a fourfold symmetry. These observations on blister shape in niobium have been related (*120*) to the intersection of certain active slip planes with the surface plane of the monocrystal during blister formation.

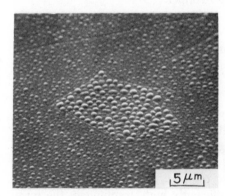

Figure 24. SEM of annealed poly-crystalline niobium surface after irradiation at room temperature with 15-keV $He^+$ ions to a dose of 3.1 $\times$ 10$^{18}$ ions cm$^{-2}$ showing different blister sizes in different grains

Surface features similar to the "crow-foot" blisters described in Figure 23 have been reported by Clausing et al. (*121*) for a polycrystalline niobium sample oxidized at 850°C for 12 hr in oxygen at 5 $\times$ 10$^{-5}$ Torr. These features were identified (*122*) as oxides formed because of internal oxidation and eruption of oxide particles through the surface of niobium samples. More recently Bauer and Thomas (*123, 124*) observed similar features in polycrystalline niobium samples implanted with 150-keV $H^+$ at temperatures of 700°–1000°C and at doses up to 3 $\times$ 10$^{19}$ ions cm$^{-2}$. They observed that the appearance of these surface features also depended on the procedure for polishing the surface. For electropolished samples, they observed these features both in the implanted area and in the unimplanted area; whereas for mechanically polished surfaces these features were observed only in the irradiated area. They initially identified these features to result from oxides and/or carbides of Nb. This is in contrast to Das and Kaminsky (*125*), who did not observe any crow-foot blisters during helium implantation in unirradiated areas of electropolished niobium samples. Moreover, they observed (*125*) holes on the surface after a quick etching of the blistered surface, and the thickness of the blister skin measured from the holes was close to the projected range of the ions. More recently Thomas and Bauer

(126) have identified the surface features observed during their 150-keV H$^+$ implantation to be niobium carbide precipitate, Nb$_2$C, and not from oxides. These different observations imply that surface features can arise by different processes even though they may display similar crystallographic features.

Monocrystalline Cu (111) surfaces show triangular-shaped oxide microcrystallites after oxidation at elevated temperatures without being exposed to any irradiation (127). Similar features have been observed on (111) copper surfaces after sputtering with He$^+$ ions (128). Similar triangular pits were observed by Kaminsky (12) from rupture of deuterium blisters on the (111) surface of copper.

### Models for Blister Formation

One of the necessary conditions for blister formation in metals and alloys is the agglomeration of implanted gas atoms to form gas bubbles in near-surface regions. Furthermore, for irradiation temperatures above liquid nitrogen temperature, a part of the lattice damage caused by the incident projectiles anneals out by recombination of interstitials and vacancies (Frenkel–pair recombinations) because of the high mobility of self interstitials. Implanted inert gas atoms such as helium are mobile at room temperature by interstitial diffusion (e.g., helium in W (129, 130) and Ni (131)) unless they are trapped. If both the vacancy and implanted gas concentrations are sufficiently high, trapping in the form of helium–vacancy complexes can occur. Experimental studies by Kornelsen et al. (130, 131) and Picraux and Vook (132) show that more than one helium atom can be trapped at a vacancy to form such helium–vacancy complexes. Theoretical calculations by Wilson et al. (133, 134, 135) for the binding energies of the helium atoms trapped at the vacancies in tungsten reasonably agree with the experimental binding energies for various postulated helium–vacancy complexes. These helium–vacancy complexes can be considered as embryos for nucleation of helium bubbles in metals. There is little theoretical work on the homogeneous nucleation of helium bubbles from these helium–vacancy complexes. Less theoretical work has been done on hydrogen implantation in metals from the diffusion and trapping point of view, than on helium in metals. As mentioned earlier, for hydrogen there is an added complication from its chemical trapping in many metals. In the following discussion of models for blister formation, we will emphasize the formation of helium blisters.

Once the gas bubbles have nucleated, depending on the various parameters mentioned earlier, they can grow by absorbing gas atoms and/or vacancies or by coalescence. Experimentally, large diameter blisters (as described earlier) have been observed even at low temperatures where vacancies are not mobile. Qualitative models for the formation of blisters from the small gas bubbles, such as helium bubbles in molybdenum with diameters of 20–40 Å, have been suggested by a number of authors (*31, 33, 40, 41, 58, 59, 100, 136–138*). These fall into three major categories:

(1) Bubble coalescence model
(2) Percolation model—based on the concepts of percolation theory
(3) Stress model—considers mainly the stresses in the implanted layer

Models based on coalescence of bubbles have been suggested independently by Blewer and Maurin (*58, 59*), Das and Kaminsky (*32, 136*), McCracken (*41*), and Evans et al. (*40, 100*). These models mostly consider the formation of helium blisters in metals. Since the diffusivity of helium in most metals at sufficiently low temperatures is very low, it is very unlikely that the small (20–40 Å diameter) bubbles will grow by diffusion of helium atoms into them. Also the mobility of helium bubbles in metals at low temperatures is very low (e.g., *see* estimates by Martin (*139, 140*) for helium bubbles in niobium). However, growth of bubbles by absorption of vacancies is possible when the vacancies become mobile. As the total dose of helium ions increases, the density of these small bubbles increases. Very high densities of small helium bubbles have been observed in many metals irradiated to high doses ($> 1 \times 10^{17}$ ions cm$^{-2}$) (*31, 38, 100*). Figure 25 shows an example in a transmission electron micrograph of molybdenum irradiated at room temperature with 18-keV He$^+$ ions to a dose of $5 \times 10^{17}$ ions cm$^{-2}$ (*40, 100*). An outstanding feature of this micrograph is that it shows blisters (as can be seen from the bend contour pattern resulting from the dome-shaped blister skin) together with the small helium bubbles. Evans et al. (*100*) observed high density of small bubbles both in the blister skin and in the areas in between the blisters.

At the critical dose for blister appearance the bubble density is high enough that the coalescence of small bubbles occurs. Figure 26a–e shows schematically various stages leading to blister formation. The transition from the stages shown in 26c–e occur very rapidly with only a relatively small increase in the helium ion dose as the in situ observations indicate (*141*). Evans (*40*) has suggested that the coalescence of bubbles becomes very rapid when the volume swelling ($\Delta V/V$) from the gas bubbles in the implant region reaches a critical value. Thus the critical dose for blister appearance $C_{bl}$ is governed by a critical value of volume swelling ($\Delta V/V$)$_c$ caused by the gas bubbles. At low temperatures, the initial coalescence of two bubbles will occur essentially at constant gas volume, thus leaving the total volume swelling in the implant region unchanged. The coalescence, once started becomes a runaway process,

*Figure 25.    Transmission electron micrograph of molybdenum irradiated at 25°C with 18-keV He⁺ to a dose of 5 × 10¹⁷ ions cm⁻² showing dome-shaped blisters [after Evans, see e.g. Refs. 40, 100].*

and an unstable high pressure cavity (26d) can form which leads to the plastic deformation of the surface layer. This dose at which the critical volume condition is reached can be considered as the critical dose for blister appearance $C_{bl}$. Using this critical volume condition Evans et al. (*100*) calculated the $C_{bl}$ values for niobium and molybdenum irradiated with He⁺ ions of different energies (30–300 keV) by assuming certain $(\Delta V/V)_c$ values and certain initial helium bubble radii. A value of $(\Delta V/V)_c \approx 50\%$ gave better agreement with the available experimental $C_{bl}$ values (*33, 37, 52*).

Now let us consider the pressure in the unstable cavity that will lead to eventual plastic deformation of the top surface. Before coalescence the gas pressure $p$ in a small bubble can be given by the well known relation:

$$p = \frac{2\gamma}{a} \tag{3}$$

where $\gamma$ is the surface energy, and $a$ is the bubble radius. If we consider a bubble of radius $\sim 20$ Å in niobium, the pressure will be $2.1 \times 10^{10}$ dynes cm$^{-2}$ (20,725 at.), taking $\gamma$ for niobium to be 2100 ergs cm$^{-2}$ (*142*). Now if two bubbles coalesce, and the new bubble is assumed to be spherical with a volume that is sum of the volume of each bubble the new equilibrium gas pressure should be lower by a factor of $2^{1/3}$, but since the number of gas atoms has not changed there will be an excess internal pressure. Thus as the coalescence progresses the excess internal pressure will increase rapidly. Once the excess pressure within the coalesced bubble (the unstable cavity in Figure 26d) exceeds a critical value, it can deform the material above it to form visible blisters (Figure 26e). An expression for this critical pressure $p_{cr}$ needed in the cavity to

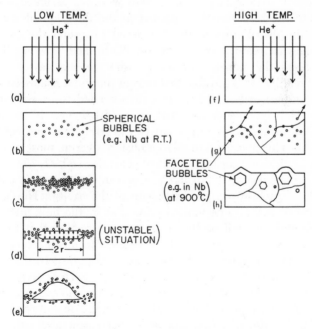

*Figure 26. Schematics of mechanisms of blister formation in metals at (a)–(e) low temperatures, (f)–(h) high temperatures*

deform the surface layer can be derived by considering the stress needed for the onset of buckling in a circular plate that is fixed along its periphery and has a uniform pressure applied to it. For a uniform pressure $p$, a plate radius $r$, and a plate thickness $t$, the maximum stress ($\sigma_r$) can be written as ($143, 144$):

$$\sigma_{r_{\max}} = \frac{3pr^2}{4t^2} \tag{4}$$

The critical pressure $p_{cr}$ in the gas filled cavity (Figure 26d) needed to deform the skin will be simply that value which makes the stress $\sigma_r$ exceed the yield strength $\sigma_y$ of the material. Thus, $p_{cr} = (4\sigma_y t^2)/3r^2$ as has been given earlier in Equation 2. This eqaution has been used by several authors ($18, 31, 32$) to estimate gas pressure needed to form the blisters. For example, if we consider the blisters formed in niobium at room temperature during irradiation with 100-keV He$^+$ ions, the blister skin thickness $t$ is 0.36 $\mu$m (Figure 8). For a gas-filled cavity with a radius of 0.1 $\mu$m, the critical pressure needed for blister formation will be 8.46 $\times$ 10$^9$ dynes cm$^{-2}$, which is larger than the pressure ($\approx 4.2 \times 10^8$ dynes cm$^{-2}$) in a spherical bubble of radius 0.1 $\mu$m estimated from Equation 3. Thus a cavity with a radius of 0.1 $\mu$m will not form a blister unless further coalescence occurs. Here we have assumed a spherical bubble, but the coalesecnce is limited in the direction of the incident beam by the distribution in the implanted ions, and the cavity has a shape somewhat like the one shown in Figure 26d. During the coalescence process the nonequilibrium pressure in the cavity cannot be estimated readily. It is possible that during the rapid coalescence process the cavity grows to larger diameters than needed to exceed the critical pressure. The coalescence can start at several regions in the irradiated area and in certain cases form a cavity extending over the entire irradiated region. This gives rise to a large blister covering most of the irradiated area as observed for high projectile energies ($60, 71, 73$), e.g., above 1.0-MeV He$^+$ ion energy in Nb and V. One can view the rapid coalescence process as a crack growing near the tip of the cavity. Thus the extent of coalescence will also be limited somewhat by the bulk microstructure of the materials. It will be shown in the next section that for identical irradiation conditions the coalescence extends to much larger diameters for the annealed targets than for the cold-worked targets. In those cases where the coalescence extends very rapidly to give cavity diameters much larger than needed for plastic deformation, the excess pressure can build up to much higher values, and the blister skin can readily rupture and exfoliate. This exfoliation process has been sometimes called "flaking" and has been distinguished from blistering; see, for example, Ref. 113. This process occurs for higher implantation energies ($> 100$ keV) in many annealed metals in certain temperature ranges. Elevated temperatures help in the rapid coalescence and in increasing gas kinetic pressure, both of which lead to increased exfoliation. The temperature range in which this occurs has been discussed earlier under "Target Temperature."

At high target temperatures (e.g., niobium at 800–900°C), helium bubbles are faceted (*38, 39*), and the average diameters are 250–1500 Å, which is larger than those observed at room temperature. In this temperature range the growth of helium bubbles to such large sizes can occur by migration and coalescence of small bubbles as shown schematically in Figure 26f–h. Some post-irradiation annealing studies on niobium implanted with $\alpha$ particles to a dose of $5 \times 10^{20}$ $\alpha$ particles cm$^{-3}$ indicate that during annealing at 950°C for 1 hr the growth of bubbles in the size range 38–80 Å in diameter occurs solely by migration and coalescence rather than by diffusion of helium between individual bubbles (*145*). It is possible that the bubbles can migrate and coalesce to large enough diameters so that the internal pressure exceeds the critical value for blister formation (26h). However, it is also possible that near the critical dose for blister appearance, a process similar to the rapid coalescence process (26d) at low temperature may occur. The diameter of a bubble formed by coalescence of smaller bubbles will be smaller at high temperatures than for low temperatures because the critical pressure needed for deforming the top layer will be lower at high temperature because of the lower yield strength $\sigma_y$ of the metal. Also the blister skin exfoliation is reduced at high temperature, as was discussed earlier.

Another model of helium clustering leading to blister formation has been considered by Thomas and Bauer (*31*) and Wilson et al. (*138*) using percolation theory (*146*). Wilson et al. (*138*) assume that the diffusion coefficient of helium atoms is concentration dependent and rapidly increases when the helium atoms become "connected" to each other. The percolation theory predicts the concentration required for the onset of an infinitely connected region of helium atoms to occur. For bcc metals at an atom fraction of 0.243, and for fcc metals at an atom fraction of 0.199, helium atoms become "connected" to each other and hence are mobile along this infinitely connected chain (*138*). Now when the gas pressure in this interconnected helium layer exceeds the critical pressure needed for surface deformation (Equation 2), blisters can form. According to this model the critical dose for blister appearance is the dose at which the concentration of helium reaches the value for the onset of percolation (i.e., an atomic fraction of 0.243 for bcc and 0.199 for fcc metals). Wilson et al. (*138*) calculated the critical dose for blister appearance $C_{bl}$ as a function of diffusion coefficient of isolated helium atoms for 300-keV He$^+$ ion irradiation of Nb and Pd for a given ion flux. From the experimental values (38) of $C_{bl}$ for 300-keV He$^+$ irradiation of Nb at 400°C ($\approx 1.2 \times 10^{18}$ ions cm$^{-2}$), they obtained an effective diffusion coefficient of helium in niobium to be $6.1 \times 10^{13}$ cm$^{-2}$ sec$^{-1}$. This value is orders of magnitude higher than the values quoted by Blow (63) for helium in niobium in the temperature range 600°– 1200°C, and the difference may be caused in part by radiation-enhanced diffusion of helium. Using such calculated effective diffusion coefficients for 15-keV H$^+$ irradiation of molybdenum and 300-keV He$^+$ irradiation of niobium and palladium, Wilson et al. (138) showed that the critical dose for blister appearance $C_{bl}$ should decrease with increasing flux, as observed by Verbeek and Eckstein (64) (Figure 16a). However, one cannot readily explain the opposite trend observed by Möller et al. (71) for 300-keV D$^+$ irradiation of nickel at low temperatures (Figure 16b). This model does not consider the yield strength of the metal which strongly depends on target temperature, and the target microstructure which influence the critical dose for blister formation. The fact that the majority of the implanted helium is in the form of bubbles at high doses does not enter into this model directly.

The mechanisms of blister formation discussed so far consider the blister formation to be caused by plastic deformation of the surface layer resulting from high gas pressure in the bubbles. More recently Behrisch et al. (76) and Roth (137) have suggested that for low energy helium ion irradiation of niobium (< 15 keV), the blister formation may be caused not by the high gas pressure but by the stresses induced in the implanted layer. The high concentration of bubbles gives rise to large lateral stresses which are responsible for the deformation of the surface layer, and the separation of the surface layer from the bulk is thought to occur near the end of the range (76). So the thickness of the blister skin is equal to the width of the implanted layer which, according to the author (137), is responsible for the observation that the blister skin thickness for niobium irradiated with low helium ion energies is larger than the calculated projected ranges. In this interpretation it is not clear why the stress distribution in the implanted layer should be at a maximum at the end of the range of ions and not near the peak in the projected range distributions. As discussed earlier under "Projectile Energy," the higher blister skin thickness for low energy implantations may in part be caused by the swelling of the skin caused by a more uniform distribution of helium bubbles over a large portion of the implant depth.

Among the three basic models described above, the coalescence model attempts to qualitatively explain many of the experimental observations, but none of the models considers all the parameters (e.g., diffusion of the atomic helium under irradiation, migration of helium bubbles, the mechanical aspects such as gas pressure needed for onset of plastic deformation, yield strength of material, microstructural factors influencing the size and distribution of helium bubbles) sufficiently to explain the various observations.

One also has to consider the influence of surface erosion from sputtering during irradiation. The depth distribution of the implanted gas atoms will change by this type of surface erosion particularly for ions with energies where the sputtering yield is near maximum ($147$).

It can be shown that the critical dose for blister appearance $C_{bl}$ has to be lower than a value given by $t\,\rho\,N/S\,A$ in order that the blisters may form, where $t$ is the blister skin thickness, $\rho$ is the density of the material, $N$ is Avogadro's number, $S$ is the sputtering yield, and $A$ is the atomic weight of the metal. In cases where the blister skin thickness is unknown, a crude estimation of $C_{bl}$ can be obtained by taking $t$ equal to $R_p$, the projected range of ions in the metal. However, as discussed earlier, $t$ can be two to three times larger than $R_p$, particularly at low projectile energies. For cases where the sputtering yields are very high and depth of penetration $R_p$ is shallow (for example, for heavier ions such as Ar, Kr, and Xe at low energies) blisters may not form at all. Using the $C_{bl}$ value given by the percolation theory, Roth ($137$) showed that for niobium implanted with Ar, blisters may form at energies above 100 keV, whereas for light ions like helium, blisters can form at energies down to 1 keV. Blisters have indeed been observed in niobium during He+ irradiation for energies down to 1 keV ($33$). More recently, however, blisters have been observed in uranium irradiated with 25-keV Ar+ ions ($148$). Blisters have also been observed in En 40 B steel under irradiation with 200-keV $N_2{}^+$ ions to a total dose of $4.0 \times 10^{17}$ ions cm$^{-2}$ ($149$).

### Reduction of Surface Damage Caused by Radiation Blistering

From the discussions so far it is clear that considerable surface damage can be caused to metals by radiation blistering, particularly during irradiation with light inert gas ions such as helium. Recently methods to reduce surface erosion caused by radiation blistering have been investigated.

**High Target Temperature.** From the earlier discussions on the effect of target temperature, it is obvious that one possible way to reduce radiation blistering is to maintain the target surface at high enough temperature so that the implanted gas is released without forming large bubbles. For many metals there is a great reduction in surface erosion from radiation blistering (Figure 19) at temperatures above 0.4–0.5 $T_m$, where $T_m$ is the melting point in K. If the irradiation temperature is sufficiently high, helium reemission during irradiation can reach almost 100% of the incoming flux, and helium agglomeration into bubbles can be prevented. For example, for vanadium and niobium irradiated at 1200°C with 300-keV He$^+$ ions, almost 100% reemission has been observed (69). Irradiation under these conditions gives rise to a porous surface structure consisting of micron-sized holes with a spacing of several microns as shown earlier for vanadium in Figure 18. Recently Bauer and Thomas (110) observed that the porous structure in niobium obtained by 300-keV He$^+$ implantation at 1200°C is rather stable. This type of surface prevented further blister formation during 300-keV He$^+$ irradiation at 400°–600°C for a dose of $\sim 4 \times 10^{18}$ ions cm$^{-2}$ where severe exfoliation of blister skin is normally observed (110). However, in many applications it may not be always possible to maintain the surface at sufficiently high temperatures. For example, in a controlled thermonuclear fusion reactor the operating temperatures of various components exposed to D, T, and He projectiles may be limited by other design criteria. In accelerator technology, maintaining components exposed to energetic projectiles at high temperatures may be a problem. A more desirable solution would be to choose materials with microstructures which minimize the formation of blisters. Only a few microstructural parameters have been investigated so far, and some of them are discussed below.

**Cold Working.** Cold working has reduced surface erosion from blistering in certain cases (32, 39). Figure 27 shows an example (39) for polycrystalline vanadium samples irradiated with 500-keV $^4$He$^+$ ions to a total dose of $6.2 \times 10^{18}$ ions cm$^{-2}$. For the annealed sample irradiated at room temperature (27a) the blister diameters are 15–350 $\mu$m, and the blister skin has ruptured and fallen off at many places. In the cold-worked sample (27b) irradiated under identical conditions most of the blisters have diameters of 5–100 $\mu$m, and no large-scale exfoliation is observed. For irradiation of the same type of samples at 900°C the blistering is similar for the annealed (27c) and the cold-worked cases (27d). Similar results have been observed for cold-worked and annealed niobium (32) samples. For irradiation at 900°C some annealing of vanadium and niobium can occur.

Thus, it appears that cold working tends to reduce blister size and blister exfoliation at room temperature, but the effect is not as pronounced above the recrystallization temperature. Several factors may be responsible for the reduction in blistering in cold-worked samples as compared with the annealed ones. The lower yield strength of the annealed material as compared with the cold-worked one will increase blister rupture and exfoliation in the annealed sample. Moreover, the increased dislocation density and the large number of subgrains in cold-worked materials can give rise to a much finer dispersion of helium bubbles and also prevent coalescence to larger diameters.

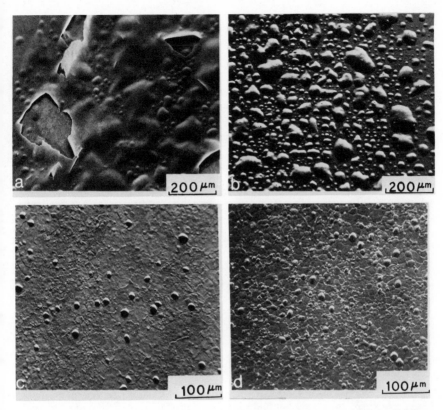

Nuclear Metallurgy

*Figure 27.    SEMs of polycrystalline vanadium surfaces after irradiation with 500-keV He⁺ ions to a total dose of $6.2 \times 10^{18}$ ions $cm^{-2}$ (a) an annealed sample at room temperature, (b) a 40% cold-worked sample at room temperature, (c) an annealed sample at 900°C, and (d) a 40% cold-worked sample at 900°C (39)*

**Grain Size and Dispersion of Second Phase.** There are no systematic studies of the effect of grain size and dispersion of second phase particles on the reduction of blistering. Some recent preliminary studies on sintered aluminum powder (SAP) containing a nominal 10.5 wt % $Al_2O_3$ (SAP 895) show a reduction in erosion from blistering as compared with annealed aluminum under identical irradiation conditions (150, 151). Figure 28 illustrates this for SAP 895 as compared with annealed aluminum for 100-keV He⁺ ion irradiation to a dose of $6.2 \times 10^{18}$ ions $cm^{-2}$ at room temperature and at 400°C. Figure 28a shows an enlarged view of a portion of an aluminum surface irradiated at room temperature. One can see that four exfoliated layers have been removed. Figure 28b shows a portion of an aluminum surface which had been irradiated at 400°C under otherwise identical conditions. The exfoliation of blister skin is

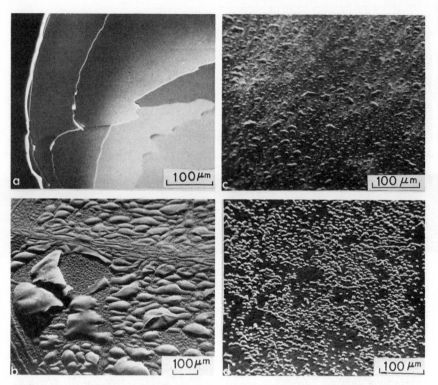

*Figure 28. SEMs of surfaces of (a) annealed aluminum irradiated at room temperature, (b) annealed aluminum irradiated at 400°C, (c) SAP 895 irradiated at room temperature, (d) SAP 895 irradiated at 400°C with 100-keV $^4He^+$ ions to a total dose of $6.2 \times 10^{18}$ ions cm$^{-2}$*

reduced at 400°C as compared with room temperature, and there is only one exfoliated layer in several areas. Figures 28c and d show typical examples of blisters formed on SAP 895 irradiated at room temperature and at 400°C, respectively. In contrast to the aluminum, where extensive exfoliation is observed, only a few blisters were ruptured in the case of SAP 895. The erosion rates estimated for the room temperature case from the ruptured and lost skins for aluminum and SAP are 1.75 ± 0.25 and 0.001 atoms per helium ion, respectively (*150*). The results for annealed aluminum held at 400°C give a value of 0.12 ± 0.05 atom per helium ion, whereas for the SAP 895 sample no exfoliation of the blisters could be observed.

The drastic reduction in erosion yield in sintered aluminum powder as compared with annealed aluminum was attributed (*151*) to the dispersion of trapped helium in the large Al–Al$_2$O$_3$ interfaces at the large grain boundaries in SAP. Formation of helium bubbles at the Al$_2$O$_3$ and aluminum interface in aluminum alloys containing dispersed Al$_2$O$_3$

particles has been observed. The average grain size in SAP 895 was ~ 0.5 μm as compared with ~ 300 μm for annealed aluminum. This large grain boundary area in SAP gives a much finer dispersion of helium bubbles than in aluminum and may prevent helium bubble coalescence to very large diameters. Furthermore, the fact that the yield strength of SAP 895, for example, is much higher ( ~ 35,600 psi) than for annealed aluminum ( ~ 1,700 psi) helps to reduce the blister rupture and exfoliation rate in SAP from that observed in aluminum.

Somewhat similar results have been recently observed for vacuum cast and sintered beryllium irradiated with 100-keV He$^+$ ions (*153*). For vacuum cast beryllium irradiated at room temperature to a dose of 6.2 × $10^{18}$ ions cm$^{-2}$, blisters with average diameters of 5–35 μm with some exfoliation were observed, whereas for sintered beryllium the blisters were smaller (average diameter of 5–15 μm), and there was no exfoliation. For irradiation at 600°C with 100-keV He$^+$ ions to a dose of 3.1 × $10^{18}$ ions cm$^{-2}$, the vacuum cast beryllium showed considerable blister exfoliation (erosion yield ~ 0.3 ± 0.1 Be atoms per incident helium ion), whereas the sintered beryllium showed greatly reduced blister exfoliation even for a higher dose of 6.2 × $10^{18}$ ions cm$^{-2}$ (erosion yield ~ 0.02 ± 0.01 Be atoms per incident helium ion).

### Future Directions

Even though a great deal of information has been generated on radiation blistering in metals, complete understanding of this phenomenon has not yet been obtained. A unified quantitative theory is lacking. It appears that considerable amount of additional experimental information is still needed on many parameters, such as critical dose for bubble and blister appearance on targets with different microstructures, grain sizes, yield strengths, and target temperatures, before such a theory can be developed.

Since radiation blistering has been identified as an important erosion process for surfaces exposed to plasmas, it is important to study this process for the radiation environments expected in controlled thermonuclear fusion reactors. While, at present, most of the data have been obtained for monoenergetic ion irradiations in a fusion reactor, one expects to have ions with a broad energy distribution impinging on the exposed surfaces. Therefore, in order to assess the severity of surface erosion by radiation blistering in thermonuclear devices, studies are needed with ions having a broad energy spectrum.

### Acknowledgments

We would like to thank J. Evans, W. Möller, and H. Verbeek for providing us with some of the original micrographs used in this review.

We are grateful to Plenum Press, New York; North-Holland Publishing Co., Amsterdam; and American Institute of Physics, New York for their permission to use certain illustrations in this manuscript. We are thankful to T. Rossing for a critical reading of the manuscript and to P. Dusza for his help in obtaining some of the experimental results used in this review.

## Literature Cited

1. Kaminsky, M., "Atomic and Ionic Impact Phenomena on Metal Surfaces," Springer-Verlag, New York, 1965.
2. Carter, G., Colligon, J., "Ion Bombardment of Solids," American Elsevier, New York, 1968.
3. Dearnaley, G., Freeman, J. H., Nelson, R. S., Stephen, J., "Ion Implantation," North-Holland, Amsterdam, 1973.
4. McCracken, G. M., *Rep. Prog. Phys.* (1975) **38**, 241.
5. Yonts, O. C., Strehlow, R. A., *J. Appl. Phys.* (1962) **33**, 2903.
6. Sheft, I., Reis, Jr., A. H., Gruen, D. M., Peterson, S. W., *Trans. Am. Nucl. Soc.* (1975) **22**, 166.
7. Sheft, I., Reis, Jr., A. H., Gruen, D. M., *J. Nucl. Mater.* (1976) **59**, 1.
8. Finn, P. A., Gruen, D. M., Page, D. L., ADV. CHEM. SER. (1976) **158**, 30.
9. Barnes, R. S., *Philos. Mag.* (1960) **5**, 635.
10. Barnes, R. S., Mazey, D. J., *Philos. Mag.* (1960) **5**, 1247.
11. Barnes, R. S., Mazey, D. J., *Proc. R. Soc.* (1963) **A275**, 47.
12. Kaminsky, M., *Adv. Mass Spectrom.* (1964) **3**, 69.
13. Kaminsky, M., *Bull. Am. Phys. Soc.* (1963) **8**, 428 (*see* Ref. *14* in Ref. *14*).
14. Primak, W., Luthra, J., *J. Appl. Phys.* (1966) **37**, 2287.
15. Stark, J., Wendt, G., *Ann. Phys.* (1912) **38**, 921.
16. Primak, W., *J. Appl. Phys.* (1963) **34**, 3630.
17. Primak, W., Dyal, Y., Edwards, E., *J. Appl. Phys.* (1963) **34**, 827.
18. Milacek, L. H., Daniels, R. D., Cooley, J. A., *J. Appl. Phys.* (1968) **39**, 2803.
19. Kaminsky, M., *IEEE Trans. Nucl. Sci.* (1971) **18**, 208.
20. Kaminsky, M., "Proceedings of the International Working Sessions on Fusion Reactor Technology," Oak Ridge National Laboratory, U. S. Atomic Energy Commission, **CONF-710624**, p. 86, 1971.
21. Kaminsky, M., *Plasma Phys. Controlled Nucl. Fusion Res. Proc. Conf.* (1975) **II**, 287.
22. Behrisch, R., Kadomstev, B. B., *Plasma Phys. Controlled Nucl. Fusion Res. Proc. Conf.* (1975) **II**, 229.
23. McDonell, W. R., *Trans. Am. Nucl. Soc.* (1975) **21**, 135.
24. Das, K. B., Roberts, E. C., Bassett, R. G., "Hydrogen in Metals," I. M. Bernstein and A. W. Thompson, Eds., p. 289, American Society for Metals, Metals Park, Ohio, 1973.
25. Hess, P. D., Turnbull, G. K., "Hydrogen in Metals," I. M. Bernstein and A. W. Thompson, Eds., p. 277, American Society for Metals, Metals Park, Ohio, 1973.
26. Kleuh, R. L., Mullins, W. W., *Trans. Metall. Soc. AIME* (1968) **242**, 237.
27. Fisher, R. M., *Electron Microsc. Struct. Mater. Proc. Int. Mat. Symp.* (1972).
28. Metals Handbook," Vol. **10**, p. 230, American Society for Metals, Metals Park, Ohio, 1975.
29. Thomas, G. J., Bauer, W., Mattern, P. L., Granoff, B., ADV. CHEM. SER. (1976) **158**, 97.
30. Martel, J. G., St. Jacques, R., Terrault, B., Veilleux, G., *J. Nucl. Mater.* (1974) **53**, 142.

31. Thomas, G. J., Bauer, W., *Radiat. Eff.* (1973) **17**, 221.
32. Das, S. K., Kaminsky, M., *J. Appl. Phys.* (1973) **44**, 25.
33. Roth, J., Behrisch, R., Scherzer, B. M. U., *J. Nucl. Mater.* (1974) **53**, 147.
34. Daniels, R. D., *J. Appl. Phys.* (1971) **42**, 417.
35. Bauer, W., Morse, D., *J. Nucl. Mater.* (1972) **44**, 337.
36. Holt, J. B., Bauer, W., Thomas, G. J., *Radiat. Eff.* (1971) **7**, 269.
37. Erents, S. K., McCracken, G. M., *Radiat. Eff.* (1973) **18**, 191.
38. Bauer, W., Thomas, G. J., *Nucl. Metall.* (1973) **18**, 255.
39. Das, S. K., Kaminsky, M., *Nucl. Metall.* (1973) **18**, 240.
40. Evans, J. H., *Nature* (1975) **256**, 299.
41. McCracken, G. M., *Jpn. J. Appl. Phys. Suppl. 2, Pt. 1* (1974) 269.
42. Blewer, R. S., Maurin, J. K., "Proceedings of Thirtieth Annual Electron Microscope Society of American Meeting," C. J. Arcenaux, Ed., p. 44, Claitor's Publishing Division, Baton Rouge, 1972.
43. Thomas, G. J., Bauer, W., "Proceedings of Thirty-third Annual EMSA Meeting," G. W. Bailey, Ed., p. 262, Claitor's Publishing Division, Baton Rouge, 1975.
44. Roth, J., Behrisch, R., Scherzer, B. M. U., *Appl. Ion Beams Met. Int. Conf.* (1974) 573.
45. Blewer, R. S., *Appl. Phys. Lett.* (1973) **23**, 593.
46. Blewer, R. S., *Appl. Ion Beams Met. Int. Conf.* (1974) 557.
47. Blewer, R. S., *J. Nucl. Mater.* (1974) **53**.
48. Roth, J., Behrisch, R., Scherzer, B. M. U., *Appl. Phys. Lett.* (1974) **25**, 643.
49. Langley, R. A., in "Ion Beam Surface Layer Analysis," O. Meyer, G. Linker, and F. Käppeler, Ed., Vol. I, p. 201, Plenum, New York, 1976.
50. Blewer, R. S., ADV. CHEM. SER. (1976) **158**, 262.
51. Pronko, P., *J. Nucl. Mater.* (1974) **53**, 252.
52. Behrisch, R., Bøttiger, J., Eckstein, W., Roth, J., Scherzer, B. M. U., *J. Nucl. Mater.* (1975) **56**, 365.
53. Langley, R. A., Picraux, S. T., Vook, F. L., *J. Nucl. Mater.* (1974) **53**, 257.
54. Hufschmidt, M., Möller, W., Heintz, V., Kamke, D., in "Ion Beam Surface Layer Analysis," O. Meyer, G. Linker, and F. Käppeler, Eds., Vol. II, p. 831, Plenum, New York, 1976.
55. Overley, J. C., Lefevre, H. W., ADV. CHEM. SER. (1976) **158**, Chap. 12.
56. Terreault, B., et al., ADV. CHEM. SER. (1976) **158**, Chap. 13.
57. Eckstein, W., Behrisch, R., Roth, J., in "Ion Beam Surface Layer Analysis," O. Meyer, G. Linker, and F. Käppeler, Eds., Vol. II, p. 821, Plenum, New York, 1976.
58. Blewer, R. S., Maurin, J. K., *J. Nucl. Mater.* (1972) **44**, 260.
59. Blewer, R. S., *Radiat. Eff.* (1973) **19**, 243.
60. Das, S. K., Kaminsky, M., *J. Nucl. Mater.* (1974) **53**, 115.
61. Kaminsky, M., Das, S. K., *Radiat. Eff.* (1973) **18**, 245.
62. Schaumann, G., Völkl, J., Alefeld, G., *Phys. Statis. Solidi.* (1970) **42**, 401.
63. Blow, S., *J. Br. Nucl. Energy Soc.* (1972) **11**, 371.
64. Verbeek, H., Eckstein, W., *Appl. Ion Beams Met. Int. Conf.* (1974) 597.
65. Thomas, G. J., Bauer, W., *J. Nucl. Mater.* (1974) **53**, 134.
66. Bauer, W., Thomas, G. J., *J. Nucl. Mater.* (1974) **53**, 127.
67. Das, S. K., Kaminsky, M., Fenske, G., in "Applications of Ion Beams to Materials," G. Carter, J. S. Colligon, and W. A. Grant, Eds., Conference Series No. **28**, p. 293, Institute of Physics, London, 1976.
68. Kaminsky, M., Das, S. K., Fenske, G., *J. Nucl. Mater.* (1976) **59**, 86.
69. Bauer, W., Thomas, G. J., *Appl. Ion Beams Met. Int. Conf.* (1974) 533.
70. Kaminsky, M., Das, S. K., *Nucl. Technol.* (1974) **22**, 373.
71. Möller, W., Pfeiffer, T., Kamke, D., "Ion Beam Surface Layer Analysis," O. Meyer, G. Linker, and F. Käppeler, Eds., Vol. II, p. 841, Plenum, New York, 1976.

72. Sinha, M. K., Das, S. K., Kaminsky, M., *Bull. Am. Phys. Soc.* (1975) **20**, 45.
73. Das, S. K., Kaminsky, M., *J. Appl. Phys.* (1973) **44**, 2520.
74. Hill, R., *Philos. Mag.* (1950) **41**, 1133.
75. Kaminsky, M., Das, S. K., Fenske, G., *Appl. Phys. Lett.* (1975) **27**, 521.
76. Behrisch, R., Bøttiger, J., Eckstein, W., Littmark, U., Roth, J., Scherzer, B. M. U., *Appl. Phys. Lett.* (1975) **27**, 199.
77. St.-Jacques, R. G., Martel, J. G., Terreault, B., Veilleux, G., Das, S. K., Kaminsky, M., Fenske, G., *J. Nucl. Mater.*, in press.
78. Schiøtt, H. E., *Mat. Fys. Medd. Dan. Vid. Selsk* (1966) **35**, no. 9.
79. Lindhard, J., Scharff, M., Schiøtt, H. E., *Mat. Fys. Medd. Dan. Vid. Selsk.* (1963) **33**, no. 14.
80. Brice, D. K., *Phys. Rev.* (1972) **A6**, 1791.
81. Ziegler, J. F., Chu, W. K., *Thin Solid Films* (1973) **19**, 281.
82. Brice, D. K., "Ion Implantation Range and Energy Deposition Distributions," Vol. 1, Plenum Data Co., New York, 1975.
83. Brice, D. K., *J. Appl. Phys.* (1975) **46**, 3385.
84. Vogl, G., *J. Phys. Paris* (1974) **35**, Supplement C-6, C6-165.
85. "Positron Annihilation," P. Hautojäroi, and A. Seeger, Eds., Lange and Springer, Berlin, 1975.
86. Brandt, W., ADV. CHEM. SER. (1976) **158**, Chap. 9.
87. Nelson, R. S., Mazey, D. J., Barnes, R. S., *Philos. Mag.* (1965) **11**, 291.
88. Chen, K. Y., Cost, J. R., *J. Nucl. Mater.* (1974) **52**, 59.
89. Boltax, A., *J. Nucl. Mater.* (1962) **7**, 1.
90. Katyal, O. P., Keesom, P. H., Cost. J. R., in "Radiation-Induced Voids in Metals," J. W. Corbett, L. C. Ianniello, Eds., CONF-710601, p. 248, National Technical Information Service, U. S. Department of Commerce, Springfield, Va., 1972.
91. Heerschap, M., Schüller, E., Langevin, B., Trapani, A., *J. Nucl. Mater.* (1973) **46**, 207.
92. Smidt, Jr., F. A., Pieper, A. G., *J. Nucl. Mater.* (1974) **51**, 351.
93. Ehrlich, K., Kaletta, D., "Proceedings of International Conference on Radiation Effects and Tritium Technology for Fusion Reactors," CONF-750989, vol. II, p. 289, National Technical Information Service, U.S. Dept. of Commerce, Springfield, Va., 1976.
94. Walker, G. K., *J. Nucl. Mater.* (1970) **37**, 171.
95. Smidt, Jr., F. A., Pieper, A. G., *Trans. Am. Nucl. Soc.* (1972) **15**, 722.
96. Serpan, Jr., C. Z., Smith, H. H., Smidt, Jr., F. A., Pieper, A. G., *AEC Symp. Ser.* (1974) **31**, 993.
97. Goland, A. N., *Philos. Mag.* (1961) **6**, 189.
98. Willertz, L. E., Shewmon, P. G., *Metall. Trans.* (1970) **1**, 2217.
99. Johnson, D. L., Cost, J. R., *Nucl. Metall.* (1973) **18**, 279.
100. Evans, J. H., Mazey, D. J., Eyre, B. L., Erents, S. K., McCracken, G. M., "Applications of Ion Beams to Materials," G. Carter, J. S. Colligon, W. A. Grant, Eds., Conference Series No. **28**, p. 299, Institute of Physics, London, 1976.
101. Brebec, G., Levy, V., Leteuirte, J., Adda, V., in "Ionic Bombardment: Theory and Applications, p. 219, Gordon and Breach, New York, 1964.
102. Rühle, M. R., in "Radiation-Induced Voids in Metals," J. W. Corbett, L. C. Ianniello, Eds., CONF-710601, p. 255, National Technical Information Service, U. S. Department of Commerce, Springfield, Va., 1972.
103. Brown, R. D., Rao, P., Ho, P. S., *Radiat. Eff.* (1973) **18**, 149.
104. Evans, J. H., *Radiat. Eff.* (1971) **10**, 55.
105. Hertel, B., Diehl, J., Botthardt, R., Sultze, H., *Appl. Ion Beams Met. Int. Conf.* (1974) 507.
106. Das, S. K., Kaminsky, M., unpublished data.
107. Das, S. K., Kaminsky, M., *Appl. Ion Beams Met. Int. Conf.* (1974) 543.
108. Roth, J., Behrisch, R., Scherzer, B. M. U., *J. Nucl. Mater.* (1975) **57**, 365.

109. St-Jacques, R. G., Veilleux, G., Terreault, B., Martel, J. G., in "Applications of Ion Beams to Materials," G. Carter, J. C. Colligan, W. A. Grant, Eds., Conference Series No. **28**, p. 313, Institute of Physics, London, 1973.
109a.Evans, *J. Nucl. Mat.* (1976) **61**, 117.
110. Bauer, W., Thomas, G. J., "Ion Beam Surface Layer Analysis," O. Meyer, G. Linker, and F. Käppeler, Eds., vol. II, p. 575, Plenum, New York, 1976.
111. Das, S. K., Kaminsky, M., *AEC Symp. Ser.* (1974) **31**, 1019.
112. Das, S. K., Kaminsky, M., *Proc. Symp. Eng. Probl. Fusion Res.*, 5th (974) IEEE Pub. No. **73**, CHO 843-3-NPS, 31.
113. Bauer, W., Thomas, G. J., *J. Nucl. Mater.* (1973) **47**, 241.
114. Das, S. K., Kaminsky, M., Rossing, T. D., unpublished data.
115. Smith, K. F., in "Reactor Handbook," Vol. 1, p. 563, Interscience, New York, 1960.
116. Keeler, J. R., Smith, K. F., in "Reactor Handbook," Vol. 1, p. 697, Interscience, New York, 1960.
117. Gemmell, D. S., *Rev. Mod. Phys.* (1974) **46**, 129.
118. Kaminsky, M., Das, S. K., *Appl. Phys. Lett.* (1972) **21**, 443.
119. Milacek, L. H., Daniels, R. D., *J. Appl. Phys.* (1968) **39**, 5714.
120. Kaminsky, M., Das, S. K., *Appl. Phys. Lett.* (1973) **23**, 293.
121. Clausing, R. E., McHargue, C. J., Spruiell, J. E., *J. Nucl. Mater.* (1974) **53**, 123.
122. Spruiell, J. E., Thesis, M.S., University of Tennessee, Knoxville, Tenn., 1960.
123. Bauer, W., Thomas, G. J., *Radiat. Eff.* (1974) **23**, 211.
124. Bauer, W., Thomas, G. J., in "Proceedings of the First Topical Meeting on the Technology of Controlled Nuclear Fusion," Vol. II, p. 517, CONF-740402-P2, National Technical Information Service, U. S. Department of Commerce, Springfield, Va., 1974.
125. Das, S. K., Kaminsky, M., *J. Nucl. Mater.* (1974) **53**, 125.
126. Thomas, G. J., Bauer, W., *J. Vac. Sci. Technol.* (1975) **12**, 490.
127. Lawless, K. R., *Rep. Prog. Phys.* (1974) **37**, 231.
128. Mazey, D. J., Nelson, R. S., Thackery, P. A., *J. Mater. Sci.* (1968) **3**, 26.
129. Kornelsen, E., *Can. J. Phys.* (1970) **48**, 2812.
130. Kornelsen, E., *Radiat. Eff.* (1972) **13**, 227.
131. Kornelsen, E., Edwards, Jr., D. E., *Appl. Ion Beams Met. Int. Conf.* (1974) 407.
132. Picraux, S. T., Vook, F. L., *Appl. Ion Beams Met. Int. Conf.* (1974) 407.
133. Wilson, W. D., Bisson, C. L., *Radiat. Eff.* (1974) **22**, 63.
134. Bisson, C. L., Wilson, W. D., *Appl. Ion Beams Met. Int. Conf.* (1974) 423.
135. Wilson, W. D., "Proceedings of the International Conference on Fundamental Aspects of Radiation Damage in Metals," **CONF-751006-P2**, vol. II, p. 1025, National Technical Information Service, U.S. Dept. of Commerce, Springfield, Va., 1976.
136. Das, S. K., Kaminsky, M., "Proceedings of the Third Conference on Application of Small Accelerators," I. L. Morgan, J. L. Duggan, Eds., Vol. I, p. 278, CONF-741040-P1, National Technical Information Service, U. S. Department of Commerce, Springfield, Va., 1974.
137. Roth, J., in "Applications of Ion Beams to Materials," G. Carter, J. S. Colligan, W. A. Grant, Eds., Conference Series No. **28**, p. 280, Institute of Physics, London, 1976.
138. Wilson, W. D., Bisson, C. L., Amos, D. E., *J. Nucl. Mater.* (1974) **53**, 154.
139. Martin, D. G., Culham Laboratory Report **CLM-R103**, UKAERE, Culham Laboratory, Abingdon, Berks, England, 1970.
140. Martin, D. G., *J. Nucl. Mater.* (1969) **33**, 23.
141. Thomas, G., Bauer, W., *J. Nucl. Mater.*, in press.

142. Radcliffe, S. V., *J. Less-Common Met.* (1961) **3**, 360.
143. Timoshenko, S., "Theory of Plates and Shells," p. 59, McGraw-Hill, New York, 1940.
144. Turner, C. E., "Introduction to Plate and Shell Theory," p. 75, Longmans, Green and Co., London, 1965.
145. Aitken, D., Goodhew, P. J., Waldron, M. B., *Nature* (1973) **244**, 15.
146. Shante, V. K. S., Kirkpatrick, S., *Adv. Phys.* (1971) **20**, 325.
147. Biersack, J. P., *Radiat. Eff.* (1973) **19**, 249.
148. Vasiliu, F., Teodorescu, I. A., in "Applications of Ion Beams to Materials," G. Carter, J. S. Colligan, W. A. Grant, Eds., Conference Series No. 28, p. 323, Institute of Physics, London, 1976.
149. Hartley, N. E. W., *J. Vac. Sc. Technol.* (1975) **12**, 485.
150. Das, S. K., Kaminsky, M., Rossing, T., *Appl. Phys. Lett.* (1975) **27**, 197.
151. Das, S. K., Kaminsky, M., Rossing, T., in "Ion Beam Surface Layer Studies," O. Meyer, G. Linker, F. Käppeler, Eds., Vol. **II**, p. 567, Plenum, New York, 1976.
152. Ruedl, E., "Irradiation Effects on Structural Alloys for Nuclear Reactor Applications," ASTM STP **484**, p. 300, American Society for Testing Materials, Philadelphia, 1970.
153. Das, S. K., Kaminsky, M., *Proc. Symp. Eng. Probl. Fusion Res., 6th,* IEEE Publ. **No. 75CH 1097-5-NPS**, p. 1151, Institute of Electrical and Electronics Engineers, New York, 1976.
154. Weissmann, R., Behrisch, R., *Radiat. Eff.* (1973) **19**, 69.
155. Wilson, K. L., Thomas, G. J., Bauer, W., *Trans. Am. Nucl. Soc.* (1975) **22**, 36.
156. Behrisch, R., Heiland, W., *Proc. Symp. Fusion Technol. 6th* (1970) 461.
157. Mikkelson, R. C., Miller, J. W., Holland, R. E., Gemmell, D. S., *J. Appl. Phys.* (1973) **44**, 935.
158. Kaminsky, M., "Atomic and Ionic Impact Phenomena on Metal Surfaces," p. 214, Springer-Verlag, New York, 1965.
159. Donhowe, J. M., Kulcinski, G. L., "Fusion Reactor First Wall Materials," L. C. Ianiello, Ed., p. 75, U. S. Atomic Energy Commission Report No. **WASH-1206**, 1972.
160. Donhowe, J. M., Klarstorm, D. L., Sundquist, M. L., Weber, W. J., *Nucl. Technol.* (1973) **18**, 63.
161. Donhowe, J. M., Klarstorm, D. L., Sundquist, M. L., Weber, W. J., *Trans Am. Nucl. Soc.* (1972) **15**, 36.
162. Roth, J., Behrisch, R., Scherzer, B. M. U., Pohl, F., *Proc. Symp. Fusion Technol., 8th* (1974) 841.
163. Bauer, W., Thomas, G. J., *J. Nucl. Mater.* (1972) **42**, 96.
164. Thomas, G. J., Bauer, W., Holt, J. B., *Radiat. Eff.* (1971) **8**, 27.
165. Stals, L., *Proc. Symp. Fusion Technol., 8th* (1974) 819.
166. Das, S. K., Kaminsky, M., Fenske, G., unpublished data.

RECEIVED May 17, 1976.

# 6

# Electron- and Photon-Induced Desorption

M. J. DRINKWINE, Y. SHAPIRA, and D. LICHTMAN

Physics Department and Laboratory for Surface Studies,
University of Wisconsin—Milwaukee, Milwaukee, Wis. 53201

*Electrons and photons can cause the desorption of adsorbed gases as a result of a quantum interaction, as opposed to thermal effects. Electron-induced desorption is fairly independent of the substrate material. Both neutrals and ions can desorb. Maximum cross sections are similar to those for dissociative reactions in electron-bombarded gas phase molecules (i.e., $10^{-16}$–$10^{-19}$ cm$^2$). Photodesorption is an entirely different process which is very substrate dependent. Photodesorption has been detected primarily from semiconductor substrates where the threshold for the reaction is the band gap energy of the semiconductor, usually 2–5 eV. Photodesorption gives rise only to neutral particle emission with cross sections in the range of $10^{-17}$–$10^{-19}$ cm$^2$ from appropriate active substrates.*

Until the advent of readily available sensitive detection equipment in the last few decades, evidence of electron- or photon-induced gas desorption as a quantum process was generally masked by thermal effects caused by the need to use relatively high power density beams. Experiments in recent years have now clearly established both the electron- and the photon-induced desorption quantum process. Since electrons and photons of similar energy produce similar effects when interacting with gas phase atoms and molecules, it was anticipated that they would show the same effect with gases adsorbed on solid substrates. It has now been shown that the processes are quite different. Each process will therefore be discussed separately.

## Electron-Induced Desorption

Electron-induced desorption has now been studied in fair detail for almost 15 yr (1, 2). Both ions and neutral particles are desorbed by the electron beam. However, because the detection efficiency of ions is much greater than that of neutrals, virtually all data have been obtained on electron-induced ion desorption. The problem in obtaining electron-induced neutral desorption is one of signal-to-noise and has as yet not been satisfactorily solved.

It is pretty well accepted that electron-induced desorption involves, basically, the interaction of electrons with the adsorbed gas components. Transitions occur to dissociative states giving rise to desorbing fragments. For example, in Figure 1 there are some of the energy levels of the hydrogen molecule. Electron-induced excitation to the various allowed states leads to the reactions as noted. If the molecule is adsorbed on a solid surface, one would expect some modifications of the energy levels. These modifications would not be comparable with the overall energy level values and, therefore, would not change the basic molecular energy level scheme very drastically. Electrons interacting with molecules adsorbed on surfaces should produce the same types of transitions leading to the same resulting components. One would expect the difference to be a relatively small shift in the energy required for the various transitions compared with that occurring in the gas phase. If the molecule is raised to an excited or ionized state, there is no reason why it should not remain

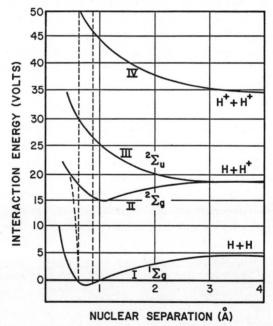

*Figure 1.  Some potential energy curves for the hydrogen molecule. (– – –) indicate the region of Frank–Condon transitions. Curves III and IV indicate states which lead to fragment production.*

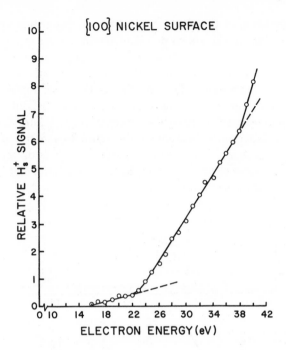

*Figure 2. Relative surface H⁺ ion current vs. electron probe energy near threshold. Electron current density is $10^{-7}$ A · cm⁻². Surface temperature is 300K.*

bound to the surface as in the normal neutral state. Should the transition take the molecule to the dissociative state, then some of the fragments will have sufficient excess kinetic energy to be able to leave the surface. This process is generally referred to as electron-induced desorption.

It is expected that the cross section for electron interaction with adsorbed gas species is comparable with the value for a similar interaction in the gas phase. Thus, cross sections for production of desorbing ion fragments range from a maximum of $10^{-18}$ cm², and indirectly determined cross sections for neutral desorption range from a maximum of approximately $10^{-16}$ cm² (2).

If one measures the cross section vs. electron energy for the production of ion fragments of adsorbed hydrogen on, for example, a nickel surface, one obtains data as seen in Figure 2 (3). The desorbing ion fragment is H⁺, and the curve is extremely similar to that obtained in electron–gas phase interactions. Thus, the threshold and various breaks in the curve undoubtedly relate to the various allowed transitions to dissociative states of the adsorbed hydrogen molecule which contribute to the desorbing H⁺ signal.

The interaction of the electrons causing ion desorption appear to be fairly independent of the substrate. This factor is seen in a collection of recent experiments by various scientists studying carbon monoxide adsorption on different metals. Electron bombardment of adsorbed CO generally gives rise to two ion species—$O^+$ and $CO^+$. In this case, a parent molecule ion $CO^+$ is seen. The explanation for its appearance is still in question. The data obtained for CO adsorbed on tungsten, rhenium, and niobium, all in different laboratories, are shown in Figure 3 (4, 5, 6). Except for a slight shift in the temperature scale caused by slightly different binding energies, one can see clear similarities in the behavior of the $O^+$ and $CO^+$ signals from all three quite different substrates.

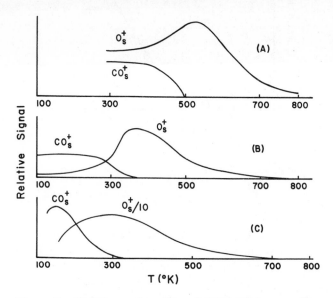

Figure 3. Relative surface $O^+$ and $CO^+$ signals caused by electron-induced desorption as the surface temperature is varied. (A) rhenium surface (adapted from Ref. 4), (B) tungsten surface (adapted from Ref. 5), (C) niobium surface (adapted from Ref. 6).

Thus, not only is the general behavior of CO on these different metals quite similar, but the interaction of the electrons causing desorption produce almost identical results, even though the substrate metals are significantly different. Additional data obtained by these experimenters indicate no significant difference when using either single crystal or polycrystalline samples.

Almost all experiments to date have concentrated on using electrons in the energy range near 100 eV. Some work is done using higher fixed electron energies (500, 1000, 2000 eV, etc.), but these energies are used primarily only because of convenience and compatibility of the electron source in conjunction with other techniques of surface analysis. Until recently there has been very little investigation on what effect increasing the bombarding electron energy above 100 eV has on the desorbing surface ion signal. Now, however, a system has been developed which ultimately can study desorbing surface ions using electron energies ranging from 0 to 20 KeV. Some preliminary data using electron energies up to 4000 eV have been obtained.

The design of the analyzer–detector is essentially the same as that described by Lichtman and McQuistan (*1*). A 6-in. radius, 60-degree magnetic sector mass spectrometer combined with an appropriate set of grids at the entrance region (which help focus and direct the incoming desorbing ions) was used on this system. An extra grid in the entrance region enables us to determine the kinetic energy distribution of the desorbing ions. Added features include an ion gun, which can sputter clean the target and perform ISS (ion scattering spectroscopy) and SIMS (secondary ion mass spectroscopy) experiments, and a very sensitive magnetic sector RGA, (residual gas analyzer), which can monitor the gas phase composition and changes in partial pressures of gas phase components during EID (electron induced desorption) experiments. However, the most important feature of the system is the manner in which the electron gun has been incorporated into the overall system set-up.

Care has been taken in the design and construction of the system so that electron energy changes are produced by a varying potential on the cathode of the electron source alone. Thus, ion collection efficiencies remain unchanged for all electron energies, and secondary electron production in the electron multiplier detector remains constant since the incoming ion energy is fixed irrespective of the bombarding electron energy. Consequently, the "overall ion optics" of the system remains fixed and any changes in the relative ion signal with changing electron energy reflect a true change between the interaction of the bombarding electrons and the surface adsorbates. With the various features incorporated into this system, many interesting and well controlled experiments can be easily done, many of which have never been attempted previously. This particularly applies to experiments involving simultaneous combinations of various surface analysis techniques.

Preliminary data have been obtained from a degreased 304 stainless steel sample at room temperature using electron bombarding energies varying from 0 to 4000 eV. The most common ion signals occur at $m/e$ ratios 1, 16, 19, 35, and 37 indicating the respective ions $H^+$, $O_{16}^+$, $F_{19}^+$, $Cl_{35}^+$, and $Cl_{37}^+$. Spectra similar to this have been obtained from similarly prepared 304 stainless steel surfaces by other researchers. The $H^+$ and $F^+$ are by far the largest surface signals. The identification of the 19 signal as $F^+$ is substantiated by the presence and similar behavior of surface signals at $m/e$ 35 and 37, which are identified as $Cl_{35}^+$ and $Cl_{37}^+$ by the 3-to-1 signal height characteristic ratio of the natural isotopic composition of chlorine. The fluorine contaminant, although generally not detectable by other means, shows up quite readily on nearly any stainless steel sample during an EID experiment. An interesting aspect of these $H^+$ and $F^+$ surface ion signals is that they often appear in highly sensitive magnetic sector RGA's and are referred to as satellite peaks. The $H^+$ and $F^+$ surface ions are produced by bombarding stainless steel surfaces inside the ionization chamber of the RGA by stray electrons from the ionization beam. Since these ions are produced at a surface, they may have more or less energy than the ions produced in the gas phase, depending on the potential of the surface at which the surface ions were produced. Consequently, the position of the peak, caused by surface $H^+$ or $F^+$, compared with where it should be in the spectrum is shifted, giving the term "satellite peaks."

Figures 4 and 5 show the relative ion signal behavior of $F^+$ and $H^+$ with electron energies of 0–4000 eV. The general shape of the curves closely resembles cross section curves for ion production by electron bombardment of gas phase molecules. Figure 6 compares the curve of Figure 5 with cross section data obtained for gas phase dissociative ionization of $H_2$. The three immediate striking features of this comparison are that (1) the general shape of both curves is essentially the same, (2) the surface ion signal reaches a maximum at a higher electron energy, and (3) the surface curve is much broadened with respect to the gas phase curve. The most probable explanation is that the electron–molecule

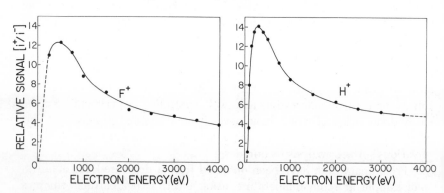

Figure 4. *Relative ion desorption efficiency for EID $F^+$ from room temperature 304 stainless steel as a function of bombarding electron energy using an electron beam current density.* $i^- = 2 \times 10^{-7}$ A/cm².

Figure 5. *Relative ion desorption efficiency of EID $H^+$ from room temperature 304 stainless steel as a function of bombarding electron energy using an electron beam current density.* $i^- = 2 \times 10^{-7}$ A/cm².

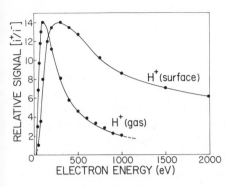

Figure 6.   *Comparison of relative EID H⁺ signal from Figure 5 with relative gas phase H⁺ production from $H_2$ as a function of bombarding electron energy. The H⁺ surface curve has been arbitrarily normalized to match the peak height of the gas phase curve. The general shape of the curves appear similar, but the surface H⁺ curve is broadened compared with the gas phase H⁺ curve.*

interaction resulting in ion production for both gas phase ions and surface ions is essentially the same. However, the surface H⁺ curve indicates that reflected primary and secondary electrons which are produced by electrons passing the adsorbed hydrogen and striking the substrate may also contribute to the ion signal. These reflected primary and secondary electrons can make a second pass at the adsorbed species producing a twofold result—the ion signal reaches a maximum at an electron energy greater than that shown for the gas phase curve and the ion signal falls off more slowly with increasing electron energy. Both results are logical in view of the fact that the second pass secondary electrons range in energy from $0-E_p$ with an energy distribution depending to a certain degree on the initial bombarding electron energy.

These results are encouraging. They are what one would intuitively expect using the assumption that the adsorbed molecules at the surface behave much the same as they do in the gas phase when bombarded by electrons. This assumption is substantiated by a considerable amount of data (Figures 2 and 3). Another aspect of these results is that it may be possible to determine absolute cross sections for EID, something normally very difficult to measure. If one assumes that the second-pass electrons essentially do not contribute to the ion signal for primary electron energies below, for example, 75 eV, then by a point–slope matching method of the surface data to gas phase data for electron energies below 75 eV, the surface ionization curve can be scaled with respect to the gas phase ionization curve. Further experiments should, therefore, hopefully provide quantitative data on both electron-induced ion and neutral particle desorption.

### Photon-Induced Desorption

Considerably greater difficulty has been encountered in attempting to measure photodesorption (7–12). This difficulty was related to two basic problems. The first is the much greater difficulty in producing controlled beams of photons whose energy is sufficient to cause desorption—that is, the uv and extreme uv range—without simultaneously causing thermal desorption. The second problem is that most early experiments were attempted on clean simple metals. In all cases, the data was very difficult to obtain and erratic at best. The increased availability of uv sources and ultra-sensitive detection systems have led to further experiments. Recent results (8) indicate that it is difficult to obtain data from clean simple metals because the cross section for photodesorption of gases adsorbed on these surfaces is very small or perhaps even zero.

Significant results were obtained when studying photodesorption from materials such as 304 stainless steel whose surface is a semiconductor, chrome oxide. Some results are shown in Figure 7. The values obtained for the various metal elements may be real or very possibly caused by scattered uv radiation striking uncovered stainless steel or similar type surfaces. At any rate, it seems quite clear that in the photodesorption process the substrate plays a significant role.

The importance of investigating semiconductor surfaces in relation to photodesorption has thus been clearly demonstrated by the results of photodesorption from the chrome oxide surface of stainless steel. To understand the interaction between photons, adsorbed gases, and semiconductor substrates, an extensively investigated semiconductor, ZnO, (13, 14) was chosen for experiments on photodesorption in our laboratory.

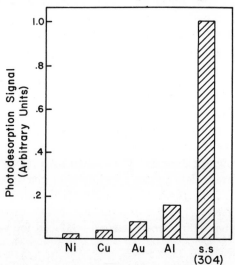

*Figure 7.    Relative photodesorption signal obtained from several metal surfaces irradiated with the full spectrum of a low pressure mercury lamp.  Flux density = 200 μW/cm²* at all wavelengths tested.

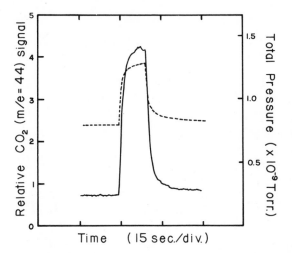

*Figure 8.   Transient change in partial pressure of
$CO_2$ (——) and in total pressure (– – –) during a
10-sec uv illumination interval. ZnO sample is at
125°C.*

Many years ago, photodesorption of oxygen from ZnO was postulated
(*15, 16*) when it was observed that huge persistent changes in the conductivity of ZnO could be brought about by band gap radiation in vacuum
while the initial conductivity could be restored almost completely by
admitting oxygen into the system.

New insight into the ZnO/oxygen system has been obtained by recent
experiments (*17*) using careful mass spectrometry of the photodesorption
products under controlled irradiation.  Short flashes of band gap radiation
incident on ZnO surfaces in a UHV ($1 \times 10^{-9}$ torr) system produced
pressure changes such as the one shown in Figure 8.  The 10-sec flash
also caused a very fast and large change in the $CO_2$ ($m/e = 44$) partial
pressure.  No such change was found in any of the other masses (between
1 and 50) that could not be accounted for by the cracking pattern of
$CO_2$ in the ionization chamber.

To establish the true photodesorptive characteristics of the experiment, measurements of the photodesorption signal were carried out as a
function of illumination intensity.  The linear dependence observed indicates the quantum features of this effect.  Furthermore, the photodesorption spectral response shown in Figure 9 indicates clearly the substrate
dependence of this effect ($E_g(ZnO) = 3.2$ eV) as well as its quantum
(and not thermal) characteristics.  These results have also been confirmed
by measurements on stainless steel (chrome oxide), CdS, and $TiO_2$ and
support the accepted model of neutralization of the chemisorbed species,
which proves to be $CO_2$, by photogenerated holes.

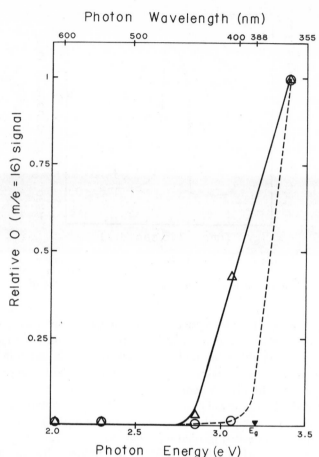

Figure 9. *Relative CO₂ photodesorption signal as a function of the energy of the incident illumination for a single crystal sample (– – –) and for a powder sample (——)*

A measurement of the photodesorbed $CO_2$ yield as a function of illumination time is shown in Figure 10. As the measurement was done by 0.25-sec flashes, Figure 10 actually shows the derivative of the $CO_2$ photodesorption yield with respect to time and indicates the gradual decay of the photodesorption rate. A recent calculation based on these experimental data suggests a value of $10^{-17}$ cm² as the cross section for neutralization of the chemisorbed $CO_2$ species by the photogenerated holes. A measurement of the photodesorption signal as a function of temperature yields an activation energy for this process of approximately 0.25 eV. There is a one-to-one correspondence between the $CO_2$ photodesorption rate and the simultaneous rate of the surface conductivity increase. Both rates are fast at the beginning of the illumination period and decay later to a much slower rate. These data support the accepted model of recombination of chemisorbed species by photogenerated holes migrating to the surface. Our measurements show, however, that this chemisorbed species is $CO_2$, and its photodesorption is directly responsible for the observed surface conductivity changes.

To understand the source of the photodesorbed $CO_2$ species, chemisorption experiments were done with several different gases such as $CO_2$, CO, and $O_2$. It was found, however, that only $O_2$ chemisorption could destroy the accumulation layer created during photodesorption. Therefore, these oxygen molecules must oxidize impurity carbon atoms on the surface and, upon electron capture, create a $CO_2^-$ ion molecule. Indeed, AES (Auger electron spectroscopy) measurements showed considerably less carbon impurities on ZnO surfaces which had been subject to prolonged illumination (19).

The observations of the impurity carbon signal by AES are in accordance with the fact that after some cycles of photodesorption–chemisorption runs, the photo-activity of the surface diminishes considerably. This activity cannot be rejuvenated by further exposure to oxygen. In other words, cleaned and carbon-depleted ZnO surfaces become inert to photodesorption and chemisorption processes. This was observed by other workers as well (18). Deposition of monolayer amounts of carbon was enough to rejuvenate the surface activity for photodesorption and conductivity to its initial level.

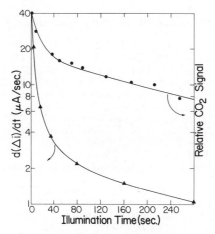

*Figure 10. Semi-log plot of the rate of surface conductivity changes (△) and of relative CO₂ photodesorption (○) as a function of net illumination time as obtained from a ZnO powder sample at 300 K*

Thus, the process of photodesorption, in many cases, seems to proceed according to the following sequence—molecular oxygen is adsorbed on surface impurity carbon atoms; capture of a substrate electron leads to the formation of a chemisorbed "$CO_2^-$" ion molecule complex; irradiation by band gap light produces electron hole pairs; some of the holes migrate to the surface, combining with the "$CO_2^-$" ion molecule leaving a physisorbed $CO_2$ molecule; and this molecule then desorbs thermally from the substrate.

*Summary*

Electron-induced desorption involves primarily the direct interaction of the incoming electron with the adsorbed gas molecule complex. Transitions to dissociative states lead to the desorption of molecular fragments as neutrals or ions. Interactions are pretty much independent of the substrates. Photodesorption seems to be primarily a substrate-dependent semiconductor process. Photons of band gap energy and above produce electron hole pairs. The holes recombine with the chemisorbed complex to produce a physisorbed molecule, generally $CO_2$, which then thermally desorbs. In photodesorption, only desorbing neutrals are detected. In both EID and photodesorption, the maximum cross sections are typical of atomic processes, i.e., $10^{-16}$–$10^{-19}$ cm$^2$.

*Literature Cited*

1. Lichtman, D., McQuistan, R. B., *Prog. Nucl. Energy Ser.* (1965) **IX**, 4 (pt. 2) 95.
2. Madey, T. E., Yates, J. T., Jr., *J. Vac. Sci. Technol.* (1971) **8**, 525.
3. Lichtman, D., Simon, F. N., Kirst, T. R., *Surf. Sci.* (1968) **9**, 325.
4. Ford, R. R., Lichtman, D., *Surf. Sci.* (1971) **26**, 365.
5. Menzel, D., *Ber. Bunsenges. Physik. Chem.* (1968) **72**, 591.
6. Davis, P. R., Donaldson, E. E., Sandstrom, D. R., *Surf. Sci.* (1973) **34**, 177.
7. Peavey, J., Lichtman, D., *Surf. Sci.* (1971) **27**, 649.
8. Fabel, G. W., Cox, S. M., Lichtman, D., *Surf. Sci.* (1973) **40**, 571.
9. Adams, R. O., Donaldson, E. E., *J. Chem. Phys.* (1965) **42**, 770.
10. Lange, W. J., *J. Vac. Sci. Technol.* (1965) **2**, 74.
11. Genequand, P., *Surf. Sci.* (1971) **25**, 643.
12. Kronauer, P., Menzel D., *Adsorption-Desorption Phenom., Proc. Int. Conf. 2nd* (1972) 313.
13. Heiland, G., Kunstman, P., *Surf. Sci.* (1969) **13**, 72.
14. Many, A., *Crit. Rev. Solid State Sci.* (1974) **4**, 515.
15. Morrison, S. R., *Adv. Catal.* (1955) **VII**, 259.
16. Melnick, D. A., *J. Chem. Phys.* (1957) **26**, 1136.
17. Shapira, Y., Cox, S. M., Lichtman, D., *Surf. Sci.* (1975) **50**, 503.
18. Levine, J. D., Willis, A., Bottoms, W. R., Mark, P., *Surf. Sci.* (1972) **29**, 165.
19. Shapira, Y., Cox, S. M., Lichtman, D., *Surf. Sci.* (1976) **54**, 43.

RECEIVED January 5, 1976. Work supported by ERDA grant No. E(11-1)-2425 and by NSF grant No. DMR 74-03947.

7

# Gas Release from Surfaces under X-Ray Impact: Photodesorption, Photocatalysis

S. BRUMBACH and M. KAMINSKY

Argonne National Laboratory, Argonne, Ill. 60439

*Experimental data on x-ray-induced gas adsorption and gas release from surfaces are reviewed with results for stainless steel and $Al_2O_3$ surfaces. $CO_2$ and $O_2$ are the major species which desorb from such surfaces under x-ray irradiation. Mean quantum yields are given for these systems. Results from experiments using photons of lower (e.g., visible or uv range, synchrotron radiation) or higher ($\gamma$-ray) energy help to interpret the data obtained with photons in the x-ray range. X-ray-induced photodesorption needs to be considered in attaining high vacuum in plasma containment devices.*

The impact of energetic photons on solid surfaces can cause the release of gases by such processes as photodesorption, photodecomposition, and photocatalysis. The release of gases from surfaces under photon impact is of current interest in such applications as accelerator technology (e.g., the attainment of ultrahigh vacuum in electron storage rings) and in plasma containment experiments in connection with the controlled thermonuclear research program. For example, in the latter application the release of gaseous impurities from irradiated components can lead to serious plasma contamination. This in turn can lead to an enhanced plasma resistivity, to a decrease in the plasma confinement time, and to plasma power losses via bremsstrahlung, line, and recombination radiation (1). In both present plasma containment devices and proposed future fusion reactors the surfaces of such major components as container walls, beam limiters, and divertor walls will be exposed to photons with a broad energy spectrum. Such photons will appear as synchrotron radiation, line and recombination radiation, bremsstrahlung, x-rays, and $\gamma$-rays.

A review of experimental data on x-ray-induced gas adsorption and gas release from solid surfaces is given below. Results from experiments using lower (eV range) or higher energy (MeV range) photons is included to help to interpret the results obtained with photons in the x-ray energy range (keV range). This review does not include data which have been reported after September 1975.

### Experiments and Results

**Photon-Induced Gas Adsorption.** While this paper deals with x-ray-induced gas release phenomena, it seems appropriate to point out that for certain gas–solid interface systems, an enhancement of gas adsorption has been observed under high energy (x-ray and $\gamma$-ray) photon impact. Since we will report gas release results from $Al_2O_3$ surfaces under x-ray irradiation, it is of interest that photon-induced adsorption was found for various gases on finely divided $Al_2O_3$ catalyst support materials. Coekelbergs et al. (2) observed photoadsorption by $\kappa$-$Al_2O_3$ of $O_2$ and CO under x-ray irradiation at 100 torr partial pressure of these species using a total pressure change technique. Gulyaev, Kolbanovskii, and Roketashvili (3) also used a total pressure change technique to observe $Co^{60}$ $\gamma$-ray-induced adsorption of CO on $\gamma$-$Al_2O_3$ in the CO partial pressure range of 0.1–1.5 torr. Kolbanovskii and Shaburkina (4) also observed a small photoadsorption effect with $CO_2$ on $\gamma$-$Al_2O_3$ under $Co^{60}$ $\gamma$-ray irradiation. The adsorption characteristics of $Al_2O_3$ surfaces for various gases at high gas pressures (0.1 torr–1 atm) and the effect of radiation on these characteristics have been reviewed by Norfolk (5). Photoadsorption in other gas–solid systems has been discussed by Molinari (6).

**Photon-Induced Gas Release.** LOW-ENERGY PHOTODESORPTION. For low energy photons in the 2–7-eV range (visible and uv radiation) the release of gases from surfaces of such solids as ZnO (7), Fe, Ni, Zr (8, 9, 10, 11), W (8, 12), stainless steel (13, 14), CdS (15), TiO (16), and $SnO_2$ (16) have been reported. Since a review of low energy photon-induced desorption has been presented in the previous paper (17), we discuss here only those experiments which help to understand the results obtained with photons in the x-ray range, i.e., photons more than three orders of magnitude higher in energy.

Fable, Cox, and Lichtman (14) used as photon sources various lines from a mercury lamp in the photon energy range 2–7 eV to induce desorption from stainless steel targets under ultrahigh vacuum conditions and used a mass spectrometer to detect the desorbing species. They found that the dominant species desorbed by photon irradiation was $CO_2$, although CO, $CH_4$, $H_2O$, and $H_2$ also were observed. Typical mass scans obtained with and without uv photon irradiation are shown for a stainless steel surface (chamber wall) in Figure 1. Quantum yields were determined for the release of $CO_2$, and a value of $5 \times 10^{-3}$ molecule/photon was observed for nearly monochromatic 6.7-eV photons. The quantum yield decreased with decreasing photon energy. Photon-induced desorption of $CO_2$ from ZnO was observed by Drinkwine, Shapira, and Lichtman (*17*) with 2–3.5-eV photons, again under ultrahigh vacuum conditions and using mass spectrometric detection. No desorption was detected for

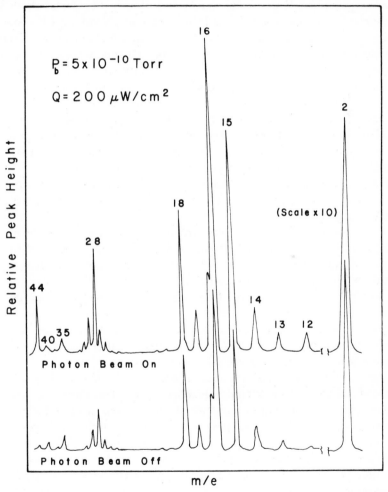

*Figure 1.   Typical gas phase mass spectra of 304 stainless steel UHV system prior to and during uv irradiation of chamber walls with the full spectrum of a low pressure mercury lamp. $P_b$, equilibrium base pressure of system with photon beam off. (14)*

any other species which could not be accounted for by the cracking of $CO_2$. The measured photodesorption cross section increased with increasing photon energy, rising sharply near the ZnO band gap. In addition to $CO_2$, Baidyaroy, Bottoms, and Mark (15) detected $O_2$, $N_2O$, $H_2S$, and Ar with their mass spectrometer while irradiating CdS with 2–3.5-eV photons at a background pressure of $1.5 \times 10^{-8}$ torr.

Many of the early low energy photon-induced desorption experiments studied the desorption of preadsorbed molecules. One example of such an experiment is that of Kronauer and Menzel (12), who studied the desorption of preadsorbed CO from tungsten. Desorbing CO molecules were detected with a mass spectrometer, and measurements of the total pressure increase were also conducted. Quantum yields for CO release were determined for photon energies between 3 and 5 eV, with a value of $4 \times 10^{-7}$ molecule/photon at 5 eV (12). Such low quantum yields are typical for desorption from clean metal surfaces (8, 9, 10, 11). Again, the quantum yields increased with increasing photon energy.

HIGH-ENERGY PHOTODESORPTION. Synchrotron Radiation. The problem of photon-induced outgassing contributing to vacuum system contamination in electron storage rings was considered by Fischer and Mack (18) and by Bernardini and Malter (19). The synchrotron radiation produced in such storage rings has an energy spectrum ranging from a few eV to approximately 10 keV.

Fischer and Mack (18) and Bernardini and Malter (19) carried out electron desorption experiments from stainless steel and copper surfaces in separate experimental systems. For example, for a copper surface and for electron energies varying from 100 eV to approximately 3000 eV, the CO desorption yield increased from about 1 to $3 \times 10^{-6}$ molecule/ electron (18). The extent to which photoelectron desorption would contribute to the gas release expected for storage rings was then estimated. Both teams suggested that the gas release observed in storage rings was caused by photoelectron desorption.

Reactor Radiations, γ-Rays. High energy photon-induced degassing has also been studied for vacuum chambers exposed to radiations from fission reactors. Dobrozemsky (20) irradiated stainless steel and aluminum vacuum chambers (base pressure $\sim 6 \times 10^{-9}$ torr) with fission reactor irradiations and used a mass spectrometer to identify the desorbing species. Partial pressure increases of approximately two orders of magnitude were reported for $CO_2$ and CO, with somewhat smaller increases for $H_2$ and $H_2O$ during the irradiation of an aluminum chamber. Partial pressure increases were also observed for CO, $CO_2$, $H_2$, and $H_2O$ during the irradiation of a stainless steel chamber but were smaller than in the aluminum chamber. The γ-ray flux and energy distribution were not given. In an earlier paper, Baltacis, Dobrozemsky, and Kubischta

(*21*) studied degassing of an aluminum chamber by fission reactor radiations at a base pressure of $6 \times 10^{-9}$ torr and observed a twofold order of magnitude increase in the total pressure during irradiation. Mass spectrometric residual gas analysis indicated that the desorbing species was mostly $H_2$ with contributions from $H_2O$, CO, and $CO_2$. From their estimated $\gamma$-ray flux value of $4 \times 10^{13}$ photons $cm^{-2}$ $sec^{-1}$ and the measured molecular flux from the surface, the authors estimated an upper limit quantum yield of $4.7 \times 10^{-3}$ molecule/photon. Muelhause, Ganoczy, and Kupiec (*22*) irradiated an aluminum chamber held at a background pressure of $5 \times 10^{-9}$ torr with the radiations from a fission reactor and measured the outgassing rates with a total pressure gauge. They also measured outgassing rates from the same chamber surface using a $Co^{60}$ $\gamma$-ray dose rate equal to the $\gamma$-ray dose rate for the fission reactor and found the outgassing rates to be equal. They therefore concluded that the outgassing in the fission reactor case was caused by $\gamma$-ray irradiation.

**X-Rays.** Until very recently, only Coekelbergs et al. (*23*) had reported any gas release from surfaces using photon irradiation in the x-ray range. They observed desorption of $CO_2$ at a partial pressure of 0.4 torr $CO_2$ from finely divided $\kappa\text{-}Al_2O_3$ under irradiation by 55-keV x-rays. The desorption was detected by measuring the change in total pressure over the solid.

More recently, x-ray-induced desorption results for mean photon energies in the 18–24-keV range obtained under ultrahigh vacuum conditions and mass spectrometric detection of desorbing species were reported by Brumbach and Kaminsky (*24, 25*). Although details of these experiments have been reported (*24, 25*) they are summarized here briefly. Figure 2 shows the experimental arrangement schematically. The x-ray spectrum was a tungsten bremsstrahlung spectrum from an x-ray tube, filtered by beryllium and characteristic for a particular electron energy. For example, for an electron energy of 50 keV, an x-ray spectrum was produced with energies ranging from about 50 keV to about 10 keV where the absorption by the two beryllium windows of $\sim 1.7$ mm thickness becomes large. A typical total photon flux of $7 \times 10^{12}$ photons $cm^{-2}$ $sec^{-1}$ on target was obtained for a 50 keV, 40 mA electron current. Pressure in the stainless steel vacuum chamber was typically $1 \times 10^{-9}$ torr during experiments. The x-rays entered the chamber through a beryllium window, and a leak valve allowed the partial pressure of various gases to be varied. A push–pull feedthrough allowed different targets to be irradiated. It was also possible to remove all the targets from the x-ray beam. The $Al_2O_3$ and stainless steel target surfaces studied by Brumbach and Kaminsky (*24, 25*) were degreased prior to irradiation, were in the chamber during a 150°C bakeout, and were held at $1 \times 10^{-9}$

torr for at least 24 hr before irradiation. Thus, the results of Brumbach and Kaminsky (24, 25) represent x-ray-induced gas release from surfaces which have not been extensively degassed. Also, Auger electron spectroscopy revealed the presence of carbon contamination on both target materials before irradiation (25). The flux of desorbing molecules was measured for a particular species using a quadrupole mass spectrometer. Calibration experiments were used to determine the mass spectrometer response to the various particle fluxes. In practice, mass spectrometer signal changes for a particular gas species released by x-ray irradiation were determined for two cases—the x-ray beam striking the target and the x-ray beam striking a gold beam stop 30 cm from the ionizer region of

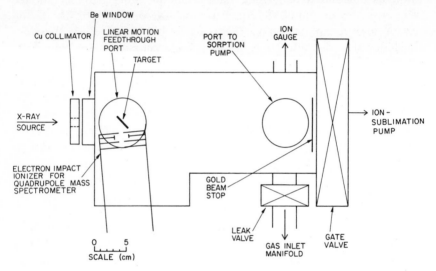

Journal of Applied Physics

*Figure 2.    Schematic of apparatus for studying photon-induced gas release (25)*

the mass spectrometer (the target had been moved out of the beam path). The no-target signal change was substracted from the signal change obtained with the target in place, and this difference attributed to the desorption from the target alone.

The results obtained by Brumbach and Kaminsky (24, 25) showed that the greatest relative signal increases were for $O_2$ and $CO_2$ when a degreased stainless steel target was irradiated with x-rays. For $Al_2O_3$ targets the greatest relative signal increase was for $CO_2$. Only small relative changes in the intensities of $H_2$, $CH_4$, and $H_2O$ were observed. Small signal changes were observed for CO for the case of $Al_2O_3$ targets and slightly larger changes in the CO signal for degreased stainless steel targets, although for the latter type targets the CO signal change was still less than for $CO_2$ or $O_2$. This is in contrast to the large relative increases in signals for $H_2$, $H_2O$, CO, and $CO_2$ when stainless steel and $Al_2O_3$ targets are heated.

Carbon dioxide release under photon impact has been observed by several authors for distinctly different regions of photon energy and for various materials. As mentioned previously, Fable, Cox, and Lichtman (*14*) and Kronauer and Menzel (*12*) observed $CO_2$ release from stainless steel surfaces with photons in the energy range 4–7 eV (uv range). Ultraviolet photons also resulted in $CO_2$ release from the surfaces of ZnO (*7, 17*) and CdS (*15*). Dobrozemsky (*20*) and Baltacis, Dobrozemsky, and Kubischta (*21*) found $CO_2$ to be desorbed by fission reactor irradiations from aluminum and stainless steel chambers. Fable, Cox, and Lichtman (*14*) also observed $H_2$, $H_2O$, $CH_4$, and CO desorption from degreased stainless steel to a greater extent than that observed by Brumbach and Kaminsky (*24, 25*), who used photons more than three orders of magnitude higher in energy. In addition to the experiment of Brumbach and Kaminsky (*25*) the only other known observation of photon-induced desorption from $Al_2O_3$ was the high pressure range (0.4 torr) experiment of Coekelbergs et al. (*23*), where the desorption of $CO_2$ was from a $CO_2$-covered $Al_2O_3$ catalyst support material using x-ray irradiation.

In photodesorption experiments it is necessary to consider that the observed gas species may have been released from the target by a thermal desorption process caused by the power deposition by the photons in near-surface regions. McAllister and White (*26*), for example, resistively heated CO-covered Ni ribbons and observed CO desorption at input powers equivalent to only 7 mW $cm^{-2}$ power density from a photon beam. Also, Lange (*8*) reported a thermal desorption contribution to his observed CO pressure increase when irradiating a CO-covered tungsten surface with uv photons. Fable, Cox, and Lichtman (*14*) estimated the temperature rise expected for their desorption experiments with degreased stainless steel and concluded that the low power available (300 $\mu W$ $cm^{-2}$) in their photon beam could not result in thermal desorption. Brumbach and Kaminsky (*24, 25*) estimated that the power deposited by the incident x-ray photons in their experiments should not be sufficient to heat the target surfaces more than 1°C and consequently could not contribute significantly to desorption. One experimental observation which they consider as evidence that a thermal desorption mechanism can be neglected has been the linear dependence of desorption rate (signal intensity) on photon flux. Figure 3 shows the observed dependence of the signal intensity for $CO_2$ released from $Al_2O_3$ under x-ray impact on the photon flux for a 50-keV bremsstrahlung spectrum. Genequand (*11*) already had pointed out that a linear relationship

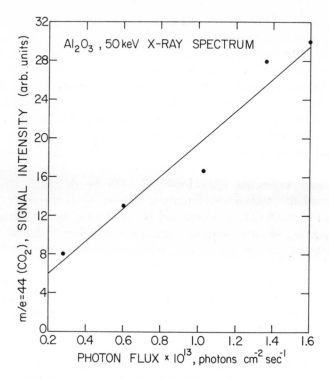

*Figure 3.   Carbon dioxide signal intensity as a function*
*of photon flux for a 50-keV x-ray spectrum and a $Al_2O_3$*
*target*

between desorption rate and photon intensity is a good indication that
the observed desorption is not thermally induced. Since the thermal
desorption rate is an exponential function of temperature, the observed
desorption rate from photon-induced surface heating should also depend
nonlinearly on photon intensity (power deposition). McAllister and
White (26), however, pointed out that over a very small temperature
range this difference in the temperature dependence would be difficult
to detect. Baidyaroy, Bottoms, and Mark (15) and Shapira, Cox, and
Lichtman (7) both found a linear dependence of the desorption rate for
$CO_2$ release from CdS (15) and ZnO (7) as a function of photon flux.
Both groups of authors concluded that the observed desorption was not
caused by surface heating.

Another piece of experimental evidence which can be used to
distinguish thermal desorption from true photodesorption is the time
response of the desorption signal. If the absorption of incident photons
caused target heating, the desorption signal would be expected to
increase in time after the photon beam has been turned on until the

surface reaches an equilibrium temperature. In turn, since the targets are in vacuum, the decrease in temperature might be expected to be slow after the photon beam has been turned off. This would result in a slow decrease in the desorption signal. Thus, if the time response is rapid with both the initiation and termination of photon irradiation, it might be concluded that thermal effects can be neglected. Lange (*8*) observed that his pressure gage signals decreased slowly when the photon irradiation was turned off, indicating some thermal desorption contribution. The sharp decrease in signal intensity when the photon source was turned off observed by Shapira, Cox, and Lichtman (*7*) indicates that thermal effects were not important in their experiments. Brumbach and Kaminsky (*24, 25*) also saw a rapid decrease in the photon-induced signal when the photon beam was turned off. This is illustrated for the case of $O_2$ released from degreased stainless steel in Figure 4. The authors considered this case also as evidence that thermal desorption effects do not play a significant role in the $O_2$ release.

If both the flux of monoenergetic photons incident on a surface and the resulting flux of molecules leaving the surface are known, a quantum yield for gas release can be obtained from the ratio of these fluxes. For example, for photon energies in the 3–7-eV range such quantum yields were obtained for the release of preadsorbed CO from W (*8, 12*) and Ni (*10*) surfaces which had been cleaned prior to CO adsorption. As mentioned earlier, the estimated quantum yields were low, in the range $10^{-9}$–$10^{-6}$ molecule/photon. A more detailed discussion of the quantum

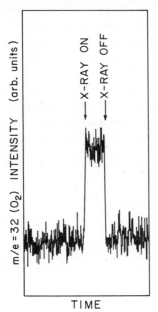

Journal of Applied Physics

*Figure 4.    X-ray-induced signal change for $O_2$, degreased stainless steel target, $7 \times 10^{12}$ photons $cm^{-2}$ $sec^{-1}$, 50-keV bremsstrahlung spectrum.*

yields typical for low energy photon impact can be found in the preceding chapter by Lichtman et al. (17). If the incident photons are not monoenergetic but have some rather broad energy distribution, then the ratio of molecular flux to total photon flux takes on a slightly different meaning. Since the quantum yield is, in general, a function of photon energy, the distribution of photon energies will result in a measured mean quantum yield. Such mean yields were determined by Brumbach and Kaminsky (24, 25) for the release of $CO_2$, CO, and $O_2$ from surfaces of degreased stainless steel and $Al_2O_3$ (not previously degassed) using the x-ray energy spectrum discussed above. Estimated values for the mean yield of $CO_2$ release from degreased stainless steel were 3 to 4 $\times$ $10^{-4}$ molecule/photon in the x-ray spectrum. The only other known quantum yield for $CO_2$ from degreased stainless steel is that reported by Fable, Cox, and Lichtman (14), who obtained a value of 5.5 $\times$ $10^{-3}$ molecule/photon using nearly monochromatic 6.7-eV photons. The mean quantum yield results of Brumbach and Kaminsky (24, 25) for degreased stainless steel and $Al_2O_3$ are summarized in Table I. The results shown in Table I for the two targets of either material represent the lower and upper limit values, since the values obtained in additional runs fell within these limits. The results for target 2 ($Al_2O_3$) and for target 1 (stainless steel) were considered the more typical cases. All targets were prepared identically (25). The only other report of a quantum yield for a broad spectrum of photon energies is that of Baltacis, Dobrozemsky, and Kubischta (21), who quoted a maximum yield of 4.7 $\times$ $10^{-3}$ molecule/photon for fission reactor radiations incident on surfaces of an aluminum chamber. This value is the ratio of the total number of all molecules released (determined by the increase in total pressure) and the total estimated $\gamma$-ray flux (no energy distribution for the $\gamma$-rays has been stated).

The dependence of the quantum yield on photon energy helps to explain the gas release process. For example, the dependence of the quantum yield on the photon energy has been studied for the uv range

### Table I.  Mean Quantum Yields for Gas Release in Molecules Released per Photon[a]

|  | CO | $O_2$ | $CO_2$ |
| --- | --- | --- | --- |
| Stainless steel, target 1 | $2 \times 10^{-4}$ | $9 \times 10^{-4}$ | $4 \times 10^{-4}$ |
| Stainless steel, target 2 | $< 6 \times 10^{-5}$ | $5 \times 10^{-4}$ | $2 \times 10^{-4}$ |
| $Al_2O_3$, target 1 | $1 \times 10^{-4}$ | $1 \times 10^{-4}$ | $2 \times 10^{-3}$ |
| $Al_2O_3$, target 2 | $< 6 \times 10^{-5}$ | $< 3 \times 10^{-5}$ | $8 \times 10^{-5}$ |
| Detection limit | $6 \times 10^{-5}$ | $3 \times 10^{-5}$ | $3 \times 10^{-5}$ |

[a] For a bremsstrahlung spectrum at 50-keV electron energy and 7 $\times$ $10^{12}$ photons $cm^{-2}$ sec $^{-1}$.

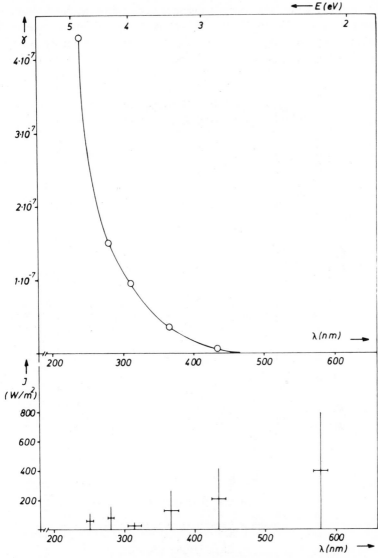

*Figure 5. Above, photodesorption probability $\gamma$ (in this paper also called quantum yield) of CO from W as a function of wavelength or energy. Below, incident power density of the irradiation used in these experiments. The horizontal bars indicate the half-widths of the interference filters used (12).*

by many authors (7, 8, 10, 11, 12, 14). Typical of the results for this energy range are those of Kronauer and Menzel (12) for the release of preadsorbed CO from tungsten, shown in Figure 5. The photon source was a mercury lamp with various discrete lines using interference filters. The upper part of Figure 5 gives the photon energy dependence of the quantum yield while the lower part shows the photon intensity (incident power density) at these various energies. The authors argued that the

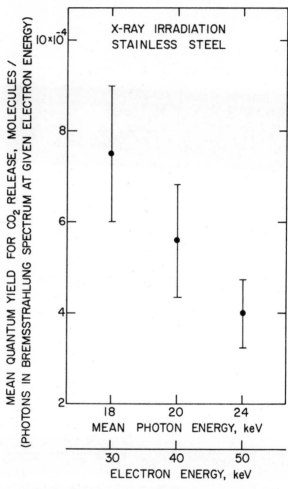

*Figure 6.   Dependence of mean quantum yield on mean photon energy for $CO_2$ released from degreased stainless steel (25)*

opposite trends of power density and quantum yield with photon energy are strong evidence against a thermal desorption process. The trend of increasing quantum yields with increasing photon energies appears to be typical for the energy range 2–7 eV. The results of Fable, Cox, and

Lichtman (*14*) for $CO_2$ release from degreased stainless steel and by Shapira, Cox, and Lichtman (*7*) for $CO_2$ release from ZnO both indicate increasing quantum yields with increasing photon energies in the visible and uv regions.

The dependence of mean quantum yields on mean photon energies was studied by Brumbach and Kaminsky (*24, 25*) in the x-ray range for $CO_2$ release from degreased stainless steel and $Al_2O_3$ surfaces. Their results for degreased stainless steel are shown in Figure 6. The abscissa shows electron energy on one scale and the corresponding mean photon energy on another scale. Experimentally, observations were made at different electron voltages. The error bars shown represent the uncertainties with which each mean yield value could be reproduced, not the total error in the absolute value of the mean quantum yield. As seen in Figure 6, the mean quantum yield decreased with increasing mean photon energy. This is opposite to the trend observed by Fable, Cox, and Lichtman (*14*) with photons more than three orders of magnitude lower in energy. The difference in the quantum yield energy dependence at low energy (*14*) and the mean quantum yields at high energy (*25*) for the case of $CO_2$ released from degreased stainless steel is shown in Figure 7. The difference in the observed photon energy dependence of the gas release yields for the two different energy ranges appears plausible if one correlates this dependence with the corresponding one for the photon absorption coefficients of the irradiated target materials as suggested by Brumbach and Kaminsky (*24, 25*). The photon absorption coefficients for both metals and metal oxides decrease with increasing photon energy in the x-ray region (*27, 28*). For the case of metal oxides, the photon absorption coefficient increases with increasing energy in the visible and ultraviolet range (*29*). This correlation between the trends in photon desorption coefficient and gas release quantum yield also supports the view of Drinkwine, Shapira, and Lichtman (*17*) that the $CO_2$ release process from oxide surfaces is substrate dependent.

Many of the surfaces from which $CO_2$ release has been observed under photon irradiation have been metal oxide surfaces. Metal oxides are known to be good catalysts for oxidation reactions, and the radiation-induced catalytic oxidation of CO to $CO_2$ on $Al_2O_3$ has been studied (*30*). Brumbach and Kaminsky (*25*) considered that the $CO_2$ released under x-ray irradiation from $Al_2O_3$ and stainless steel may have been produced by photocatalytic oxidation of CO, since CO was a major component in the background pressure of the chamber used in their experiments. In order to explore the possibility of CO oxidation, Brumbach and Kaminsky (*25*) studied the intensity of the photon-induced $CO_2$ signal at various CO partial pressures. Their results are shown in Figure 8 for both degreased stainless steel and $Al_2O_3$. No distinct trend was observed over

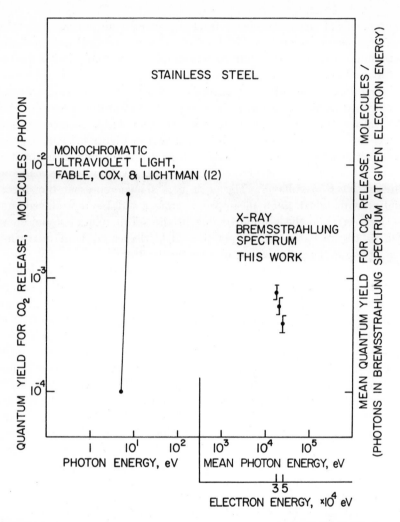

*Figure 7. Dependence of quantum yield on photon energy in the uv and x-ray regions for release of $CO_2$ from degreased stainless steel*
(25)

the range of CO partial pressures studied, within experimental uncertainty. The authors concluded that either $CO_2$ is not produced in a photocatalytic CO oxidation reaction or that the active sites on the surface are already saturated for the surface conditions used and at partial pressures less than $8 \times 10^{-11}$ torr, the lowest background CO partial pressure achieved in the chamber. Drinkwine, Shapira, and Lichtman (17) chemisorbed CO, $O_2$, and $CO_2$ on ZnO to study the effect of these gases on $CO_2$ photodesorption. They found that only $O_2$ leads to an enhancement of photodesorption. Drinkwine et al. (*17*) concluded that their $CO_2$ arose from oxidation of surface carbon and were able to increase the amount of $CO_2$ released by depositing carbon on their ZnO surfaces. Brumbach and Kaminsky also detected carbon impurities on their degreased stainless steel and $Al_2O_3$ surfaces (*25*).

A dependence of the mean quantum yield for $CO_2$ release on the x-ray dose was observed for the $Al_2O_3$ and degreased stainless steel target 1 (*25*). For a total x-ray dose of $2 \times 10^{17}$ photons cm$^{-2}$ after about 500 min irradiation time, the mean quantum yield decreased from $2 \times 10^{-3}$ to $2 \times 10^{-4}$ molecule/photon for the $Al_2O_3$ target 1 and from $4 \times 10^{-4}$ to $2 \times 10^{-4}$ molecule/photon for the degreased stainless steel target 1. For mean quantum yield values for $CO_2$ release in the range from $10^{-3}$ to $10^{-4}$ molecule/photon a total dose of $2 \times 10^{17}$ photons cm$^{-2}$ will remove only the equivalent of a fraction of one monolayer of adsorbed $CO_2$. Recently obtained results of gas release from discharge-cleaned stainless steel surfaces under x-ray impact will be published soon (*31*).

## Application: X-Ray Induced Desorption in Plasma Devices

An estimate of the $CO_2$ release rates from degreased stainless steel surfaces has been made recently (*24*) for a tokamak fusion reactor, conceptually designed by the University of Wisconsin–Madison team (UWMAK-I) (*32*). According to this design a bremsstrahlung power loading of 28.2 W cm$^{-2}$ may be expected on the stainless steel surfaces

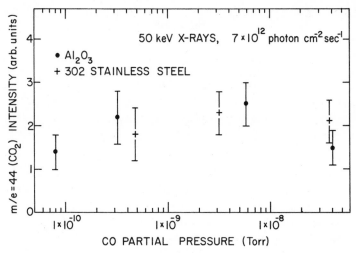

*Figure 8.    Intensity of $CO_2$ gas release signal as a function of CO partial pressure for $Al_2O_3$ and degreased stainless steel (25)*

exposed to bremsstrahlung radiations. If one arbitrarily assumes that the bremsstrahlung is typical for a tungsten bremsstrahlung at 30-keV electron energy (resulting in a mean photon energy of 18 keV) and takes a mean quantum yield for $CO_2$ release of $4 \times 10^{-4}$ atom/photon for a degreased stainless steel surface, a release rate will be $5.3 \times 10^{12}$ $CO_2$ molecules $cm^{-2}$ $sec^{-1}$. If in actuality the mean photon energy is shifted to lower energies, e.g., to the 1–5-keV range, even higher gas release yields can be expected on the basis of an extrapolation of our data to energies in the 1–5-keV range (*see* Figure 7). The $CO_2$ release rates mentioned above will contribute to the plasma contamination of fusion devices for as long as the x-ray–surface interactions release $CO_2$ with the high mean quantum yields quoted above. With increasing x-ray dose, a depletion of those adsorbed gases which contribute to the $CO_2$-release is expected to result in decreased $CO_2$ release rates.

## Conclusions

The relatively scarce experimental data on x-ray-induced gas release from non-discharge cleaned surfaces of stainless steel and $Al_2O_3$ reveal the following trends.

(1) The dominant gas species released are $O_2$ and $CO_2$ and not $H_2$, $H_2O$, or CO, which are typically observed in thermal desorption.

(2) The mean quantum yields for $O_2$ and $CO_2$ release are in the range $8 \times 10^{-5}$–$2 \times 10^{-3}$ molecule/photon in the tungsten bremsstrahlung spectrum at 50-keV electron energy.

(3) The energy dependence of the mean quantum yield for $CO_2$ release reveals an increase in the yield with decreasing mean photon energy.

(4) The observed energy dependence of the quantum yields for $CO_2$ release from stainless steel at both low photon energies (2–7-eV range) and high photon energies (18–24-keV range) has been correlated with the energy dependence of the photon absorption coefficient.

(5) The amount of $CO_2$ released was found not to be a function of the CO partial pressure in the $10^{-10}$–$10^{-8}$ torr range.

The mechanism for the observed dominant $O_2$ and $CO_2$ release under x-ray impact from stainless steel and $Al_2O_3$ surfaces is not yet understood. More experimental studies are needed to provide the needed data base to establish the dominant mechanism for the x-ray impact-induced gas release from surfaces. Data are also needed to test if the suggested correlation between the observed energy dependence of the quantum yields for gas release and the one for the photon adsorption coefficients hold true for the photon energy region from $\sim 10$ eV to 10 keV, a region for which reliable data are practically nonexistent.

*Acknowledgments*

We are grateful to Academic Press, Inc., London; Journal of Applied Physics, American Institute of Physics, New York; Surface Science, MIT, Cambridge, Mass. for their permission to use certain illustrations in this manuscript.

## Literature Cited

1. Hogan, J. T., Clarke, J. F., in "Surface Effects in Controlled Fusion," H. Wiedersich, M. Kaminsky, K. M. Zwilsky, Eds., North-Holland, Amsterdam, 1974.
2. Coekelbergs, R., Crucq, A., Decot, J., Degols, L., Randoux, M., Timmerman, L., *J. Phys. Chem. Solids* (1965) **26**, 1973.
3. Gulyaev, G. V., Kolbanovskii, Y. A., Roketashvili, E. G., *High Energy Chem.* (1971) **5**, 391.
4. Kolbanovskii, Y. A., Shaburkina, V. I., *High Energy Chem.* (1967) **1**, 510.
5. Norfolk, D., *Radiat. Res. Rev.* (1974) **5**, 373.
6. Molinari, E., *Symp. Electron. Phenom. Chemisorpt. Catal. Semiconduct.* (1969).
7. Shapira, Y., Cox, S. M., Lichtman, D., *Surf. Sci.* (1975) **50**, 503.
8. Lange, W. J., *J. Vac. Sci. Technol.* (1965) **2**, 74.
9. Lange, W. J., Riemersma, H., *Am. Vac. Soc. Trans.* (1961) **8**, 167.
10. Adams, R. O., Donaldson, E. E., *J. Chem. Phys.* (1965) **42**, 770.
11. Genequand, P., *Surf. Sci.* (1971) **25**, 643.
12. Kronauer, P., Menzel, D., "Proc. of 2nd Intl. Conf. on Adsorption-Desorption Phenomena," F. Ricca, Ed., p. 313, Academic, London, 1972.
13. Peavy, J., Lichtman, D., *Surf. Sci.* (1971) **27**, 649.
14. Fabel, G. W., Cox, S. M., Lichtman, D., *Surf. Sci.* (1973) **40**, 571.
15. Baidyaroy, S., Bottoms, W. R., Mark, P., *Surf. Sci.* (1971) **28**, 517.
16. Petrera, M., Trifiro, F., Benedek, G., *Jpn. J. Appl. Phys. Suppl. 2* (1974) Pt. 2, 315.
17. Drinkwine, M. J., Shapira, Y., Lichtman, D., ADV. CHEM. SER. (1976) **158**, 171.
18. Fischer, G. E., Mack, R. A., *J. Vac. Sci. Technol.* (1965) **2**, 123.
19. Bernardini, M., Malter, L., *J. Vac. Sci. Technol.* (1965) **2**, 130.
20. Dobrozemsky, R., *Nucl. Instrum. Methods* (1974) **118**, 1.
21. Baltacis, E., Dobrozemsky, R., Kubischta, W., *Proc. Int. Vac. Congr.* (1968) 767.
22. Muehlhause, C. O., Ganoczy, M., Kupiec, C., *IEEE Trans. Nucl. Sci.* (1965) **NS-12**, 478.
23. Coekelbergs, R., Crucq, A., Decot, J., Degols, L., Randoux, M., Timmerman, L., *J. Chim. Phys.* (1966) **63**, 218.
24. Brumbach, S., Kaminsky, M., *Symp. Eng. Prob. Fusion Res., 6th* (1976) IEEE Pub. No. 75CH1097-5-NPS, 1135.
25. Brumbach, S., Kaminsky, M., *J. Appl. Phys.* (1976) **47**, 2844.
26. McAllister, J. W., White, J. M., *J. Chem. Phys.* (1973) **58**, 1496.
27. Henke, B. L., Tester, M. A., *Adv. X-ray Anal.* (1975) **18**, 76.
28. Siegbahn, K., *Alpha-, Beta-, Gamma-Ray Spectros.*, American Elsevier (1968).
29. Hagemann, H. J., Gudat, W., Kunz, C. (1974) DESY SR-74/7.
30. Coekelbergs, R., Collin, R., Crucq, A., Decot, J., Degols, L., Timmerman, L., *J. Catal.* (1967) **7**, 85.
31. Brumbach, S., Kaminsky, M. J., *Nucl. Mater.* (1976) **63**, in press.
32. Wisconsin Tokamak Reactor Design, UWFDM-68 (1974).

RECEIVED June 21, 1976. Work performed under the auspices of the U.S. Energy Research and Development Administration.

# 8

# X-Ray Emission from the Surface Region of Solids

ROBERT L. PARK

Department of Physics and Astronomy and Center of Materials Research,
University of Maryland, College Park, Md. 20742

*A fast electron incident on a surface may conserve energy
by emitting a bremsstrahlung photon. Near the short wavelength limit the bremsstrahlung spectrum measures a density
of unoccupied conduction states averaged over all atomic
species on the surface region. The characteristic soft x-ray
band spectrum, by contrast, is considered a bulk probe of
the filled valence states. By using electrons with energies
just above the threshold for core level scattering, however,
the probe depth is reduced to the inelastic mean free path
for electrons. In soft x-ray appearance potential spectroscopy, thresholds are detected in the derivative of the total
x-ray yield. The yield is related to the density of conduction
band states. The core level probes examine separately the
states associated with different elements on the same surface.*

In the decay of an excited state, energy may be conserved by the emission of photons or electrons. Electron spectroscopy is popular because
emitted electrons can be manipulated easily by electrostatic or magnetic
fields. For many purposes, however, the indifference of photons to these
fields is an advantage. The escape depth of a soft x-ray photon is orders
of magnitude greater than for an electron of the same energy, and the
classic method of soft x-ray band spectroscopy is generally regarded as a
way to examine the bulk properties of materials. Indeed, in the early
1930's soft x-ray band spectra of the alkali metals (1) were thought to
represent a major triumph of the free electron theory of metals.

If the energy of the electrons used to create the excited state is
reduced to near threshold values, however, the sampling depth is determined by the electron mean-free path. Therefore, under conditions of

threshold excitation, the surface sensitivity of x-ray techniques is essentially the same as for electron spectroscopies involving the same levels. Such measurements were originally undertaken to avoid the distortion introduced by self-absorption of the emitted x-rays, but eliminating self-absorption is equivalent to probing only the outermost layers of the material. Similar surface sensitivities can be achieved using high energy electrons at grazing incidence to create the excited states or by using protons or heavier ions, which have relatively short stopping lengths, as the exciting particles. In this chapter the various schemes for obtaining x-rays from the surface region of solids are reviewed briefly, and the kinds of information conveyed by these x-rays are discussed.

## Sampling Depth

The surface region of a solid can reasonably be defined as the layer in which the atomic potentials differ from those of atoms still deeper within the solid (2). This would of course include adsorbed foreign atoms, but it also includes those substrate atoms that sense the altered chemical environment imposed by the loss of translational symmetry (3). For clean metals the surface region may include no more than the outermost two atomic layers (4). The extent to which a measurement is specific to this region is frequently referred to as the "surface sensitivity." In most surface spectroscopies this sensitivity is determined by the in-

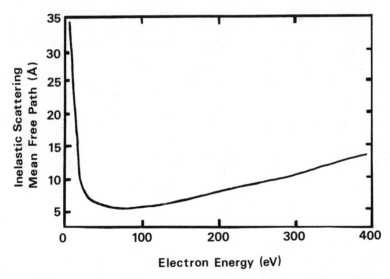

*Figure 1.    The energy dependence of the mean free path for inelastic scattering of an electron in aluminum from Ref. 6 based on a parameterization of Ref. 5's calculations. The sampling depth is a minimum for electron energies of 50–100 eV for various materials.*

elastic mean-free path of electrons—that is, by the range over which an electron will travel before losing energy.

The mechanisms by which a particle loses energy in passing through a solid strongly depend on its energy. For electrons in the energy range below a few keV, collisions with nuclei are rare, and the principal source of energy loss is the creation of plasmons by electron–electron scattering. The inelastic mean-free path for plasmon excitation has been treated theoretically by Quinn (5). A parameterization of Quinn's calculation for aluminum by Duke et al. (6) is shown in Figure 1. Coupling to plasmon excitations is less efficient at high energies, resulting in a steady increase in the mean-free path above approximately 50 eV. Since Quinn's calculation considers only plasmon losses, the inelastic mean-free path becomes infinite below the plasmon energy, which for aluminum is 15.4 eV. Indeed, even before the wave nature of the electron had been established, Farnsworth (7) reported that electrons of a few eV energy are reflected from surfaces without measurable energy loss. There are of course excitations other than plasmons by which electrons can lose energy in this low energy range, such as the creation of photons, but these losses can only be detected with very high resolution electron spectrometers (8, 9). Thus, although the inelastic mean-free path does not become infinite at low energies, as this model would seem to suggest, one would expect that electron spectroscopies would be most sensitive to the surface of aluminum at 50–100 eV. A great body of experimental evidence indicates that this is true, not only for aluminum but for a wide variety of other materials as well (10). This assumes that electrons which have lost appreciable energy are excluded from measurement. This is a reasonable assumption within one plasmon energy of a threshold.

When a solid stops an electron, the lattice is left essentially undisturbed. For ions however the situation may be quite different. At energies of a few kilovolts or less, the displacement of lattice atoms, including sputtering, accounts for most of the energy loss (11). At very high energies, however, the cross section for displacement is relatively low, and the creation of excited electronic states is the principal energy loss mechanism. Under these conditions ions penetrate deeply into the solid. Near the end of its path, of course, the destruction of the lattice by an impinging ion is always quite severe.

## The Bremsstrahlung Spectrum

Classically, the deceleration of the electron in the solid must result in the emission of electromagnetic radiation or "bremsstrahlung." For the case of a slow electron ($v << c$) decelerated in the direction of its motion, the radiation field is formally identical to that of a radiating electric dipole. The angular distribution of radiated energy is thus the familiar $\sin^2\theta$ distribution. In practice, however, both the angular distribution and the polarization of the outgoing radiation are greatly modified by scattering of the electrons in the material.

It is clear from Figure 1 that the stopping length of the electron will be small compared with the wavelength of the emitted bremsstrahlung. Thus, for the purpose of estimating the expected spectral distribution, the deceleration can be expressed in terms of a delta function. The wave field is given by:

$$E(t) = \frac{e \sin\theta}{4\pi\epsilon_0 c^2 r} \sqrt{\frac{2eV}{m}} \, \delta(t - t_0) \tag{1}$$

where $V$ is the potential through which the incident electron is accelerated to the sample, and $t_0$ is the time at which the pulse of radiation is emitted. Since the radiation field is represented by a delta function, its Fourier components are independent of frequency except for phase, and the energy radiated in a unit frequency interval is:

$$W_\omega = -\frac{e^2}{3\pi\epsilon_0 c} \frac{2eV}{mc^2} \tag{2}$$

Thus, the simple classical approximation predicts a power spectrum for bremsstrahlung which is independent of frequency. Even classically, of course, the high frequency components of the spectrum will be cut off if the stopping time of the electron is made finite.

In fact, however, the high frequency cutoff corresponds to the conversion of the total kinetic energy of the electron into a single photon, i.e., $\hbar\omega_{max} = eV$. This so-called "Duane–Hunt limit" was the basis for an early determination by Duane and Hunt ($12$) of Planck's constant. In terms of the number of quanta emitted during the deceleration, the spectrum below the cutoff as shown in Figure 2 is given by:

$$N(\omega) = \frac{e^2}{4\pi\epsilon_0 \hbar c} \frac{2}{3\pi} \frac{2eV}{mc^2} \frac{1}{\omega} \tag{3}$$

At low frequencies the bremsstrahlung spectrum appears to follow the $1/\omega$ behavior predicted by Equation 3. At very low frequencies, this results in the so-called "infrared catastrophe," with the emission of an infinite number of zero energy photons. Near the cutoff frequency, however, the bremsstrahlung spectrum of most materials exhibits very distinct structure. This structure can be understood in terms of the band theory of solids. Let us rederive the spectrum from this point of view.

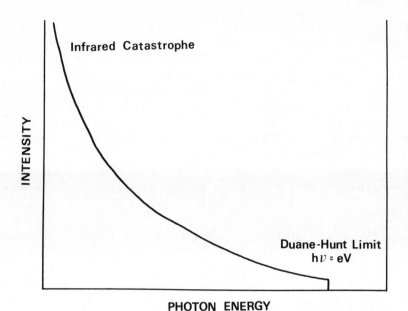

**PHOTON ENERGY**

*Figure 2. Semiclassical prediction of the bremsstrahlung spectrum produced by uniformly stopping a beam of electrons of energy eV.*
The infrared catastrophe refers to the $1/\nu$ increase in the number of photons at low energies. The Duane–Hunt limit corresponds to the conversion of the kinetic energy of an incident electron into a single bremsstrahlung photon. Uniform stopping of an electron is equal to a constant density of states above the Fermi energy.

If a potential is applied between a thermionic emitter and an anode, electrons will arrive at the anode surface with an energy $eV + e\phi_c + kT$ relative to the Fermi energy of the anode, where $e\phi_c$ is the emitter work function, and $kT$ is the average energy of the thermionically emitted electrons just outside the cathode surface (Figure 3). If an incident electron is captured by a state $\epsilon_1$ above the Fermi level, energy may be conserved by the direct emission of a bremsstrahlung photon of energy:

$$h\nu_1 = eV + e\phi_c + kT - \epsilon_1 \qquad (4)$$

The number of photons of energy $h\nu_1$ is thus a measure of the density of states available at $\epsilon_1$. The quantum mechanical equivalent of a uniformly decelerated electron is just a constant density of states above the Fermi energy. To a first approximation, the bremsstrahlung spectrum near the cutoff reflects the density of unoccupied states of the solid, with the cutoff corresponding to Fermi energy.

Experimentally, it is inconvenient to actually scan the soft x-ray spectrum for a fixed accelerating potential. In practice therefore the pass band of the x-ray spectrometer is set at some arbitrary value, removed from any characteristic lines, and the electron energy is varied to move the Duane–Hunt limit across the pass band. Near the cutoff, this is entirely equivalent to scanning with the x-ray spectrometer (*13*).

This so-called "bremsstrahlung isochromat" technique has been developed largely by Ulmer and his co-workers at Karlsruhe, and the results are described well by the density-of-states model (*14*). The agreement of the bremsstrahlung isochromats with a rigid band model of the filling of the $3d$ band for the elements iron through zinc is evident in Figure 4 taken from the work of Turtle and Liefeld (*15*). The pronounced maximum above the threshold in Fe, Co, and Ni corresponds to the unfilled portion of the $d$ band.

Ulmer (*16*) has stressed the complementary nature of bremsstrahlung isochromat spectroscopy, which examines the unfilled (conduction) states, and photoelectron spectroscopy which probes the filled (valence) states. Curiously, however, photoelectron spectroscopy has been em-

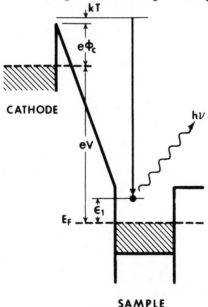

*Figure 3.    Energy level diagram of the bremsstrahlung process.*

*If the incident electron is captured in a state above the Fermi level $E_F$, energy may be conserved by the emission of a bremsstrahlung photon of energy $h\nu = eV + e\phi_c + kT - \epsilon_1$, where $e\phi_c$ is the work function of the electron emitter, and $kT$ is the average energy of a thermionically emitted electron. Near the cutoff at $\epsilon_1 = 0$, the bremsstrahlung spectrum directly probes the density of unfilled states.*

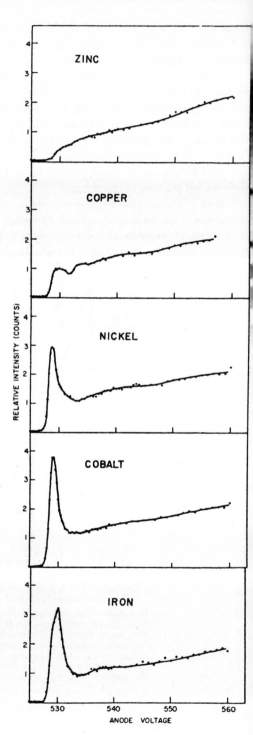

Physical Review B

*Figure 4. X-ray bremsstrahlung isochromats for the elements iron through zinc obtained by Ref. 15.*

*The x-ray analyzer window was fixed at $hv = 530$ eV. The large peak at threshold in iron, cobalt, and nickel corresponds to the unfilled portion of the 3d band. As the 3d states are filled in going from iron to copper, the peak narrows and then disappears with only the 4s states remaining unfilled.*

braced by surface scientists, who have exploited its sensitivity to the surface region. By contrast, bremsstrahlung isochromat spectroscopy, which probes essentially the same region, has been largely ignored by surface scientists. Thus, although its sensitivity to the surface region and hence to contamination has been noted by the Karlsruhe group (*17, 18*), its applications have been concerned mostly with a comparison of metal and alloy spectra with theories appropriate to bulk structure (*16–22*). An example of the systematic changes in the isochromat spectrum with alloying is shown in Figure 5.

The only deliberate exploitation of the surface sensitivity of bremsstrahlung isochromat spectroscopy has been its application by Merz to the study of work functions (*23*). One possible reason for this neglect by surface scientists has been the high currents required to obtain spectra in a reasonable time. This may rule out the technique for such important surface problems as chemisorption.

### Characteristic Excitation

Long counting times are required to obtain the bremsstrahlung isochromat because of the small cross section for radiative capture of an

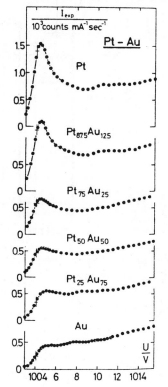

Zeitung der Physik

*Figure 5.   X-ray bremsstrahlung isochromats of the alloy system platinum-gold obtained by Ref. 22. The results are not consistent with a rigid band model of the filling of the 5d band.*

electron. Energy may also be conserved in the capture of an incident electron by the excitation of a core electron as shown in Figure 6. If the incident electron is captured in a state $\epsilon_1$ above the Fermi energy, the core electron will be excited to a state $\epsilon_2$ given by:

$$\epsilon_2 = eV + e\phi_c + kT - \epsilon_1 - E_B \qquad (5)$$

where $E_B$ is the core electron binding energy relative to the Fermi energy. The recombination of the core hole may occur radiatively with the emis-

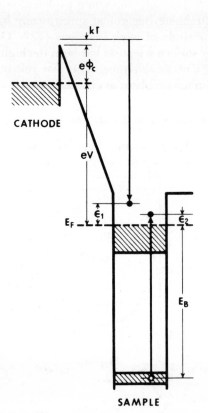

*Figure 6. Energy level diagram of the characteristic excitation process. If the energy of the incident electron eV + e$\phi_c$ + kT exceeds the binding energy of a core electron, it may be scattered from the core electron into a state above the Fermi energy $\epsilon_1$ and excite the core electron into a second state $\epsilon_2 = eV + e\phi_c + kT - \epsilon_1 - E_B$. The core hole may subsequently recombine by the emission of a characteristic x-ray photon. The excitation probability, measured by the number of characteristic x-rays emitted, is a two-electron probe of the density of unfilled states.*

sion of an x-ray photon, but unlike the production of bremsstrahlung, the characteristic process has a distinct threshold at:

$$eV_{\text{crit}} = E_B - e\phi_c - kT \qquad (6)$$

Above this threshold the excitation probability is determined by the density of unoccupied states, but the view of these conduction states is very different from that provided by the bremsstrahlung spectrum since the excited core electron, as well as the incident electron must be fitted into available states above the Fermi level.

The density of states for the two electrons is determined by all the combinations of $\epsilon_1$ and $\epsilon_2$ allowed by the conservation of energy as expressed by Equation 5. The two-electron density-of-states $N_{2c}(E)$ is thus given by the self-convolution of the density of conduction band states for one electron $N_c(E)$, i.e.,

$$N_{2c}(E) = \int_0^E N_c(E') N_c(E - E') dE' \tag{7}$$

Assuming constant oscillator strengths (as we tacitly did in the bremsstrahlung case), the excitation probability is proportional to the integral product of the final state distribution given by Equation 7 and an initial state distribution $N(E)$ corresponding to the filled core level. The width of the core level is a consequence of the finite lifetime $\tau$ of the hole that must be created to verify its existence. $N(E)$ is thus represented by a Lorentzian of width $h/\tau$. According to this simple picture, the excitation edge should have the shape of the self-convolution of the bremsstrahlung isochromat broadened by the lifetime of the core level. To observe the core level excitation spectrum, it must first be distinguished from a background of bremsstrahlung.

Conventionally this has been accomplished by using a soft x-ray spectrometer adjusted to pass a single characteristic emission line associated with the recombination of a particular core hole (*24, 25, 26*). The electron energy is then varied as in the bremsstrahlung isochromat technique. The spectrum in the "characteristic isochromat," however, is determined almost entirely by the chracteristic emission in the pass band, which is far more intense than the bremsstralung in the same band.

The spectral window for the characteristic isochromat is the atomic core level in which the initial hole was created. The comparatively broad pass band of the analyzer is only intended to reduce the bremsstrahlung background. Thus, the instrumental resolution is determined solely by the energy spread of the incident electrons. Since this can, without much difficulty, be kept as low as .25 eV, the resolution is substantially better than for other core level spectroscopies.

The detail with which the conduction band can be examined, however, is limited by the lifetime broadening of the core level. To study the density of states, therefore, it is necessary to measure only those excited core levels with relatively long lifetimes. For example, the density of states viewed by transitions involving a $2s$ core hole shows less detail than the view involving a $2p$ hole because of the strong $2p \rightarrow 2s$ Coster–Kronig transition (*27*).

The smearing of the density-of-states information, however, is the price imposed by the uncertainty principle for a local view of the density of states. Thus, in contrast to the bremsstrahlung edge, a characteristic edge is specific to a particular element. This makes it possible to examine separately the states accessible to core electrons of different elements on the same surface. This very local view of the density of states can provide information on questions such as the extent to which atoms retain their own states or share them with other elements in a common band. The position of the edge, as well as its shape, can be influenced by changes in chemical configuration. These "chemical shifts" have received a great deal of attention in ESCA studies of surfaces, but their interpreation in terms of charge transfer requires a knowledge of bond lengths and angles which is often unavailable.

Despite the specificity of the excitation edges to particular elements, conventional characteristic isochromat spectroscopy is poorly suited to an analysis of the elemental composition of the surface region. The position of an x-ray line must be known in advance in order to adjust the window of the spectrometer. To attempt to survey a wide energy range in search of edges in a sample of unknown composition would be hopelessly tedious.

### Appearance Potential Spectroscopy

In 1954 Shinoda, Suzuki, and Kato (28) demonstrated that it is possible to suppress the bremsstrahlung background without a dispersive analyzer by electronic differentiation of the total x-ray yield as a function of the incident electron energy. Differentiation weights the Fourier components of the spectrum by their frequency, thus enhancing the abrupt characteristic thresholds relative to the smoothly varying bremsstrahlung background. This approach was rediscovered by Park, Houston, and Schreiner (29) in 1970 who used potential modulation differentiation and synchronous detection to extract the excitation thresholds. Under the name "soft x-ray appearance potential spectroscopy," this has since become an important technique for studying the composition and electronic structure of solid surfaces. It is, moreover, a remarkably simple technique.

An appearance potential spectrometer in its simplest form (30) is shown in Figure 7. The sample is bombarded by electrons from a bare filament emitter. Photons produced by electron impact pass through a grid and are absorbed by a cylindrical metal photocathode. A bias voltage between the filament and grid prevents filament electrons from entering the photocathode can. Electrons from the photocathode are collected on a positively biased coaxial collector wire. The work function of the

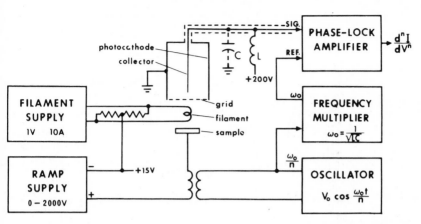

*Figure 7.   An appearance potential spectrometer using photoelectron detection. The sample is bombarded by electrons from a bare filament emitter. X-rays produced by electron impact pass through a grid and are absorbed by a metal photocathode. Electrons from the photocathode are collected on a positively biased coaxial collector wire. To obtain the $n^{th}$ derivative of the x-ray yield, a small sinusoidal oscillator is superimposed on the sample potential. The $n^{th}$ harmonic of the collector signal is detected.*

photocathode, which is usually gold, is high enough to discriminate against low energy bremsstrahlung and most filament incandescence.

The collector current is a measure of the total x-ray yield. To obtain the first derivative spectrum, a small sinusoidal oscillation is superimposed on the potential of the sample. That portion of the collector current that varies at the frequency of the oscillation is selected by a high-$Q$ resonant L-C circuit and further filtered and detected by a phase-lock amplifier. The $n^{th}$ harmonic of the modulation frequency can be shown to correspond to the $n^{th}$ derivative of the total yield, broadened by an appropriate instrument response function (*31*).

The second derivative appearance potential spectrum of a uranium surface obtained with a simple spectrometer of the type shown in Figure 7 is shown in Figure 8. Several impurities are evident on the surface including carbon, calcium, oxygen, and iron. As a technique for analyzing surface composition, however, appearance potential spectroscopy has some interesting limitations. At least two elements, palladium and gold, do not appear to exhibit detectable thresholds in the energy range of interest for surface studies (*32*). The magnitude of the matrix elements connecting the core levels to states near the Fermi level are just too small to produce detectable edges. For most elements, however, is is possible to find easily detectable levels somewhere in the region below 2 keV. Often, as in the case of uranium (*33*), all the levels in this energy range can be detected. A more serious objection to the use of the simple spectrometer shown above is the large primary current ($\sim$ 1–10 mA)

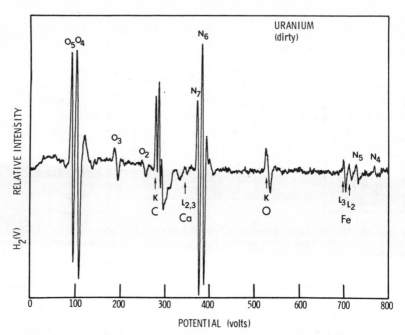

*Figure 8. Second derivative soft x-ray appearance potential spectrum of a contaminated uranium surface. The complex carbon K spectrum is characteristic of graphite. All of the uranium levels can be detected except the $O_1$ which is obscurred by the graphite spectrum.*

required to obtain spectra in a reasonable period. Not only do such currents induce desorption of weakly bound adsorbates by direct excitation into anti-bonding states, but at higher voltages the sample heating can be substantial. It seemed therefore that appearance potential spectroscopy would find little application in the important area of chemisorption.

The high currents used in appearance potential spectroscopy were largely to compensate for the low efficiency of the photoelectric detection scheme. In the soft x-ray region photoemitters have a quantum efficiency no better than 1%. The field was revolutionized, therefore, when Andersson, Hammarqvist, and Nyberg (34) developed a soft x-ray appearance potential spectrometer using a solid state detector with a quantum efficiency near 100% in the soft x-ray region. This allowed them to operate at incident electron currents 100 times lower than those conventionally used in appearance potential spectroscopy. Additional improvement resulted from the use of a thin aluminum foil filter to remove low energy bremsstrahlung (35). Since the appearance potential technique does not depend on a focused electron beam for its resolution, the current density is at least 100 times lower than is customary in electron-excited Auger electron spectroscopy. This has enabled Andersson and Nyberg to successfully apply appearance potential spectroscopy to the study of chemisorbed layers of oxygen, carbon, nitrogen, and sulfur on several transition metal surfaces (*36, 37, 38*) without appreciable sample heating or electron-induced desorption (Figure 9).

The importance of these developments is not the ability of appearance potential spectroscopy to analyze the elemental composition of the surface region; the number of techniques which can serve this purpose threatens to exceed our capacity to devise new acronyms by which to label them (*39*). The unique advantage of appearance potential spectroscopy is the resolution with which the local density of unoccupied states can be examined.

Initially, however, the interpretation of appearance potential spectra in terms of the simple one-electron density-of-states model described by Equation 7 seemed highly suspect. Among the first spectra obtained were the *K* edges of boron (*40*) and graphite (*41*). The derivative spectrum in both cases consisted not of a single *K* edge, but of a series of six nearly equally spaced peaks. It seemed quite improbable that such

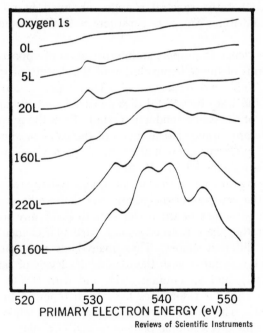

Reviews of Scientific Instruments

*Figure 9.   Soft x-ray appearance potential spectrum of the oxygen K level as a function of oxygen exposure of a nickel surface (L = 1 × 10⁻⁶ Torr-sec) (34). The single-peaked oxygen spectrum corresponds to a chemisorbed state. The multipeaked spectrum corresponds to the formation of NiO.*

peculiar spectra could be explained by simple density-of-states considerations. It was, moreover, fashionable at that time to attribute every slightly suspect bump in core level spectroscopies to many-body singularities (42). This fashion has abated somewhat (see, for example, Ref. 43), and we now have a great deal more data from which to form a judgment.

The first systematic attempt to compare appearance potential spectra with the predictions of a one-electron density-of-states model was a comparative study of the $L$-shell spectra of the $3d$ transition metals from scandium to nickel (27). The results agreed remarkably well with a simple rigid band model, with no suggestion of any severe complications from many-body effects. A detailed comparison of the appearance potential spectrum of the aluminum $L_{2,3}$ edge agreed quantitatively with density-of-states calculations (44) except for a suggestion of a threshold peak which might be the result of many-body effects (45).

Ertl and Wandelt (46) found good agreement between the appearance potential spectrum of Ni–Cu alloys and density-of-states calculations based on the coherent potential approximation. Later these researchers demonstrated that the spectral changes observed in the oxidation of nickel (47) and iron (48) were consistent with the predictions of band theory.

Direct evidence to support an interpretation of appearance potential spectra in terms of the self-convolution of the density of states has been provided by Webb and Williams (49). They report a double-peaked $L_3$ threshold in the transition metal dichalcogenides attributable to crystal field splitting of the $d$-like conduction band. Even the graphite spectrum whose bizarre appearance led to much of the concern about many-body anomalies in appearance potential spectroscopy now seems interpretable by the same simple model (45, 50, 51).

Only among the rare earth elements do many-body resonances appear to dominate the structure of the edges. Not that we know enough about the electronic structure of these elements to make any sort of detailed comparison; rather the various edges of a particular element are observed to have very different shapes. Thus, many-body resonances have been reported to be associated with the $M_5$ and $N_5$ levels of samarium (52), lanthanum (53), and gadolinium (53), and with the $O_5$, $O_4$ levels of thorium (53). Similar resonances have been reported in the spectrum of barium (54). The resonances seem to persist even when the rare earth is present as only a few atomic percent in a Ni–Cr alloy (55), thus indicating the local character of the interaction.

### Soft X-Ray Emission Spectra

In 1967 Weber and Peria (56) published a description of the use of a low energy electron diffraction system to obtain an electron-excited

Auger electron spectrum. Just one month later, Sewell and Cohen (57) reported that they had used a modified reflection high energy electron diffraction system to obtain the soft x-ray emission spectrum of the surface region. The surface sensitivity in this case was a consequence of using grazing incidence electrons to excite the spectrum.

With this apparatus, Sewell and Cohen were able to study the surface crystallography and composition simultaneously. Despite the necessity of using a window to isolate the specimen vacuum from the x-ray spectrometer, they observed elements down to carbon. It is a measure of the relative difficulty of detecting and analyzing soft x-rays as compared with low energy electrons that, while the Auger spectrometer of Weber and Peria was widely copied, it seems no one adopted the method of Sewell and Cohen.

At about this time, there were other efforts to limit the excitation of x-rays to the surface region. The objective, however, was not to study the surface but simply to eliminate line shape distortions produced by self-absorption of the emitted x-rays. Liefeld (58) demonstrated that as the energy of exciting electrons is reduced the shape of the soft x-ray emission band of a metal undergoes a drastic change. To demonstrate that this change is caused by reduced self-absorption because of the shallower sampling depth at low electron energies, Liefeld plotted the ratio of intensities for the nickel $L$-emission spectrum taken at high (15 keV) and low (2 keV) electron energies. The resulting "self absorption spectrum" was nearly identical to the known absorption spectrum for nickel foil. Using this approach Liefeld and Hanzély obtained $L_\alpha$ spectra of the $3d$ transition series which are essentially free of satellites and self-absorption effects (59). These spectra probably represent a much more accurate view of the valence band states in these materials than that provided by photoelectron spectroscopy since the latter have never been properly corrected for inelastic tailing (2).

To the x-ray spectroscopist, however, surface effects are generally regarded as additional artifacts with which to cope. Thus, although soft x-ray emission spectra, taken under conditions of near-threshold excitation, offer the most accurate means of studying the valence band electronic structure in the surface region, they have not been widely exploited for this purpose. Surface scientists have shown little inclination to undertake such demanding studies. This is not because surface scientists are inherently indolent; soft x-ray spectrometers do not lend themselves to the ultra-high vacuum techniques which are an essential characteristic of modern surface science (69), and the electron bombardment currents are excessive for studying fragile surfaces such as chemisorbed layers.

*Ion-Induced X-Rays*

It is of course possible to excite soft x-rays with any incident projectile of energy greater than the binding energy of the core electrons. In contrast to electron excitation case, the collision of protons or heavier ions with a solid results in negligible bremsstrahlung (*61*). Unfortunately, the characteristic x-ray yield is also relatively small, and it is no longer practical to use dispersive analyzers to obtain the spectrum. Measurements are generally made instead with solid state detectors, using ions of 0.1–5 MeV to excite the x-rays.

There have been great strides recently in solid state energy analysis. Unlike dispersive analyzers, which actually discriminate on the basis of momentum, solid state detectors are not confined to studying point sources, and the solid collection angle can be relatively large. Unfortunately, a state-of-the-art resolution of 140 eV is unlikely to impress soft x-ray spectroscopists accustomed to thinking in terms of tenths of an electron volt. What is sacrificed of course is any sensitivity to line shape and hence to chemical environment. The technologist often regards the dependence of spectral-line shape on chemical environment as more of a hindrance than a help in any case. The use of solid state detectors in electron-excited x-ray analysis, however, has the effect of letting too much bremsstrahlung through. The resulting high background makes quantitative measurement difficult.

Although it is generally used to analyze thin films on solid substrates (*62*), Van der Weg et al. (*63*) report that the monolayer sensitivity of proton-induced x-rays is at least comparable with that of Auger spectroscopy. Although most of the effort in proton-induced x-ray analysis has been in studying the elemental composition of the surface (*64*). Nagel has suggested (*64*) that the strong satellite structure produced by high energy ion impact may help identify satellite structure in conventional x-ray band structure studies.

Aside from the convenience of proton-induced x-ray analysis as an accessory in Rutherford backscattering studies (*66*), however, it is difficult to see any great advantage over electron excitations that would justify investment in a Van de Graaff accelerator for this purpose. We should not encourage innovation in cases of doubtful improvement.

### Conclusions

Voltaire observed that "the rude beginnings of every art acquire a greater celebrity than the art in perfection." This is evident in the rush to embrace each newly discovered or rediscovered spectroscopy. It seems, however, that greater progress might be made by more effective use of existing techniques.

In particular, it seems a pity that bremsstrahlung isochromat spectroscopy and threshold-excited soft x-ray band emission have thus far been largely ignored by surface scientists. Together they provide a view of both the conduction and valence band density-of-states in the surface region of a solid, which could help us to understand the complexities of surface phenomena. Thus far, only appearance potential spectroscopy, among the several surface sensitive x-ray techniques, has been at all widely used by surface scientists.

*Literature Cited*

1. O'Bryan, H. M., Skinner, H. W. P., *Phys. Rev.* (1934) **45**, 370.
2. Park, R. L., *Phys. Today* (April 1975) **28**, 52.
3. Houston, J. E., Park, R. L., Laramore, G. E., *Phys. Rev. Lett.* (1973) **30**, 846.
4. Laramore, G. E., Camp, W. J., *Phys. Rev. B* (1974) **9**, 3270.
5. Quinn, J. J., *Phys. Rev.* (1962) **126**, 1453.
6. Duke, C. B., Anderson, J. R., Tucker, Jr., C. W., *Surf. Sci.* (1970) **19**, 117.
7. Farnsworth, H. E., *Phys. Rev.* (1925) **25**, 41.
8. Propst, F. M., Piper, T. C., *J. Vac. Sci. Technol.* (1967) **4**, 53.
9. Ibach, H., *J. Vac. Sci. Technol.* (1972) **9**, 713.
10. Powell, C. J., *Surf. Sci.* (1974) **44**, 29.
11. Lindhard, J., Nielson, V., Scharff, M., Thomson, P. V., *Fys. Medel. Dan. Vid. Selsk.* (1963) **33** (10).
12. Duane, W., Hunt, F. L., *Phys. Rev.* (1915) **6**, 166.
13. Ulmer, K., in "X-Ray Spectra and Electronic Structure of Matter," Vol. **II**, p. 213, Kiev Institute of Physics of Metals, Kiev, 1969.
14. Merz, H., Ulmer, K., *Z. Phys.* (1968) **212**, 435.
15. Turtle, R. R., Liefield, R. J., *Phys. Rev. B* (1973) **7**, 3411.
16. Ulmer, K., in "Band Structure Spectroscopy of Metals and Alloys," p. 205, Fabian and Watson, Eds., Academic, New York, 1973.
17. Merz, H., Ulmer, K., *Phys. Lett.* (1966) **22**, 251.
18. Rempp, H., *Z. Physik* (1974) **267**, 181.
19. Eggs, J., Ulmer, K., *Phys. Lett.* (1968) **26A**, 246.
20. Merz, H., Ulmer, K., *Z. Phys.* (1968) **210**, 92.
21. Edelmann, F., Eggs, J., Ulmer, K., in "X-Ray Spectra and Electronic Structure of Matter," Vol. **I**, p. 166, Kiev Institute of Physics of Metals, Kiev, 1969.
22. Rempp, H., *Z. Phys.* (1974) **267**, 187.
23. Merz, H., *Phys. Status Solidi* (1970) **A1**, 707.
24. Liefeld, R. J., *Soft X-Ray Band Spectra* (1968) 133.
25. Dev, B., Brinkman, H., *Ned. Tidgschr. Vacuumtech.* (1970) **8**, 176.
26. Burr, A. F., *Adv. X-ray Anal.* (1970) **13**, 426.
27. Park, R. L., Houston, J. E., *Phys. Rev. B* (1972) **6**, 1073.
28. Shinoda, G., Suzuki, T., Kato, S., *Phys. Rev.* (1954) **95**, 840.
29. Park, R. L., Houston, J. E., Schreiner, E. G., *Rev. Sci. Instrum.* (1970) **41**, 1810.
30. Park, R. L., Houston, J. E., *Surf. Sci.* (1971) **26**, 664.
31. Houston, J. E., Park, R. L., *Rev. Sci. Instrum.* (1973) **29**, 571.
32. Park, R. L., *Surf. Sci.* (1975) **48**, 80.
33. Park, R. L., Houston, J. E., *Phys. Rev. A* (1973) **7**, 1147.
34. Andersson, S., Hammarqvist, H., Nyberg, C., *Rev. Sci. Instrum.* (1974) **45**, 877.
35. Baun, W. L., Chamberlain, M. B., Solomon, J. S., *Rev. Sci. Instrum.* (1973) **44**, 1419.

36. Andersson, S., Nyberg, C., *Solid State Commun.* (1974) **15**, 1145.
37. Nyberg, C., *Surf. Sci.* (1975) **52**, 1.
38. Andersson, S., Nyberg, C., *Surf. Sci.* (1975) **52**, 489.
39. Park, R. L., "Surface Physics of Crystalline Solids," Vol. II, Chapter 8, p. 377, Academic, New York, 1975.
40. Houston, J. E., Park, R. L., *J. Vac. Sci. Technol.* (1971) **8**, 91.
41. Houston, J. E., Park, R. L., *Solid State Commun.* (1972) **10**, 91.
42. Mahan, G. D., *Phys. Rev.* (1967) **163**, 612.
43. Robinson, J. E., Dow, J. D., *Phys. Rev.* (1975) **B 11**, 5203.
44. Nilsson, P. O., Kanski, J., *Surf. Sci.* (1973) **37**, 700.
45. Park, R. L., Houston, J. E., Laramore, G. E., *Jpn. J. Appl. Phys. Suppl. 2* (1974) **2**, 757.
46. Ertl, G., Wandelt, K., *Phys. Rev. Lett.* (1972) **29**, 218.
47. Ertl, G., Wandelt, K., *Z. Naturforsch.* (1974) **29a**, 768.
48. Ertl, G., Wandelt, K., *Surf. Sci.* (1975) **50**, 479.
49. Webb, C., Williams, P. M., *Phys. Rev. B* (1975) **11**, 2082.
50. Webb, C., Williams, P. M., *Surf. Sci.* (1975) **53**, 110.
51. Houston, J. E., *Solid State Commun.* (1975) **17**, 1165.
52. Chamberlain, M. B., Baun, W. L., *J. Vac. Sci. Technol.* (1974) **11**, 441.
53. Murthy, M. S., Redhead, P. A., *J. Vac. Sci. Technol.* (1974) **11**, 837.
54. Nilsson, P. O., Kanski, J., Wendin, G., *Solid State Commun.* (1974) **15**, 287.
55. Harte, W. E., Szczepanek, P. S., Leyendeker, A. J., *Phys. Rev. Lett.* (1974) **33**, 86.
56. Weber, R. E., Peria, W. T., *J. Appl. Phys.* (1967) **38**, 4355.
57. Sewell, P. B., Cohen, M., *Appl. Phys. Lett.* (1967) **11**, 298.
58. Liefeld, R. J., *Soft X-Ray Band Spectra* (1968) 133.
59. Hanzély, S., Liefeld, R. J., "Electronic Density of States," L. H. Bennett, Ed., National Bureau of Standards Publication **323**, p. 319, 1971.
60. Duke, C. B., Park, R. L., *Phys. Today* (August, 1972) **25**, 23.
61. Merzbacher, E., Lewis, H. W., in "Encyclopedia of Physics," Vol. **34**, p. 166, S. Flügge, Ed., Springer-Verlag, Berlin, 1958.
62. Hart, R. R., Olson, N. T., Smith, H. P., Khan, J. M., *J. Appl. Phys.* (1968) **39**, 5538.
63. Van der Weg, W. F., Kool, W. H., Roosendaal, H. E., *Surf. Sci.* (1973) **35**, 413.
64. Musket, R. G., Bauer, W., *Appl. Phys. Lett.* (1972) **20**, 411.
65. Nagel, D. J., in "Band Structure Spectroscopy of Metals and Alloys," D. J. Fabian and L. M. Watson, Eds., p. 77, Academic, New York, 1973.
66. Bauer, W., Musket, R. G., *J. Appl. Phys.* (1973) **44**, 2606.

RECEIVED January 5, 1976. This work was supported by the Office of Naval Research under grant N00014-75-C-0292 and the National Science Foundation under grant DMR72-03021-A04.

9

# Positron Interaction with Solid Surfaces

WERNER BRANDT

Department of Physics, New York University, New York, N.Y. 10003

*Current knowledge of positron–surface interactions is summarized with applications to surface science. The review interlinks four areas of positron physics which have developed almost independently, but which are possibly equally relevant to the study of surfaces: (1) the trapping at surfaces or the escape as positronium of positrons implanted in small solids, (2) the interaction of positrons with crystal defects and voids in solids, (3) the emergence of positrons from positron beam moderators, and (4) the formation of positronium in ground and excited states by positrons backscattered from solid surfaces. The positron method offers a potentially very sensitive technique for the systematic study of surfaces.*

## The Positron Method

Positrons are the antiparticles of electrons in the sense of Dirac's theory of the electron. They annihilate with an electron into gamma quanta. In some insulators, positrons form bound states with an electron called para positronium (para Ps; positron–electron spin state $S = 0$) and ortho positronium (ortho Ps; positron–electron spin state $S = 1$), normally in the ratio 1:3. The annihilation rate, measured as indicated in Figure 1, is proportional to the electron density at the site of the positron. Practically all positrons, when implanted into solids, annihilate with an electron into two gamma quanta, each of energy $mc^2 = 0.511$ MeV. The energy distribution of the electrons in the overlap region of the electron and positron wave functions can be measured by the small Doppler shifts, relative to $mc^2$, of the gamma rays emitted by positron–electron pairs annihilating in motion relative to the laboratory frame of reference. The two gamma rays emerge from the annihilation site in opposite directions to conserve momentum. Small deviations from 180° of the two-gamma angular correlation are detected in coincidence experiments as indicated

219

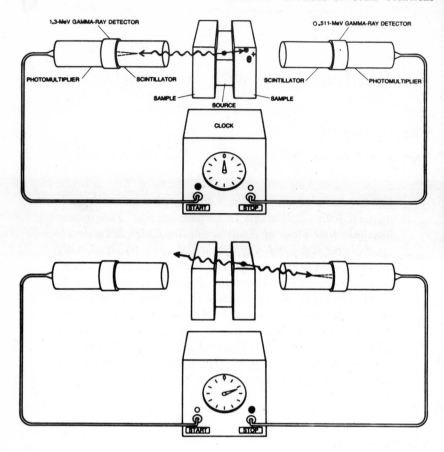

*Figure 1. In a typical positron lifetime apparatus, $^{22}$Na emits a 1.28-MeV gamma ray whenever a positron is injected into the sample. The time delay between it (upper figure) and one of the two 0.511-MeV annihilation gamma rays (lower figure) is recorded (3).*

The distribution of such time delays, i.e., the lifetime spectrum is analyzed in terms of lifetime components, each characterized by a mean lifetime $\tau$ and a relative intensity I which is a measure of the probability for positron disappearance in this channel. The mean disappearance rate in a positron state $\tau^{-1}$ is proportional to the electron density at the site of annihilation and the trapping rate into other states.

in Figure 2. They measure the momentum distribution of the annihilating pairs and give equivalent information to Doppler shifts with higher accuracy. In short, the "positron method" through the annihilation characteristics provides measurements of the density, energy, and momentum distributions of the electrons in the domain probed by the positron wavefunction. Several reviews of the positron method have appeared recently (1, 2, 3).

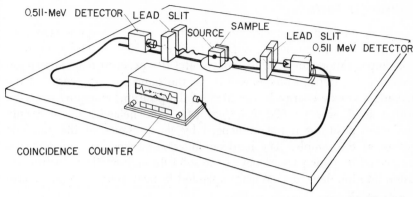

Scientific American

*Figure 2. In a typical angular correlation apparatus, the angle between the two 0.511-MeV annihilation gamma rays is measured by the dependence of the coincidence count rate on the angular deviation from $180° = \pi$ radian (3). The width of the distribution is proportional to the mean momentum of the electrons encountered by the positrons in the sample. The count rate in a $\pi$-radian coincidence apparatus (PICA) is often recorded for expediency to study changes in the angular correlation induced by sample treatments.*

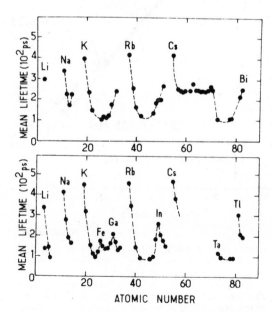

Physical Review Letters

*Figure 3. Mean positron lifetimes in metals as a function of their atomic number. Upper graph gives the measured values (5), the lower graph the values calculated by current theory for the annihilation in a conduction-electron gas (6). The dashed lines are to aid the eye.*

## Positrons at Surfaces

Most positrons from radioactive sources are so energetic that they come to rest at mean depths ranging from 10 to 100$\mu$m. The mean positron implantation range $R_+$, or conversely, the positron mass-absorption coefficient $\alpha_+$, depend on the density of the sample $d$ in g/cm$^3$ and the maximum kinetic energy $E_{+M}$ in MeV of the positrons emerging from the source as $R_+ \equiv \alpha_+^{-1} \simeq 60d^{-1}E_{+M}^{3/2}\mu m$  (4). The behavior of so deeply implanted positrons is not influenced by the presence of the entrance surface of the sample. The hard annihilation gamma rays are virtually unaffected by their passage out of the solid through ancillary instrumentation into the detectors. When implanted in homogeneous solids, there-

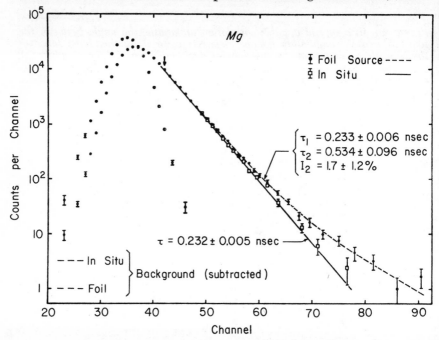

Physical Review

*Figure 4. Positron lifetime spectra in Mg metal in the form of the coincidence count rate per channel.*

*16.5 channels correspond to a 1-nsec time delay. The solid curve was obtained with an Mg sample which was activated by the $^{25}Mg(p,\alpha)$ $^{22}Na$ reaction and thus had the positron source imbedded in the bulk. The dashed curve was obtained with the same, but unirradiated metal, sandwiched as indicated in Figure 1 around a $^{22}Na$ positron source. A long-lived component of low intensity is apparent in this spectrum. The prompt time resolution spectrum of the instrument is also shown (14).*

fore, positrons have annihilation characteristics that are well defined and indicate the many-body properties of the bulk of the material.

The screening of the Coulomb interaction between a positron and an electron by the valence electrons prevents formation of positronium (Ps) in metals, and only one positron lifetime determined mainly by the conduction–electron density should be observed. Figure 3 shows the most recent compilation of experimental positron lifetimes in metals (upper graph) (5) which is in detailed agreement with the predictions of current theory (lower graph) (6). Gerholm (7) was the first to resolve a second component in some metals with an intensity of the order of 1% of all annihilations and a lifetime some two times longer than that in the bulk. Similar components were subsequently reported by many investigators with intensities that appeared to vary from sample to sample. Positron backscattering from surfaces can also have an effect (8, 9). Systematic studies (10, 11, 12) demonstrated that this "anomalous" component is linked to positrons that annihilate in the entrance surface and that it is affected by the conditions of the surface. When the positron-emitting isotope is imbedded in the sample bulk, no such component is observed (13, 14). An example is given in Figure 4.

Positrons in solids diffuse over mean distances $\Lambda_+ = (2D_+\tau_c)^{1/2} \simeq$ $10^2$ to $10^3$Å until they annihilate, $D_+$ being the positron diffusion constant and $\tau_c$ the mean lifetime in the bulk (15, 16). A fraction of all positrons, $\sim \alpha_+\Lambda_+$ or 0.01–1%, can reach the entrance surface, where they encounter a milieu of lower electron density and momenta than in the bulk. In consequence, positrons live longer in surface states, and the angular correlation between the two gamma rays is narrower than that from positrons annihilating in the bulk.

### Small Solids

The surface components became the subject of active research with the discovery (17) that the long-lived component acquires a high intensity in powders of insulators with grains of mean radius R so small that $\alpha_+R << 1$. The positrons are then deposited uniformly through each grain. Approximately the fraction $(1 - \Lambda_+/R)^3$ annihilates in the bulk while the rest diffuses to the grain surface before annihilation. Figure 5 shows the measured intensity $\Phi_2$ of the bulk lifetime component attributed to ortho Ps annihilating in $SiO_2$ grains, as a function of $(D_{Ps}\tau_2)^{1/2}/R$. A detailed theory (18, 19) predicts the solid curve. The fit to the data amounts to a determination of the positronium diffusion constant $D_{Ps}$ in $SiO_2$ which was found to be $(1.5 \pm 0.2) \times 10^{-5}$ cm²/sec. The properties of the longest-lived component with lifetime $\tau_2$ have been proven explicitly to be those of ortho Ps (20, 21, 22) in the interstices. The concomitant narrowing of the two-gamma anagular correlation with de-

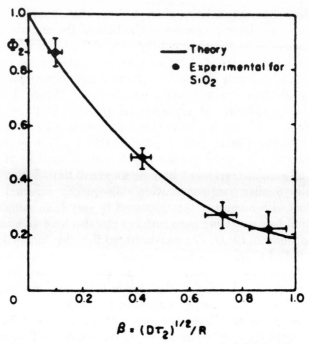

*Figure 5.   Intensity of the lifetime component attributed
to Ps annihilation in SiO$_2$ powder grains, as a function of
the inverse grain radius R$^{-1}$.*

*The theoretical curve was fitted to the data by the proper
choice of the constant $(D\tau_2)^{1/2}$. Since $\tau_2$ is known from experiment, this amounts to an experimental determination of
the Ps diffusion constant in SiO$_2$, of value (1.5 ± 0.2) × 10$^{-5}$
cm$^2$/sec (18).*

creasing SiO$_2$ grain size shown in Figure 6 (23) has confirmed this
interpretation quantitatively.

In semiconductors such as Si and Ge (24, 25), the intensity of the
longest-lived component also rises sharply when the grain size of the
solids becomes smaller than the implantation depth of the positrons and
comparable with the positron diffusion length, corresponding to diffusion
constants ~ 10 cm$^2$/sec. Many investigations have addressed the diffusion of positrons through internal surfaces into the void structure of
molecular powders, sieves, gels, and similar materials (26, 27, 28, 29).

The intensity of the surface lifetime component of metals increases
rapidly in fine metal powders as the grain size approaches $\Lambda_+$ (30). It is
convenient to characterize powders by the specific surface S in cm$^2$/g of
the sample which can be measured by various methods. The product Sd
is equal to the surface-to-volume ratio and hence a measure of a mean
R$^{-1}$. Figure 7 shows the fraction of positrons annihilating in the long-lived

component $\Phi_s$ in the form $Y \propto \ln(1 - \Phi_s)^{-1}$ vs. $Sd$, for seven metal powders of nearly the same conduction electron density (*30*). The linear rise supports the interpretation of $\Phi_s$ as a surface component and gives a diffusion length $(2D_+\tau_c)^{1/2} = 177\text{Å}$, corresponding to a positron diffusion constant $D_+ = (1.0 \pm 0.5) \times 10^{-2}$ cm²/sec in such metal grains at room temperature. Consistent with Figure 7, the two-gamma angular correlation narrows linearly with increasing $\Phi_s$. Figure 8 displays the data for the change in count rate $\Delta I_o/I_o$ in a $\pi$-radian coincidence apparatus (PICA) which increases in proportion to the narrowing of the angular correlation, as a function of $F_s \simeq \Phi_s$, the measured fraction of positrons in the long-lived component underlying Figure 7 (*31*).

The Zeeman transition of ortho Ps diffusing out of the grains of $Al_2O_3$ powder shows a shift to lower frequencies, which grows with $S$, when compared with the transition measured in a gas (*21, 22*). This confirms the suggestion (*18*) that the Zeeman resonance technique can provide a precise method for studying the interaction of positrons and positronium with solid surfaces.

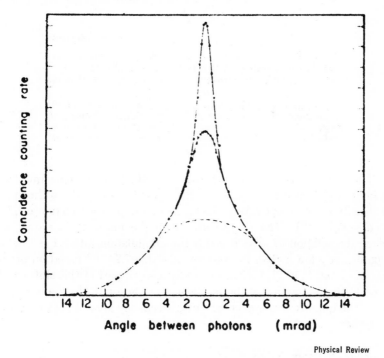

Physical Review

*Figure 6.   The effect of the grain size in $SiO_2$ powders on the correlation of the angular deviations from 180° between the two 0.511-MeV annihilation gamma rays. The upper curve is for grains of 50Å diameter; the middle curve for grains of 400Å diameter. The lower curve is measured for crystalline quartz (23).*

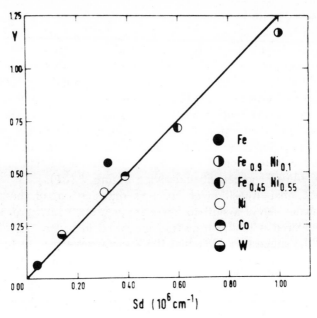

Physical Review Letters

*Figure 7. The fraction of positrons $\Phi_s$ annihilating in metal surface states, plotted in the form $Y \propto ln(1 - \Phi_s)^{-1}$, as a function of the surface-to-volume ratio Sd of metal powders.*

*The reciprocal of the abscissa values are a measure of the mean grain size. The slope is equal to $(D_+\tau_c)^{1/2}$. Since $\tau_c$ is known (Figure 3), the positron diffusion constant in these metal grains can be estimated to be $D_+ \sim 1 \times 10^{-2}$ $cm^2/sec$ (30).*

The self-annihilation rate of ortho Ps into three gamma quanta in the interstices between the grains of $SiO_2$ powders depends on the apparent density $d_p$ of the powders, shown in Figure 9 as a function of $\rho^* = d_p/(d - d_p)$ (32). The slope varies with the mean grain radius, and linear extrapolation to $\rho^* = 0$ yields the annihilation rate 7.1 $\mu s^{-1}$, which is significantly lower than the theoretical value 7.24 $\mu s^{-1}$ based on predictions of quantum electrodynamics. Only a small part of this shift can be traced to the interaction of the positronium atoms with the grain surfaces (33). The quantitative aspects of this discrepancy are not yet understood everywhere.

### Inhomogeneities in Solids

The positron method has developed into one of the most sensitive techniques for the study of point defects in crystals (2, 34). Such defects can be viewed as providing internal microsurfaces for positron absorption, leading to localized states with distinct annihilation characteristics (*15, 16, 35–41*).

The positron wavefunction in alloys distributes itself unevenly between the domains occupied by the alloyed elements. In consequence, thermal vacancy formation in microscopically different parts of the material can be studied (*42, 43, 44, 45*). Clusters in alloys can form a potential well to trap positrons, and the cluster boundary acts as an internal positron-absorbing surface. For example, the annihilation characteristics in PbIn alloys suggest that the positron wavefunction becomes progressively more localized in Pb clusters as their size grows with increasing lead concentration (*46*). This situation is radically different from that of trapping in single vacancies, where the degree of positron localization per trapping event is virtually independent of vacancy concentration. Much interesting work with positrons can be expected on the nature of boundaries between compositional fluctuations inside alloys.

### Voids in Metals

Voids form and grow in reactor construction materials after exposure to high neutron fluences of the order of $10^{22}$ neutrons/cm$^2$ (*47, 48*). The void formation is accompanied by the swelling of the material, often by

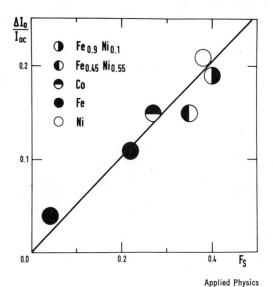

Applied Physics

*Figure 8. The narrowing of the two-gamma angular correlation, proportional to $\Delta I_o / I_{oc}$, for metal powders in which an increasing positron fraction $F_s \simeq \Phi_s$ (Figure 7) annihilates in surface states as Sd increases* (31)

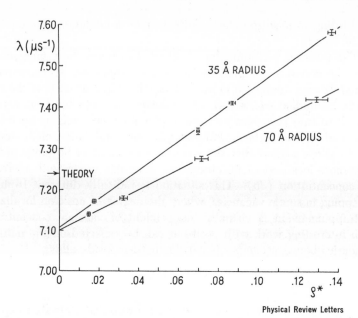

*Figure 9.   The three-gamma disappearance rate attributed to orthopositronium self-annihilations in the interstices between powder grains of two different grain sizes as a function of the free volume in the powder. The linearly extrapolated rate at infinite volume falls significantly below the vacuum value expected by theory (32).*

as much as several percent. Voids trap positrons at least as efficiently as vacancies (49). The application of the positron method to the study of voids in metals has attracted wide interest since the first reports of correlations between void formation and changes of annihilation characteristics in 1972 (50, 51).

Figure 10 compares two-gamma angular correlation curves of normal Al, of Al heated to 600°C so that thermal vacancies form which trap positrons, and of Al containing voids of 100Å diameter (as judged by electron microscopy) after neutron irradiation with fluence $4.5 \times 10^{21}$ neutrons/cm$^2$ (52). The positron lifetime spectrum of molybdenum after neutron irradiation with fluence $1.5 \times 10^{18}$ neutrons/cm$^2$ is shown in Figure 11 (53). The spectrum has two components, one with a short lifetime $\tau_1$ of intensity $I_1$, and the other with a longer lifetime $\tau_2$ of intensity $I_2 = 1 - I_1$. The trapping rate $\kappa$ into the void is given by (54):

$$\kappa = I_2(\tau_1^{-1} - \tau_2^{-1})$$

For spherical voids of radius $r_o$ and density $3/(4\pi r_1^3)$, $\kappa$ can be written in form (15, 16, 19):

$$\kappa = 3r_o D_+ / r_1^3$$

With an effective diffusion constant $D_+ \simeq 10^{-2}$ cm²/sec inferred from measurements on metal powders (30) and a void density $\sim 3 \times 10^{17}$ cm⁻³ from electron microscopic observations, one calculates from the lifetime data a void radius of 5–10Å, which is to be compared with the mean radius $\sim 20$Å of the voids detected with the electron microscope.

A consequence of the analysis of lifetime spectra in terms of trapping is that the short lifetime caused by the disappearance of positrons from the crystal bulk should become shorter when either voids and vacancies or external surfaces are present. This feature of our model (54) has not as yet emerged from any published analysis on such specimen, which leaves the method on not as safe a footing as one might wish.

### Dynamic Processes

The positron method opens the possibility of following void formation and annealing. Figure 12 (53) summarizes the changes in neutron-

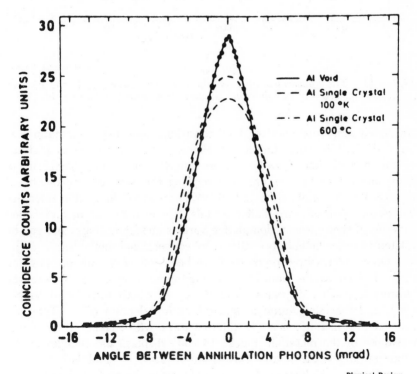

*Figure 10.    Change of angular correlation for aluminum (100°K) (lowest
peak) with the formation of vacancies (600°C) and voids (52)*

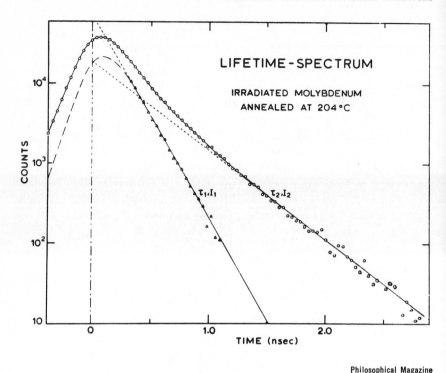

*Figure 11.   The appearance of a long-lived component $\tau_2$, $I_2$ in the positron lifetime spectrum of molybdenum because of neutron-produced voids (53)*

irradiated molybdenum after 1 hr at various annealing temperatures in terms of the following annihilation characteristics: $\tau_1$, the short lifetime; $\tau_2$, the long lifetime; $I_2 = 1 - I_1$, the intensity of the long-lived component; and $S$, the line parameter describing the Doppler broadening of the 0.511-MeV annihilation radiation. The dashed lines give reference values for a pair of unirradiated crystals. The changes can be understood in terms of the processes dominating various annealing stages and can be related to independent observations by conventional methods.

Other microscopic processes have been followed with the positron method, such as the annealing of x-ray-induced positive ion vacancy clusters in NaCl (55), vacancy clustering in electron-irradiated copper (56), and recrystallization of spherulites in selenium (57). They are correlated, if in fundamentally different ways, with changes in the global resistivity of the material. Figure 13 (56) displays the changes of the counting rates in the angular correlation of positrons annihilating in copper with valence electrons $I_v$ and with core electrons $I_c$ and of the resistivity $\Delta\rho$ after isochronal anneals at increasing temperatures. The drop in $\Delta\rho$ points to a reduction in electron scattering centers (58). The rise in

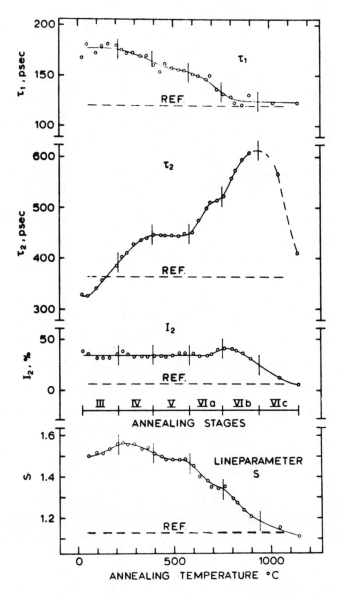

Philosophical Magazine

*Figure 12.  Changes of four positron annihilation charac-
teristics in neutron-irradiated molybdenum with isochronal
anneals. (– – –), reference values of unirradiated material
(53).*

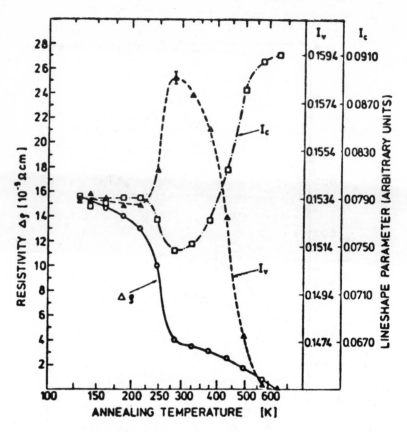

Physical Review Letters

*Figure 13.   Changes of the PICA count rate, attributed to annihilations with valence electrons $I_v$ of the large angle correlation count rate attributed to annihilations with core electrons $I_c$ and of the resistivity $\Delta\rho$ for electron-irradiated copper after isochronal anneals (56)*

$I_v$ can signify an increase in the concentration of positron trapping sites $\propto r_1^{-3}$ as well as an increase in the trap size $r_0$ because the trapping rate, $\kappa = 3r_0 D_+ r_1^{-3}$, is proportional to their product. The combined evidence suggests that radiation-produced vacancies become mobile and coalesce into clusters before they anneal.

As a new application of the positron method, we expect that similar effects should accrue with the formation of fissures and cavities under stress and in metal fatigue, which warrant experimental exploration. Simulation of neutron-induced vacancy clustering and void formation by heavy ion bombardment is now being studied with positrons (59). In this context, the positron method should detect blistering of metal surfaces and help to elucidate the processes leading to this important phenomenon, which occurs after heavy bombardment with ions ($\sim 10^{18}$ ions/cm$^2$) (*60*). Blistering warps surface layers within the range of the ions. It can become a significant cause for the erosion by surface flaking of the first wall in future fusion reactors (*61*).

### Positron–Surface Interactions

Many experimental methods have been developed to a high degree of sophistication for the analysis of surfaces (*62*), and concerted efforts are being made to advance the theory of surface structures with modern tools of atomic and many-body physics by two main approaches. One starts with the band structure of the solid and builds on calculations of the "renormalized" surface atom (*63, 64, 65*). It neglects many-body effects and is best suited for insulators, semiconductors, and perhaps some transition metals. A description of the surface in terms of molecular orbitals or "dangling bonds" emerges such that the energy levels of the surface atoms resemble more nearly those of isolated atoms than of atoms in the bulk.

The other approach builds on the electron gas model for the valence electrons in the solid, which was first applied to the metal surface by Frenkel (*66*). The ions are treated as a uniform positive background for the homogeneous conduction electron gas that fills the half space $z < 0$, of bulk density $n_o$, electron spacing $r_s$ defined by $4\pi r_s^3 n_o/3 = 1$, and Fermi momentum $p_F = (9\pi/4)^{1/3} r_s^{-1}$ all in atomic units. A self-consistent treatment of the electron density $n(z)$ near the surface, located at $z = 0$, leads to a distribution, which we call the electronic selvage of the solid, approximately given by (*15, 16*):

$$n(z) \equiv 3/4\pi r_s^3(z) = (n_o/2)\{1 - \text{signum } z \, [1 - \exp(-|z|/a_s)]\}$$

where $a_s \simeq r_s^{1/2}$ is the surface screening length. Treatments of metal vacancies in terms of the electron gas model yield the same selvage function for the surfaces of voids in metals when the void radius is large compared with $a_s$ (*15, 16, 67, 68*), and the annihilation characteristics of "chemisorbed" positrons trapped in internal or external surfaces are similar.

To illustrate the chemisorption of atoms on such surfaces (*69, 70, 71, 72, 73*), Figure 14 shows the result of a self-consistent calculation (*70*) of the binding energy of a proton, a "heavy" positron, in the selvage of a solid with the electron density of tungsten ($r_s = 1.5$). The plane for the image potential is calculated to be located at the mean distance of the electrons in the selvage for $z > 0$. As the proton approaches the surface to distances closer than 2.5Å, the interaction energy begins to deviate significantly from the image potential, and a broad minimum is reached at a distance comparable with $a_s$. The minimum gives an ionic desorption energy of 9 eV, compared with an experimental value 11 eV. The image plane is located on the vacuum side of the adsorbed proton, which implies that the effect of external fields at the chemisorption site is screened by the selvage. Indeed it is known (74) that the electric fields necessary to desorb hydrogen are essentially the same as the fields required to evaporate the substrate.

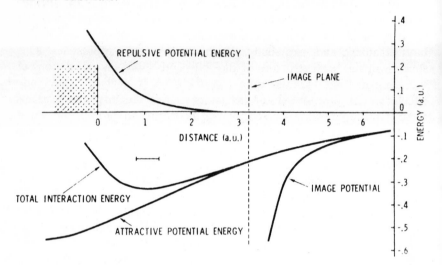

*Figure 14.    Results of a self-consistent calculation of the binding for a proton to a metal surface described by an electron gas model (70)*

The problem of chemisorbed positrons is complicated by the fact that the positron is a quantum particle. A treatment of the image force for a positron has been given by Hodges (75) who concludes that the quantum mechanical theory evolves by replacing the classical variables in the image-potential expressions with the appropriate quantum operators, with additional modifications for the self energy in the velocity operator. A method is developed for calculating quantum corrections to the classical image potential for a particle incident on the surface with a velocity. This opens the way for the formulation of a theory of chemisorbed positrons.

To guide experiments, crude estimates can be made in the statistical approximation (15, 16) of the positron binding energy in a surface state at $z_{min} \simeq a_s \simeq r_s^{1/2}$, viz., $1/2(e^2/4a_s) = 3.4r_s^{-1/2}$ eV. The lifetime of a positron $\tau(z)$ changes with the distance $z$ from the surface as:

$$\tau(z) = [r_s^3(z)/12]\{1 + f_G[r_s^3(z) + 10]/6\}^{-1} \text{ nsec}$$

**Table I.    Values of $\tau_s/\tau_c = \tau(r_s^{1/2})/\tau(r_s)$**

| $r_s$ (a.u.) | 2 | 3 | 4 | 5 |
|---|---|---|---|---|
| $\sim$ metal | Al | Ca | Na | Rb |
| $\tau_s/\tau_c$ | 2.2 | 1.4 | 1.2 | 1.1 |

where $f_G = 1$ for metals and $f_G < 1$ for materials with an energy gap. At $z_{min} = r_s^{1/2}$, the ratio of the lifetime in the surface state $\tau_s$ to that in the bulk $\tau_c$ becomes $\tau_s/\tau_c = \tau(r_s^{1/2})/\tau(r_s)$, with values given in Table I.

For metal surfaces contaminated such that $f_G < 1$, the ratios $\tau_s/\tau_c$ are larger than those for clean surfaces. The ratio of the PICA count rate $I_o$ for positrons in surfaces, $I_{os} \simeq 3/4 p_F(a_s)$, to the PICA count rate of positrons annihilating in the bulk of the conduction electron gas, $I_{ov} = 3/4 p_F$, becomes $I_{os}/I_{ov} = p_F/p_F(a_s) = 1.8$, in fair agreement with data derived from metal powders *(31)* and from metals with voids *(50)*.

Positrons are chemisorbed by some metal surfaces and escape from others *(76, 77)*. This poses the question as to the causes for such differences in surface behavior. Figure 15 *(78)* shows, in the upper part, how the selvage electrons spill into the vacuum, leaving part of the positive background unscreened. A dipole layer forms, as sketched in the lower

**Table II.    Positron and Positronium Work Functions of Metals of Atomic Number Z and Mean Valence Electron Radius $r_s$ a.u.[a]**

| Metal | Z | $r_s$ | D(eV) | $E_o$(eV) | $-E_{corr}$(eV) | $\Phi_+$(eV) | $\Phi_{Ps}$(eV) |
|---|---|---|---|---|---|---|---|
| Sc | 21 | 2.39 | 2.8 | 3.7 | 8.2 | 1.7 | −1.6 |
| Ti | 22 | 1.92 | 4.3 | 4.8 | 9.2 | 0.1 | −2.3 |
| V | 23 | 1.66 | 5.4 | 5.7 | 9.4 | −1.7 | −4.2 |
| Cr | 24 | 1.48 | 5.7 | 6.3 | 9.8 | −2.2 | −4.5 |
| Fe | 26 | 1.85 | 5.2 | 5.7 | 10.1 | −0.8 | −3.0 |
| Co | 27 | 1.81 | 5.0 | 5.7 | 9.9 | −0.8 | −2.6 |
| Ni | 28 | 1.81 | 5.0 | 5.3 | 9.9 | −0.4 | −2.0 |
| Cu | 29 | 2.12 | 3.8 | 4.9 | 9.5 | 0.8 | −1.2 |
| Y | 39 | 2.66 | 2.6 | 3.3 | 8.0 | 2.1 | −1.5 |
| Zr | 40 | 2.11 | 4.3 | 4.3 | 9.0 | 0.4 | −2.4 |
| Nb | 41 | 1.81 | 5.4 | 5.4 | 9.6 | −1.2 | −3.7 |
| Mo | 42 | 1.61 | 5.8 | 6.0 | 9.8 | −2.0 | −4.2 |
| Rh | 45 | 1.96 | 4.6 | 5.2 | 9.8 | 0.0 | −2.0 |
| Pd | 46 | 1.81 | 3.8 | 5.4 | 9.6 | 0.4 | −0.8 |
| Ag | 47 | 3.02 | 2.6 | 4.5 | 9.5 | 2.4 | −0.4 |
| Ta | 73 | 1.80 | 4.8 | 5.3 | 10.1 | 0.0 | −2.6 |
| W | 74 | 1.62 | 6.0 | 6.3 | 10.2 | −2.1 | −4.2 |
| Ir | 77 | 1.80 | 4.8 | 6.7 | 10.1 | −1.4 | −3.5 |
| Pt | 78 | 1.83 | 4.5 | 5.6 | 9.9 | −0.2 | −1.2 |
| Au | 79 | 2.09 | 3.5 | 5.3 | 9.9 | 1.1 | −5.4 |

[a] Auxiliary quantities calculated in Ref. *83*.

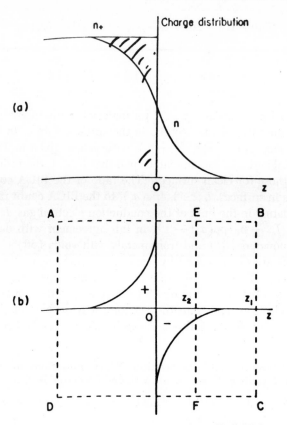

Physical Review B

*Figure 15. (a) The charge distribution in the selvage of a semi-infinite metal with surface at $z = 0$; $n_+$ is the positive background and $n$ the electron density. (b) The surface dipole layer resulting from the selvage. The field at $z_1$ is zero. At $z_2$ close to the surface, the net charge in AEFD is positive. The double layer attracts an electron but repels a positron coming from the outside. The situation is reversed for particles approaching the surface from within the metal (78).*

part, which repels an electron but attracts a positron as it enters the selvage from the sample bulk. When the positron has passed the surface at $z = 0$, it can be trapped in a surface state or ejected into the vacuum.

Electrons must overcome the work function of the metal surface to emerge from the solid, as described well by the electron gas model (79, 80, 81). Tong (78) has introduced the notion of a negative work function for the positron, if it gains energy by passing through the selvage.

Positron work functions $\Phi_+$ are difficult to predict accurately because they depend on a surface and a bulk term, $\Phi_+ = -D - \mu_+$, where $D$ is the surface–dipole barrier, and $\mu_+ = E_0 + E_{corr}$ is the chemical potential of the positron in the bulk. It is composed of the positive zero-point energy $E_0$ of the positron in the lattice of ion cores and the negative electron–positron correlation energy $E_{corr}$ (78, 82). The results of a recent calculation of $\Phi_+$ for transition metals (83) are summarized in Table II. The three metals with the most negative work functions—Cr, Mo, and W—show a sharp symmetrical peak at roughly the $\Phi_+$ values in the energy distribution of positrons emerging from the solids into the vacuum (77). This can be viewed as experimental evidence for the occurrence of negative work functions for positrons. The second, broad and nondistinct peaks may have other, possibly instrumental origins not linked to material-characteristic properties of the solid.

Figure 16 shows the dependence of the $\Phi_+$ values in Table II on $r_s$. The $r_s$ values are calculated from the density of valence electrons. If plotted against $r_s^{eff}$ based on the density of electrons participating in bulk plasmon excitations (84), the monotonic correlation between the calculated $\Phi_+$ values and $r_s$ is lost, presumably because the coupling between the conduction electrons and ion cores is reduced in the selvage. The curve represents an estimate based on the statistical model of metals according to $\Phi_+ = -D - E_0 - E_{corr}$. We derived the surface dipole

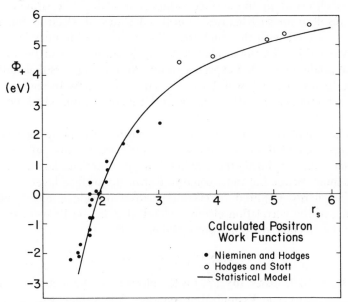

*Figure 16.    Variation of the positron work function $\Phi_+$ with $r_s$. The points, depicting the values given in Table II, and the circles are calculated for individual metals (82, 83). The curve is estimated by the statistical method with the formulae given in the text.*

energy, $D = 2\pi\epsilon^2 n_0 a_s^2 (4 - e)/e \simeq 15 r_s^{-2}$ eV, $\epsilon$ being the electronic charge and $e = 2.718 \ldots$ , by potential theory from the selvage function $n(z)$ given above; it is consistent for $a_s = 0.8 r_s^{1/2} a_0$, where $a_0 = \hbar/me^2$ with self-consistent calculations of the electronic work function (79, 80, 85). The zero-point energy, $E_0 = 14.2 r_s^{3/2}$ eV, is based on Wigner–Seitz calculations of the positron wavefunction (67, 68, 83). The correlation energy, $E_{corr} \simeq -12.4 r_s^{-1}[h(r_s) - 1]^{1/3}$ eV $= -6.8(1 + 10 r_s^{-3})^{1/3}$ eV, is given in terms of the density enhancement factor $h(r_s)$ (15, 16, 86); it approaches the positronium ionization energy, 6.8 eV, for large $r_s$. The precise $r_s$ value of the crossing point $\Phi_+ = 0$ is quite uncertain. For example, Fe($r_s = 1.85$) and Ni($r_s = 1.81$), where powder measurements point to surface trapping of positrons (30, 31), have weakly negative $\Phi_+$ values by Table II. However, the real values may well have been positive because of the passivation of the grain surfaces during the powder preparation.

Such surface treatments, generally, introduce an electronic energy gap near the surface ($f_G < 1$) (86). A gap reduces $D$ and $E_{corr}$ and, in consequence, increases or decreases $\Phi_+$ depending on the detailed quantum mechanical conditions imposed by the surface. Similarly, anions near the surface must decrease $E_0$ and thus increase $\Phi_+$. In fact, the changes observed in the surface component with Cl⁻ treatments (12) could result from such causes. Localized states of electrons on metal surfaces covered with thin dielectric films and their lifetimes have been discussed (87). The isomorphic states for positrons should be analyzed and the predictions tested experimentally. All this suggests that judicious surface treatments may well lower the curve in Figure 16 and so widen the range of materials with negative work functions from which positrons can be ejected.

When implanted in materials with positive $\Phi_+$, positrons may accumulate somewhat preferentially near surfaces, because all positrons reaching a surface during the slowing down process with kinetic energies $< \Phi_+$ cannot escape but must come to rest in the surface layer.

The energy required to take a positron outside a metal without stripping of the correlation electron cloud, that is, the Ps work function $\Phi_{Ps}$ can be calculated from a Born–Haber cycle:

$$\Phi_{Ps} = \Phi_- + \Phi_+ - E_B$$

where $\Phi_- \simeq (4.67 - 0.39 r_s)$eV (79) is the electron work function and $E_B = 1/4\, a.u. = 6.8$ eV the Ps ionization energy. The values $\Phi_{Ps}$ are always below $\Phi_+$ and, in fact, are negative for $r_s \lesssim 3$. Nevertheless, the image forces (88) still may not let thermalized positrons leave a metal as Ps atoms, and so far the evidence for Ps emergence from metals is sparse (89). The experimental evidence for Ps formation in metal voids is contradictory (51, 90) and in need of clarification. On insulators, where long range surface forces are absent, Ps atoms have to emerge through internal and external surfaces of the solids in which the positrons were stopped.

## Positron Beams from Surfaces

The first attempt by Madansky and Rasetti (91) to observe positrons leaving solids was unsuccessful. It was interpreted by Ferrell (92) to be caused, in effect, by a positive positron work function. In an unpublished thesis, Cherry (93) observed positrons from a $^{22}$Na source to emerge with energies of a few eV after transmission through mica from the surface of a chromium film. Later, Madey (94) reported the emission of slow positrons from polyethylene with energies peaked near 20 eV. The first beam experiments were begun in 1965 by McGowan and co-workers with a 1-amp peak current, 55-MeV linear accelerator (95). The positrons were created through pair production by the bremsstrahlung from 20-nsec bursts of electrons on a tantalum target and slowed down in foils of Al, mica, and CsBr with 100–200Å Au coatings. The positrons left the Au surface with an energy that peaked between 0.75 and 2.9 eV. This was attributed to a negative work function of gold, as suggested by Kohn and Callaway and elaborated by Tong (78).

In another type of study, started by Paul and his colleagues in 1966, positrons were emitted from a $^{22}$Na source and slowed down by gold-covered moderators (96). Following post-acceleration, the energy was analyzed in a spherical spectrometer. The positron energy spectra were similar to those recorded with accelerator-produced beams. The slow positron yield of $10^{-8}$ did not change under the influence of external electric fields up to 200 kV/cm.

The yield from rolled Ni foils rose by a factor $10^3$ if the exit surface was made rough. The low yields of polished Ni surfaces became comparable with those of the rough surfaces after etching. Adsorbed gas enhanced the yields by factors of the order of $10^2$ over those of degassed surfaces (96). The properties of metal oxides with regard to positron and positronium surface annihilations appear to be complicated (97), but high positron yields, $\sim 10^{-5}$, were observed with MgO surfaces (98, 99).

The production of copious beams of low energy positrons is important for atomic scattering experiments. Several groups have entered the field to produce such beams. They have begun systematic studies to establish reproducible correlations between the yields of positrons emerging from moderators and the preparation of their surfaces and to search for relations to the microscopic conditions underlying the concept of positron work functions.

### Excited-Positronium Formation on Surfaces

The first observation of Lyman-α radiation from excited positronium Ps* was reported in 1975 (*100, 101*). Positronium was formed in vacuo with high efficiency by backscattering of slow positrons incident on a solid surface, as sketched in Figure 17. Single photons of the 2430-Å Lyman-α (2P − 1S) Ps line were observed in coincidence with the gamma rays from the annihilation of groundstate Ps, as illustrated in Figure 18. Since the pioneering work of Deutsch (*102*), the existence of Ps had been inferred from annihilation characteristics and the fine structure resonance of the Ps ground state. The optical observations constitute the first classical, spectroscopic proof for the existence of the leptonic element Ps.

The intensities correspond to a yield of $10^{-4}$ Ps* atoms per incident positron. Yields of $10^{-3}$ excited species have been observed with backscattered $H^+$ and $He^+$ ions (*103*). The Ps* yields were nearly the same for targets of Ge, MgO, Au, Ti, and $SiO_2$. Carbon black produced no measurable Ps* signal. Changes in temperature from 30° to 450°C increased Ps formation by a factor of three but had no effect on the Ps* yield. In complete contrast with the situation of positrons moderated inside meals, the groundstate Ps production by the backscattering technique has efficiencies of 60–80% (*104*). Here, positrons arrive with kinetic energies of ~ 10eV, enough to excite surface plasmons and to overcome $\Phi_+$.

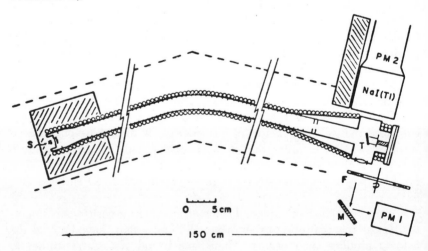

Physical Review Letters

*Figure 17. Slow positron beam apparatus. The positrons of the source S are guided by a 75G solenoid to the target T. The filtered (F) Lyman-α-radiation is detected in a photomultiplier (PM 1) in coincidence with an annihilation gamma ray recorded in the Na(Tl) crystal coupled to a photo multiplier (PM 2) (100).*

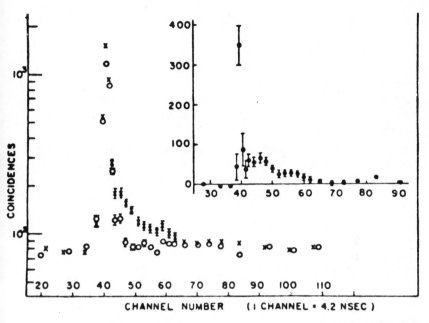

Physical Review Letters

*Figure 18.    Coincidences between the photon (start) emitted in the optical transition Ps\* → Ps and a gamma ray (stop) emitted in the Ps annihilation as a function of time delay, given in channel number. The insert shows a linear plot of the signal with background subtracted (100).*

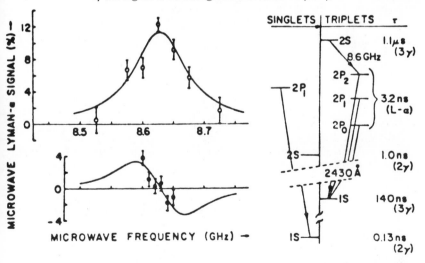

Physical Review Letters

*Figure 19.    The observed Lyman-α signal (○) and the differential signal as a function of microwave frequency. The insert is the schematic term diagram for the* n = 2 *and* n = 1 *Ps states indicating the relevant transitions and life-time τ for each level (101).*

The splitting of the $2^3S_1 - 2^3P_2$ line of Ps was subsequently measured by inducing *rf* transitions in a weak magnetic field and observing the resonant increase in the Lyman-$\alpha$ emission from Ps* formed at surfaces, as shown in Figure 19. The value $8628.4 \pm 2.8$ MHz agrees within two standard deviations with the theoretical value computed with a Lamb shift of 231 MHz included. In this sense, the experiments have proved quantum electrodynamics to be correct to third order terms in the fine structure constant.

Such sensitive methods of observing the states populated by slow positrons backscattered from surfaces will complement in important ways the methods of measuring the states of positrons that reach internal or external surfaces after moderation within solids. These intriguing possibilities for the study of solid surfaces with the positron as a quantum particle probe clearly merit further study.

## Acknowledgment

I am grateful for the aid of Temple-Jene Harris and Diane Isaacson in the preparation of these notes.

## Literature Cited

1. Proc. Third International Conference on Positron Annihilation 1973, P. Hautojärvi and E. Seeger, Ed., Springer-Verlag, Heidelberg, 1975.
2. West, R. N., *Adv. Phys.* (1973) **22**, 263.
3. Brandt, W., *Sci. Am.* (1975) **233** (1), 33.
4. Brandt, W., Paulin, R., *Phys. Rev. B*, in press.
5. MacKenzie, I. K., Jackman, T. E., Thrane, N., *Phys. Rev. Lett.* (1975) **34**, 512.
6. Brandt, W., Isaacson, D., Paulin, R., *Phys. Rev. Lett.* (1975) **35**, 1180.
7. Gerholm, T. R., *Ark. Fys.* (1956) **10**, 523.
8. Bertolaccini, M., Zappa, L., *Nuovo Cimento B* (1967) **52**, 487.
9. MacKenzie, I. K., Schulte, C. W., Jackman, T., Campbell, J. L., *Phys. Rev. A* (1973) **7**, 135.
10. Kohonen, T., *Ann. Acad. Sci. Fenn. Ser. A* (1963) **6** (130).
11. *Positron Annihilation, Proc. Conf.* (1967) 277.
12. Kohonen, T., Kugel, H. W., Funk, E. G., Mihelich, J. W., *Phys. Lett.* (1966) **20**, 364.
13. Schwarzschild, A., Brookhaven National Laboratory, private communication of unpublished results, 1963.
14. Weisberg, H., Berko, S., *Phys. Rev.* (1967) **154**, 249.
15. Brandt, W., *Appl. Phys.* (1974) **5**, 1.
16. Brandt, W., Proc. Third International Conference on Positron Annihilation 1973, P. Hautojärvi and E. Seeger, Ed., Springer-Verlag, Heidelberg, 1975.
17. Paulin, R., Ambrosino, G., *J. Phys. (Paris)* (1968) **29**, 263.
18. Brandt, W., Paulin, R., *Phys. Rev. Lett.* (1968) **21**, 193.
19. Brandt, W., Paulin, R., *Phys. Rev. B* (1973) **8**, 4125.
20. Paulin, R., Thesis, Orsay 1969 (unpublished).
21. Judd, D. J., Lee, Y. K., Madansky, L., Carlson, E. R., Hughes, V. W., Zundell, B., *Phys. Rev. Lett.* (1973) **30**, 202.
22. Egan, P. O., Carlson, E. R., Hughes, V. W., Mourino, M., Varghese, S. L., Leventhal, M., zu Putlitz, G., *Bull. Am. Phys. Soc.* (1973) **18**, 1503.

23. Steldt, F. R., Varlashkin, P. G., *Phys. Rev. B* (1972) **5**, 4265.
24. Gainotti, A., Ghezzi, C., *Phys. Rev. Lett.* (1970) **24**, 349.
25. Gainotti, A., Ghezzi, C., *J. Phys. C* (1972) **5**, 779.
26. Gol'danskii, V. I., Mokrushin, A. D., Prokop'ev, E. P., *Sov. Phys. Solid State Engl. Transl.* (1972) **13**, 2686.
27. Tatur, A. O., Mokrushin, A. D., *Sov. Phys. Solid State Engl. Transl.* (1973) **14**, 2617.
28. Chuang, S. Y., Tao, S. J., *Can. J. Phys.* (1973) **51**, 820.
29. Brandt, W., Chiba, T., *Phys. Lett.* (1976) **57A**, 395.
30. Paulin, R., Ripon, R., Brandt, W., *Phys. Rev. Lett.* (1973) **31**, 1214.
31. Paulin, R., Ripon, R., Brandt, W., *Appl. Phys.* (1974) **4**, 343.
32. Giley, D. W., Marko, K. A., Rich, A., *Phys. Rev. Lett.* (1976) **36**, 395.
33. Ford, G. W., Sauder, L. M., Witten, T. A., *Phys. Rev. Lett.* (1976) **36**, 1269.
34. Seeger, A., *J. Phys. F* (1973) **3**, 248.
35. Hodges, C. H., *Phys. Rev. Lett.* (1970) **25**, 284.
36. Brandt, W., Waung, H. F., *Phys. Rev. B* (1971) **3**, 3432.
37. Hodges, C. H., *J. Phys. F* (1974) **4**, L230.
38. Cotterill, R. M. J., Petersen, K., Trumpy, G., Träff, J., *J. Phys. F* (1972) **2**, 459.
39. McKee, B. T. A., Saimoto, S., Stewart, A. T., Stott, M. J., *Can. J. Phys.* (1974) **52**, 759.
40. Brandt, W., Paulin, R., Dauwe, C., *Phys. Lett.* (1974) **48A**, 480.
41. Manninen, M., Nieminen, R., Hautojärui, P., Arponen, J., *Phys. Rev. B* (1975).
42. Stott, M. J., Stewart, A. T., Kubica, P., *Appl. Phys.* (1974) **4**, 213.
43. Stott, M. J., Kubica, P., *Phys. Rev. B.* (1975) **11**, 1.
44. Kubica, P., McKee, B. T. A., Stewart, A. T., Stott, M. J., *Phys. Rev. B* (1975) **11**, 11.
45. Sueoka, O., *J. Phys. Soc. Jpn.* (1975) **39**, 969.
46. Lock, D. G., West, R. N., *J. Phys. F* (1974) **4**, 2179.
47. Cawthorne, C., Fulton, E. J., *Nature* (1967) **216**, 575.
48. "Radiation-Induced Voids in Metals," J. W. Corbett and L. C. Ianiello, Ed., U. S. Atomic Energy Commission **CONF-710601**, April 1972.
49. Hodges, C. H., Stott, M. J., *Solid State Commun.* (1973) **12**, 1153; **13**, vii(E).
50. Mogensen, O., Petersen, K., Cotterill, R. M. J., Hudson, B., *Nature* (1972) **239**, 98.
51. Cotterill, R. M. J., MacKenzie, I. K., Smedskjaer, L., Trumpy, G., Träff, J. H. O. L., *Nature* (1972) **239**, 99.
52. Triftshäuser, W., McGervey, J. D., Hendriks, R. W., *Phys. Rev. B* (1974) **9**, 3321.
53. Petersen, K., Thrane, N., Cotterill, R. M. J., *Philos. Mag.* (1974) **29**, 9.
54. Brandt, W., in *Positron Annihilation, Proc. Conf.* (1967) 181.
55. Brandt, W., Paulin, R., *Phys. Rev. B* (1973) **8**, 4125.
56. Mantl, S., Trifthäuser, W., *Phys. Rev. Lett.* (1975) **34**, 1554.
57. Brandt, W., Oremland, M., *Phys. Lett.* (1976) **57A**, 387.
58. Martin, J. W., Paetsch, R., *J. Phys. F* (1973) **3**, 907.
59. Engman, U., Bolmqvist, B., *Radiat. Eff.* (1975) **24**, 65.
60. Kaminsky, M., Das, S. K., *Appl. Phys. Lett.* (1972) **21**, 443.
61. Roth, J., Behrisch, R., Scherzer, B. M. U., *J. Nucl. Mat.* (1975) **57**, 365.
62. Heiland, W., *Mod. Aspects Electrochem.* (1975) 85.
63. Watson, R. E., Ehrenreich, H., Rodges, L., *Phys. Rev. Lett.* (1970) **24**, 829.
64. Hodges, L., Watson, R. E., Ehrenreich, E., *Phys. Rev. B* (1972) **5**, 3953.
65. Levin, K., Liebsch, A., Bennermann, K. H., *Phys. Rev. B* (1973) **7**, 3066.
66. Brandt, W., Fahs, J., in press.
67. Frenkl, J., *Z. Phys.* (1928) **51**, 232.

68. Fahs, J. H., Thesis, New York University, 1975, unpublished.
69. Schrieffer, J. R., *J. Vac. Sci. Technol.* (1972) **9**, 561.
70. Smith, J. R., Ying, S. C., Kohn, W., *Phys. Rev. Lett.* (1973) **30**, 610.
71. Appelbaum, J. A., Hamann, D. R., *Phys. Rev. Lett.* (1975) **34**, 806.
72. Einstein, T. L., *Phys. Rev. B* (1975) **12**, 1283.
73. Gomer, R., *Solid State Phys.* (1975) **30**, 93.
74. Müller, H. E., Krishnaswamy, S. V., McLane, S. B., *Surf. Sci.* (1970) **23**, 112.
75. Hodges, C. H., *J. Phys. C* (1975) **8**, 1849.
76. Costello, D. G., Grace, D. E., Herring, D. F., McGowan, J. Wm., *Phys. Rev. B* (1971) **5**, 1433.
77. Pendyala, S., Bartell, D., Girouard, F. E., McGowan, J. Wm., *Phys. Rev. Lett.* (1974) **33**, 1031.
78. Tong, B. Y., *Phys. Rev. B* (1972) **5**, 1436.
79. Lang, N. D., Kohn, W., *Phys. Rev. B* (1971) **3**, 1215.
80. *Ibid.*, (1973) **8**, 6010.
81. Lang, N. D., *Solid State Phys.* (1973) **28**, 255.
82. Hodges, C. H., Stott, M. J., *Phys. Rev. B* (1973) **7**, 73.
83. Nieminen, R. M., Hodges, C. H., *Solid State Commun.* (1976) **18**, 1115.
84. Isaacson, D., "Compilation of $r_s$ Values," New York University, Document No. **02698**, National Auxiliary Publication Service, New York, 1975.
85. Smith, J. R., *Phys. Rev. Lett.* (1970) **25**, 1023.
86. Brandt, W., Reinheimer, J., *Phys. Lett.* (1971) **35A**, 109.
87. Cole, M. W., *Phys. Rev. B* (1971) **3**, 4418.
88. Zaremba, E., Kohn, W., *Phys. Rev. B* (1976) **13**, 2270.
89. Tsuchiya, Y., Noguchi, S., Hasegawa, M., *Phys. Lett.* (1975) **54A**, 276.
90. Sen, P., Cheng, L. J., Kissinger, H. E., *Phys. Lett.* (1975) **53A**, 299.
91. Madansky, L., Rasetti, F., *Phys. Rev.* (1950) **79**, 397.
92. Ferrell, R. A., *Rev. Mod. Phys.* (1956) **28**, 308, footnote 53.
93. Cherry, W. H., Thesis, Princeton University, 1958, unpublished.
94. Madey, J. M. J., *Phys. Rev. Lett.* (1969) **22**, 784.
95. Costello, D. G., Grace, D. E., Herring, D. F., McGowan, J. Wm., *Phys. Rev. B* (1972) **5**, 1433.
96. Keever, W. C., Jaduszliwer, B., Paul, D. A. L., *At. Phys.* (1973) **3**, 561.
97. Sen, P., Sen, C., *Nuovo Cimento.* (1975) **29B**, 124.
98. Canter, K. F., Coleman, P. G., Griffith, T. C., Heyland, G. R., *J. Phys. B* (1972) **5**, L167.
99. Curry, S. M., Schawlow, A. L., *Phys. Lett.* (1971) **37A**, 5.
100. Canter, K. F., Mills, Jr., A. P., Berko, S., *Phys. Rev. Lett.* (1975) **34**, 177.
101. Mills, Jr., A. P., Berko, S., Canter, K. F., *Phys. Rev. Lett.* (1975) **34**, 1541.
102. Deutsch, M., *Phys. Rev.* (1951) **82**, 455.
103. Baird, W. E., Zivitz, M., Thomas, E. W., *Phys. Rev. A* (1975) **12**, 876.
104. Canter, K. F., Mills, Jr., A. P., Berko, S., *Phys. Rev. Lett.* (1974) **33**, 7.

RECEIVED March 15, 1976. Work supported by the National Science Foundation.

# 10

# Backscattering of Light Ions from Metal Surfaces

H. VERBEEK[1]

Solid State Division, Oak Ridge National Laboratory, Oak Ridge, Tenn. 37830

*Light ions backscattered from a solid target cause an energy distribution which reaches from zero to almost the primary energy. The number of the backscattered particles and their energy, angular, and charge distributions depend largely on the energy and the ion target combination. For high energies, particles are backscattered in a single collision governed by the Rutherford cross section. For lower energies, multiple collisions and the screening of the Coulomb potential have to be considered, which makes the theoretical treatment more difficult. This energy region is, however, of special interest in the field of nuclear fusion research.*

When a beam of ions impinges onto a metal target, some of the ions are implanted, and some are reflected from the surface or backscattered from deeper layers. The implanted atoms may diffuse to the surface and be released with thermal energies, or they may be trapped inside the material, e.g., at lattice defects. They also may cluster together and form gas bubbles inside the metal. The amount of trapping depends strongly on the particular ion target combination, on the temperature, and on the bombardment dose. This paper deals only with the particles which are kinetically backscattered and is restricted to light ions such as hydrogen and helium.

The particles which are backscattered cause an energy distribution which extends from zero energy to almost the primary energy. A typical example is given in Figure 1, which shows the energy distribution for 15-keV protons backscattered from an Au target. The sharp threshold at high energies is caused by ions which are backscattered from the

[1] Present address: Max Planck Institut für Plasmaphysik, D-8046 Garching, West Germany.

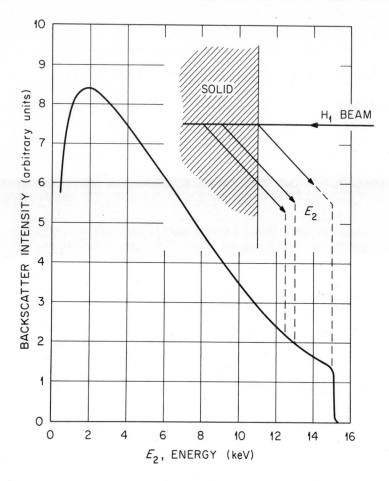

*Figure 1.   Energy distribution of hydrogen atoms backscattered from Au which is bombarded by 15-keV protons*

surface atoms of the target. The position of this edge is determined by the kinematics of a single scattering event. When a particle of mass $M_1$ and an energy $E_1$ is scattered from an atom with mass $M_2$, the energy after scattering is:

$$E_2 = k^2 E_1$$

where

$$k = [M_1 \cos \theta + (M_2{}^2 - M_1{}^2 \sin^2 \theta)^{\frac{1}{2}}]/(M_1 + M_2) \qquad (1)$$

for $M_1 < M_2$, and $\theta$ is the scattering angle in the laboratory system.

The backscattering intensity in Figure 1 at lower energies results from scattering events deeper in the solid. Along their passage through the solid, the particles lose energy in elastic nuclear collisions and quasi-continuously by excitation of target electrons (*1*, *2*). Therefore, they appear outside the target with all energies below the energy of the particles backscattered from the surface.

This paper is divided into three sections, each of which treats a certain range of primary energies. These energy ranges arise quite naturally from the different theoretical and experimental treatments of the backscattering effect they require.

For protons at energies above ≈ 50 keV the scattering can be well described by the single collision model (*3*, *4*, *5*). This model assumes that the backscattering occurs in a single large angle deflection from a certain nuclear collision, which is describable by the Rutherford cross section (*6*). The trajectories of the ions inside the material to the scattering center and back to the surface are taken as straight lines. Along their trajectories they lose energy by exciting the target electrons. This can be described by the differential energy loss $dE/dx$, i.e., the energy loss per unit path length. This makes the theoretical calculation of the energy spectra rather simple.

For this energy range the most convenient measuring method uses surface barrier detectors. These give energy proportional signals which are easily analyzed in a multi-channel analyzer system. They are equally sensitive to ions and neutral atoms.

At high energies, which depend strongly on the ion target combination, the validity of the Rutherford scattering law is limited by the occurrence of nuclear reactions and resonances. This area will be covered by R. S. Blewer in this volume. A compilation of nuclear reaction data may be found in Refs. 7, 8, and 9. Protons and helium ions which are easily available from small accelerators with energies from ≈ 100 keV to several MeV are widely used for surface layer analysis and depth profiling (*see*, for instance, Refs. *10* and *11*).

At energies below ≈ 50 keV the screening of the nuclear charge by the outer electrons becomes more important. Thus the Coulomb potential leading to the Rutherford cross section is no longer appropriate. More complicated potentials which are derived from the Thomas–Fermi theory have to be used (*2*). Also the validity of the single collision model breaks down. With decreasing energy the cross section for nuclear collisions increases, and one has to account for multiple collisions of the backscattered particles. The theoretical treatment becomes much more complicated, and an analytical form for the energy distributions can no longer be derived.

Also the experimental techniques must be modified. At energies below ≈ 20 keV surface barrier detectors no longer give energy information. Electrostatic or magnetic energy analyzers can be used, but these are only sensitive to charged particles. Also, the charged fraction of the backscattered particles decreases with decreasing energy (12, 13, 14, 15), and the spectra of the neutral and charged components may be very different (13, 14). Thus one needs a means to ionize the neutral backscattered particles in a definite manner if quantitative results are to be obtained. Another way to overcome this difficulty is to use time-of-flight techniques (16), but these too require particle detectors with known sensitivity for neutrals.

The range of primary energies from 5 to 20 keV is of special interest for fusion technology. For plasma experiments and later fusion reactors where the mean particle confinement times in the plasma are much shorter than the desired burning times, the interaction of the diffusing plasma particles with the first walls is important. Particularly with respect to the question of recycling (17), it is necessary to know the total number and angular, energy, and charge distributions of light ions (H, D, T, He) backscattered from solid surfaces.

Below the primary energy of $\approx$ 1 keV the experiments are extremely difficult. Most of the backscattered particles are neutral, and the ionization and detection methods currently in use break down at these low energies. No experiments in this energy range are known which deal with the total number of backscattered particles. Consequently, one has to rely on computer simulations. The backscattering of primary ions with energies below 1 keV is of particular interest for today's plasma experiments.

If one investigates the backscattering from single crystals, the results are largely influenced by the crystal structure which causes channeling and blocking effects. This offers a variety of measuring methods in depth profiling and lattice site determination. Recent reviews are given in Refs. 18 and 19; the present review deals only with the backscattering from amorphous or polycrystalline materials.

### High Energies

The principle of backscattering of ions with primary energies larger than $\approx$ 50 keV can be explained with the aid of Figure 2. A particle of energy $E_1$ enters a solid at an angle $\alpha$ to the surface normal. The trajectory inside the solid is a straight line until it encounters a target atom close enough to cause a large angle deflection. The outgoing ion trajectory is again a straight line, and it leaves the surface at an angle $\beta$ to the normal. In the experiment $\alpha$ and $\beta$ are determined by the primary beam direction and the position of the detector with respect to the target.

The particles lose energy along their paths through the solid. For light ions with energies above a few keV, the energy loss is primarily from ionization and excitation of target electrons (20). One can characterize this phenomenon by the differential energy loss $dE/dx$ measured in eV/Å or MeV/(mg/cm$^2$). Sometimes it is more convenient to use the stopping cross section or stopping power $\epsilon = 1/N\, dE/dx = M_2/(N_o\rho)\, dE/dx$, where $N$ is the number of target atoms per unit volume, $N_o$ is Avogadro's number, $\rho$ is the density, and $M_2$ is the mass number of the

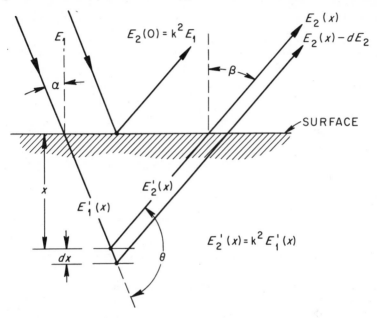

*Figure 2. Principle of the backscattering of light ions from a solid*

target atoms. Thus $\epsilon$ is measured in eV cm$^2$/atom. $dE/dx$ (or $\epsilon$) is a function of energy. Today a number of stopping power tables are available (21, 22, 23). These are semi-empirical tables based on experimental data which are inter- and extrapolated using theoretical functional dependences. As an example, Figure 3 shows $dE/dx$ curves for H$^+$ and He$^+$ in Ni from the tables of Northcliffe and Schilling (22). Sigmund has recently reviewed various energy loss mechanisms (24).

With the scheme of Figure 2 the energy of the outcoming particles $E_2$ can be related to the depth from which the backscattering occurs. Particles which are backscattered from the surface have lost energy only in the elastic collision, i.e., $E_2 = k^2 E_1$ according to Equation 1. For regions not too far from the surface, the relation between depth $x$ and the observed energy $E_2$ is:

$$\Delta E = k^2 E_1 - E_2(x) = \left[\frac{k^2}{\cos \alpha}\left(\frac{dE}{dx}\right)_{\overline{E_1}} + \frac{1}{\cos \beta}\left(\frac{dE}{dx}\right)_{\overline{E_2}}\right]x \qquad (2)$$

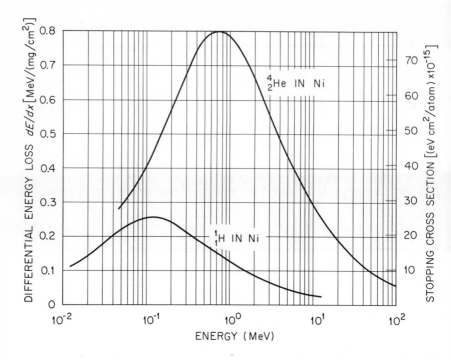

*Figure 3.   Differentital energy loss for H and He ions in Ni (22)*

where $\Delta E$ is the difference between the energies of particles backscattered from the surface and from a depth $x$, and $(dE/dx)_{\overline{E_1}}$ and $(dE/dx)_{\overline{E_2}}$ are the differential energy losses for the mean energies along incoming and outcoming trajectories. For this equation it was assumed that $dE/dx$ varies only slightly along the particle trajectories. For the analysis of thick layers the target can be divided into thin slices. For each slice the incident energy can be calculated using the $dE/dx$ curve. In many laboratories, computer programs are used for this procedure.

When composite targets—compounds or mixtures— are investigated, one obtains an overlay of the spectra of each component. These are shifted in energy relative to each other because $k^2$ depends on the target atom mass number $M_2$ (Equation 1). The stopping power for a compound is the sum of the stopping powers of the constituents weighted by the relative amounts with which they occur in the compound (Bragg's rule (25)). This relation has been proved valid in several cases (26). For instance, the stopping power in $SiO_2$ is $\epsilon_{SiO_2} = \epsilon_{Si} + 2\epsilon_O$. A rigorous treatment of the analysis of targets whose composition varies with depth was given by Brice (27).

An example of these prinicples is presented in Figure 4, which shows the spectrum of 2-MeV $He^+$ ions backscattered from a film of $Nb_3Ge$ (a compound with high superconducting $T_c$) deposited on an $Al_2O_3$ substrate (*28*). The spectra of the constituents of the film and of the substrate are clearly visible. The structure on top of the backscattering spectrum indicates that this film is not uniform. It is Ge-rich near the surface. At the high energy edge the slope of the spectrum is determined by the detector resolution. At the interface between film and substrate the slope is considerably less steep. This is from straggling in the energy loss. Energy straggling becomes increasingly important with increasing depth, and it finally limits the resolution of backscattering spectrometery. Energy straggling depends on the energy and on the ion–target combinations. It was treated theoretically in an early paper by Bohr (*1*) and by Lindhard et al. (*20*). Ref. 29 contains recent experimental determinations using backscattering techniques.

The backscattering yield, i.e., the height $H$ of the energy distributions (number of counts per channel) is related to the number density of atoms in a layer $dx$ near the surface:

$$H = Q\sigma\Omega N dx \qquad (3)$$

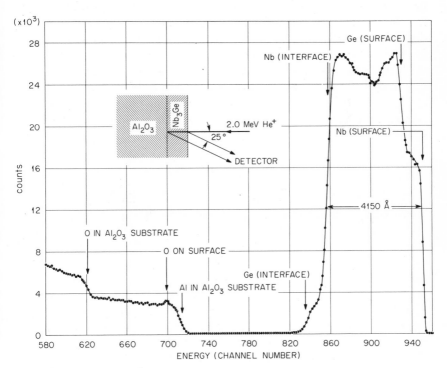

*Figure 4.  Backscattering spectrum of 2-MeV from a $Nb_3Ge$ film on an $Al_2O_3$ substrate (28)*

where $Q$ is the number of primary particles arriving on the target, $\Omega$ is the solid angle subtended by the detector, $N$ is the number of target atoms per unit volume, and $\sigma$ is the cross section averaged over the solid angle $(\sigma = 1/\Omega \int (d\sigma/d\Omega) d\Omega)$. It has been shown (5) that for protons with energies $> 50$ keV, the Rutherford cross section is valid. In laboratory coordinates this is:

$$d\sigma = \frac{Z_1 Z_2 e^2}{16E} \cdot f(\theta) \, d\Omega$$

$$f(\theta) = 4 \left[ \cos \theta + \left\{ 1 - \left( \frac{M_1}{M_2} \sin \theta \right)^2 \right\}^{\frac{1}{2}} \right]^2 \sin^{-4} \theta \left\{ 1 - \left( \frac{M_1}{M_2} \sin \theta \right)^2 \right\}^{-\frac{1}{2}}$$

$$(4)$$

where $Z_1$, $Z_2$ and $M_1$, $M_2$ are the nuclear charges and mass numbers of projectiles and target atoms, respectively, $e$ is the elementary charge, and $\theta$ is the scattering angle. (For a derivation of this formula, see Ref. 30). Using Equation 2, $dx$ in Equation 3 can easily be related to an energy interval $dE_2$ (for instance that corresponding to the width of a channel in the multichannel analyzer), when it is assumed that $dE/dx$ is constant for the energy interval under consideration. If the functional energy dependence of the stopping cross section from energy is known, formulae for the backscattering yield from thick targets can be derived (31, 32). With these the stopping cross sections can be determined from the absolute height of the energy distributions (31, 33, 34).

If the particles are backscattered from a thin film of thickness $t$, the sum of the counts in all channels containing counts from particles backscattered from this film is:

$$\int H(E_2) \, dE_2 = Q\sigma\Omega N t \qquad (5)$$

This is independent of the stopping power. Thus the number of target atoms per unit area $Nt$ can be determined directly.

As an example, in Figure 5 the backscattering of 150-keV protons from a Nb film on a Be substrate is shown (35). This film was sputtered by 5-keV deuterons. From the decrease of the number of backscattered particles after sputtering, the sputtering yield could be determined.

Backscattering is a very unlikely process in this energy range. To illustrate this, let us assume a 1-$\mu$m thick Ni foil which is bombarded by 1-MeV protons. A detector of 1-cm diameter in a distance of 10 cm from the target at $\theta = 135°$ counts only $1.1 \times 10^{-7}$ particles per incident ion. Nevertheless this is a very useful and nondestructive method for surface analysis. In some cases surface impurities of less than $10^{-4}$ monolayer have been detected (36).

*Medium Energies*

As mentioned already in the introduction, surface barrier detectors are no longer suitable to measure particle energies below 20 keV. It is necessary in this energy range to use electrostatic or magnetic spectrometers. These are, naturally, only sensitive to charged particles. There are several papers (*4, 37, 38, 39, 40*) which report on measured energy distributions of the charged component of the backscattered particles. Most of the backscattered particles are, however, neutral at energies below 40 keV (for hydrogen) (*12, 13, 14, 15*). A comparison with the theoretical values is only possible if the total number of backscattered particles is known, since the theory of the charged fraction is not yet well developed. Therefore, a direct measurement of the neutrals seemed to be desirable.

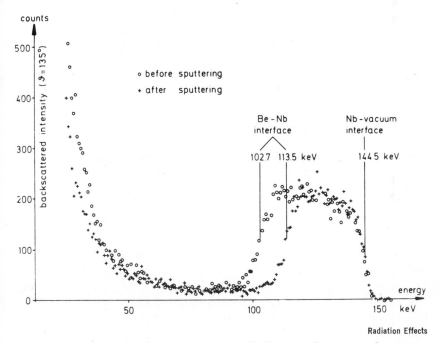

Figure 5.   *Energy distributions of protons backscattered at an angle of θ =*
*135° from a 600-Å Nb film on a Be substrate before and after sputtering with*
*5-keV D⁺ ions (35)*

Buck et al. (*16*) successfully performed time-of-flight measurements for He in the energy range of 6–32 keV. For this kind of experiment the primary beam has to be pulsed. The velocity of the particles backscattered from the target is determined by the time which elapses from the start signal given by the beam pulse and the detection of the particle in

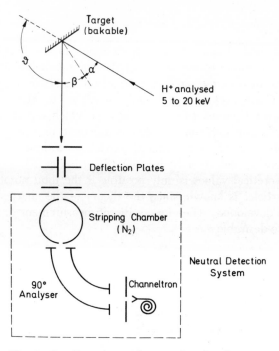

*Figure 6.   Experimental setup for the detection*
*of neutral backscattered particles*

a multiplier situated in a certain distance. This method is suitable for the spectroscopy of neutrals as long as the detection efficiency of the multiplier is known.

Another method, which has been used at the IPP in Garching, is to ionize the neutral particles by stripping in a gas cell (*12, 13, 14*). The principle is explained by Figure 6. A magnetically selected ion beam impinged onto the target, which could be rotated such that the entrance and exit angles $\alpha$ and $\beta$ could be varied. A scattered beam corresponding to a scattering angle of $\theta = 135°$ was selected. With deflection plates the charged component could be removed from the beam. In the stripping cell, which was filled with $2 \times 10^{-3}$ torr $N_2$, a part (known from a previous calibration) of the neutrals was ionized and energy was analyzed in a 90° electrostatic spectrometer which used a channeltron multiplier detector.

Representative measurements of the energy distributions of positively charged and neutral particles backscattered at $\theta = 135°$ from a Ta target bombarded with 18.5-keV protons are shown in Figure 7. The charged fraction $Q = N^+/(N^+ + N°)$, i.e., the number of positively charged to the number of neutrals plus positives, decreases from ~40% at 20 keV to ~10% at 2 keV. Both the neutral and the charged spectra show distinct maxima at low energies. Besides their relative height, the shapes of the two spectra are rather different. Consequently, a derivation of the neutral spectrum from the charged is not easily accomplished. The shapes of the backscattering spectra depend only slightly on the target material. For decreasing primary energy the maximum is more pronounced, but the charged fraction depends only on the exit energy. The charged fraction decreases slightly with increasing angle of emergence $\beta$ (*13, 14*). At low energies negative particles were also observed (*41*). In this case the negative fraction is small, but especially with target materials with low work function, the negative fraction may be much larger than the positive.

In a separate experiment an electrostatic analyzer was used which could be swiveled around the target to determine the angular distribution of the particles backscattered when a proton beam is impinging normal onto a Nb target (*42*). With this instrumentation only the charged component (including positive and negative ions) could be measured. For protons on Nb it was possible to determine the total number of backscattered particles using the charged fraction measured previously (*13, 14*).

Reference *42* shows that the angular distribution of all backscattered particles is very close to a cosine distribution for primary proton energies of 4–15 keV. Only the ions scattered with the highest energies are preferentially scattered into smaller scattering angles. These contribute, however, only very little to the total backscattering intensity. Two examples

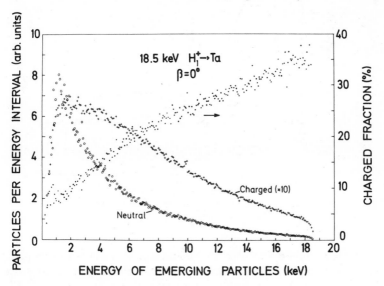

*Figure 7. Energy distributions of neutral and positively charged hydrogen atoms backscattered from Ta bombarded with 18.5-keV protons. The charged fraction $N^+/N^+ + N^\circ$ is given by dots.*

of energy distributions of all particles backscattered into the whole half-space are shown in Figure 8. Both spectra with primary energies $E_1 = 10.22$ keV and $E_1 = 4.16$ keV (measured with 8.32-keV $H_2^+$ molecular ions) show a distinct maximum at $\approx 1$ keV. This is caused by the fact that the scattering cross section increases as the ions lose energy in the solid. Below the maximum the probability of multiple scattering and removal from the beam exceeds the probability of backscattering events. By integration of the spectra in Figure 8 over all energies, the particle reflection coefficient $R_N$—the number of backscattered-to-incoming particles—can be determined. $R_N$ can also be determined from the number of backscattered particles measured at a specific angle (e.g., $\theta = 135°$), as the angular distribution is well approximated by a cosine distribution (42, 43).

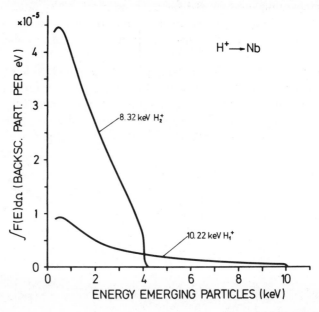

Journal of Applied Physics

Figure 8. Energy distributions of all particles backscattered into $2\pi$ solid angle when a Nb target is bombarded with 10.22-keV $H_1^+$ and 8.32-keV $H_2^+$ ions

Another method was used by Sidenius et al. (44) who mounted the targets inside a proportional counter so that all backscattered particles were absorbed and created pulses proportional to their energy. In Figure 9 from Ref. 43 results from the two experiments (43, 44) for stainless steel and Nb are compared with theoretical calculations. To compare the reflection coefficients for different materials, they were plotted vs. reduced energies (20):

$$\epsilon = E \cdot aM_2 / [(M_1 + M_2) Z_1 Z_2 e^2]$$

where $a = 0.468 \, (Z_1^{2/3} + Z_2^{2/3})^{-1/2}$ Å is the screening length in the interaction potential.

In the theoretical work of Weissmann and Sigmund (45) and of Bøttiger and Winterbon (46) the slowing down of the protons in an amorphous solid of infinite extent is calculated by the Boltzmann transport equation. This model assumes that the atoms start from a plane in the solid. All atoms which finally come to rest behind this plane are considered to be backscattered. Bøttiger and Winterbon (46) also include a surface correction. J. E. Robinson (47), O. S. Oen and M. T. Robinson (48), and Ishitani et al. (49) obtained values for $R_N$ by computer simulation. These simulations also assume amorphous materials. In all theoretical calculations an analytical approximation to the Thomas–Fermi interaction potential was used.

For stainless steel the measured values of $R_N$ agreed well with the computer simulations of Oen and Robinson (48). The fact the experimental results for Nb from two different experiments are considerably lower than those from the calculations can probably be attributed to two main differences.

(1) The Nb target had rather large crystal grains. Thus the backscattering was more from individual single crystals than from amorphous material. This leads to larger penetration depths and hence less backscattering.

(2) The Nb was very likely covered with an oxide layer which also reduces the backscattering.

A number of authors (50, 51, 52) determined reflection coefficients by measuring the amount of gas which is trapped in the target. When no gas is released thermally, the sum of the trapping and reflection coefficients is unity. However, this method is restricted to certain ion target combinations.

The knowledge of reflection coefficients as well as energy and angular distributions is very important for plasma experiments. Therefore, more experiments are necessary. Another important number in this context is the energy reflection coefficient $R_E$, the total energy carried away by the backscattered particles related to the incoming energy. It is $R_E = R\overline{E}/E_1$, where $\overline{E}$ is the mean energy of all reflected particles. $\overline{E}$ can be determined from the spectra in Figure 8. As seen from Figure 9 again good agreement of the experimental values and the calculations was achieved while the experimental data for Nb are below those expected from the

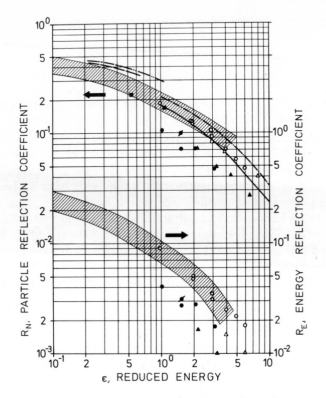

*Figure 9. Particle and energy reflection coefficients*
$R_N$ *and* $R_E$ *as function of the reduced energy. Hatched*
*area: comuuter simulation* (48); *theoretical curves:*
———— $H$, — · — $D$ (45); — — $H$, — · · — $D$ (46); *ex-*
*perimental data* △ $H \rightarrow SS$, ▲ $H \rightarrow NB$ (44); ○ $H \rightarrow$
$SS$, □ $D \rightarrow SS$, ● $H \rightarrow Nb$, ■ $D \rightarrow Nb$ (43).

computer simulations. The energy reflection coefficient $R_E$ was directly
measured by a Danish group (53) using different calorimetric methods.
They found fair agreement with the computer simulation of Oen and
Robinson (48).

### Low Energies

At very low energies the fraction of the backscattered particles which
are charged becomes very small. At energies below ≈ 200 eV, the ionization method by stripping in a gas cell breaks down and because the cross
sections for electron loss are small compared with those for scattering in
the gas, and the latter effect distorts energy and angular distributions. In
the energy range above several tens of eV the ionization by electron

impact is also impossible since the required electron densities cannot be achieved. The detection methods for neutrals also break down. All currently used methods for detecting neutrals depend on the creation of secondary electrons. At energies where the potential emission of ions dominates the kinetic emission ($\lesssim 200$ eV), there is no longer any emission of secondary electrons by neutrals. Because of the lack of detectors the time-of-flight methods are then also no longer usable at these low energies.

Therefore, one has to rely on the results of computer simulations. These are, in turn, especially suitable for low energies since it is easy to obtain sufficient statistics without too much computer time. Energy and angular distributions for protons scattered from Cu obtained recently by Oen and Robinson (48) are shown in Figure 10. The energy distribution

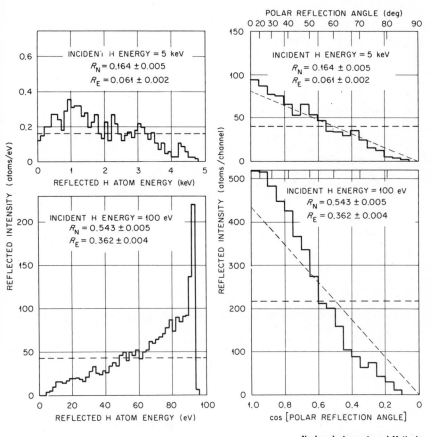

Nuclear Instruments and Methods

*Figure 10. Energy (left) and angular (right) distributions of hydrogen atoms backscattered from Cu bombarded with 100-eV and 5000-eV protons. (Computer simulation by Oen and Robinson (48).*

with a primary energy of 5 keV shows a maximum at ≈ 1 keV which agrees with the experimental observations. At a primary energy of 100 eV the spectrum is sharply peaked at high energies corresponding to backscattering from the surface. The angular distributions show remarkable deviations from a cosine distribution, which would give the dotted lines. The backscattered intensity is peaked in the entrance direction which was normal to the surface. At grazing incidence the authors found reflection coefficients close to one with the intensity highly peaked in the direction of specular reflectance.

## Acknowledgments

I am indebted to B. R. Appleton, J. W. Miller, O. S. Oen, M. T. Robinson, H. H. Anderson, and J. Bøttiger for making their results available to me prior to publication. I appreciate a critical reading of the manuscript by B. R. Appleton and O. S. Oen. Finally I want to thank the people of the Solid State Division of ORNL for their warm hospitality during my year in Oak Ridge.

## Literature Cited

1. Bohr, N., *K. Dan. Vidensk. Selsk. Mat. Fys. Medd.* (1948) **18** (8).
2. Lindhard, J., Nielsen, V., Scharff, M., *K. Dan. Vidensk. Selsk. Mat. Fys. Medd.* (1968) **36** (10).
3. Rubin, S., *Nucl. Instrum. Methods* (1959) **5**, 177.
4. McCracken, G. H., Freeman, N. J., *J. Phys. B* (1969) **2**, 661.
5. van Wijngaarden, A., Brimmer, E. J., Baylis, W. E., *Can. J. Phys.* (1970) **48**, 1835.
6. Rutherford, E., *Philos. Mag.* (1911) **21**, 669.
7. Hornyak, W. F., Lauritzen, T., Morrison, P., Fowler, W. A., *Rev. Mod. Phys.* (1950) **22**, 291.
8. McGowan, F. K., Milner, W. T., Kim, H. J., Hyatt, Wanda, *Nucl. Data Tables* (1969) **6**, 353.
9. *Ibid.* (1969) **7**, 1.
10. "Ion Beam Surface Layer Analysis," J. W. Mayer and J. F. Ziegler, Eds., Elsevier Sequoia, Lausanne, 1974.
11. "Ion Beam Surface Layer Analysis," O. Meyer, G. Linker, F. Käppler, Eds., Plenum, New York, 1976.
12. Behrisch, R., Eckstein, W., Meischner, P., Scherzer, B. M. U., Verbeek, H., "Atomic Collisions in Solids," S. Datz, B. R. Appleton, and C. D. Moak, Eds., p. 315, Plenum, New York, 1975.
13. Meischner, P., Verbeek, H., *J. Nucl. Mater.* (1974) **53**, 276.
14. Meischner, P., Verbeek, H., Report **IPP 9/18**, Max Planck Inst. f. Plasmaphysik, Garching, Germany, 1975.
15. Buck, T. M., Feldman, L. C., Wheatley, G. H., "Atomic Collisions in Solids," S. Datz, B. R. Appleton, and C. D. Moak, Eds., p. 331, Plenum, New York, 1975.
16. Buck, T. M., Chen, Y. S., Wheatley, G. H., van der Weg, W. F., *Surf. Sci.* (1975) **47**, 244–255.
17. Hinnov, E., *J. Nucl. Mater.* (1974) **53**, 9.
18. Gemmell, D. S., *Rev. Mod. Phys.* (1974) **46**, 129.

19. Morgan, D. V., "Channeling, Theory, Observation and Applications," Wiley, New York, 1973.
20. Lindhard, J., Scharff, M., Schiøtt, H. E., *K. Dan. Vidensk. Selsk. Mat. Fys. Medd.* (1963) **33** (14).
21. Whaling, W., "Encyclopedia of Physics," S. Flügge, Ed., Vol. **34**, p. 193, Springer Verlag, 1958.
22. Northcliffe, L. C., Schilling, R. F., *Nucl. Data Tables* (1970) **A 7**, 233.
23. Ziegler, J. F., Chu, W. K., *At. Data Nucl. Data Tables* (1974) **13** (5).
24. Sigmund, P., "Proceedings of the Advanced Study Institute on Radiation Damage Processes in Materials," Corsica, 1973.
25. Bragg, W. H., Kleeman, R., *Philos. Mag.* (1905) **10**, S 318.
26. Feng, J. S.-Y., Chu, W. K., Nicolet, M.-A., *Thin Solid Films* (1973) **19**, 227.
27. Brice, D. K., *Thin Solid Films* (1973) **19**, 121–135.
28. Miller, J. W., Appleton, B. R., Gavaler, J. R., private communication.
29. Harris, J. M., Chu, W. K., Nicolet, M.-A., *Thin Solid Films* (1973) **19**, 259.
30. Ziegler, J. F., Lever, R. F., *Thin Solid Films* (1973) **19**, 291.
31. Behrisch, R., Scherzer, B. M. U., *Thin Solid Films* (1973) **19**, 247.
32. Chu, W. K., Ziegler, J. F., *J. Appl. Phys.* (1975) **46**, 2768.
33. Wenzel, W. A., Whaling, W., *Phys. Rev.* (1952) **87**, 499.
34. Powers, D., Whaling, W., *Phys. Rev.* (1962) **126**, 61.
35. Eckstein, W., Scherzer, B. M. U., Verbeek, H., *Radiat. Eff.* (1973) **18**, 135.
36. Ball, J. D., Buck, T. M., Caldwell, C. W., McNair, D., Wheatley, G. H., *Surf. Sci.* (1972) **30**, 69.
37. Crawthron, E. R., Cotterel, D. L., Oliphant, M., *Proc. Roy. Soc. Lond.* (1970) **A319**, 435.
38. Morita, K., Akimune, H., Suita, T., *Jpn. J. Appl. Phys.* (1968) **7**, 916.
39. Eckstein, W., Verbeek, H., *J. Vac. Sci. Technol.* (1972) **9**, 612.
40. Mashkova, E. S., Molchanov, V. A., *Radiat. Eff.* (1972) **13**, 131.
41. Verbeek, H., Eckstein, W., Datz, S., *J. Appl. Phys.* (1976) **47**, 1785.
42. Verbeek, H., *J. Appl. Phys.* (1975) **46**, 2981.
43. Eckstein, W., Matschke, F. E. P., Verbeek, H., *J. Nucl. Mater.* (1976) **63**.
44. Sidenius, G., Lenskjaer, T., *Nucl. Instrum. Methods* (1976) **132**, 673.
45. Weissmann, R., Sigmund, P., *Radiat. Eff.* (1973) **19**, 7.
46. Bøttiger, J., Winterbon, K. B., *Radiat. Eff.* (1973) **20**, 65.
47. Robinson, J. E., *Radiat. Eff.* (1974) **23**, 29.
48. Oen, O. S., Robinson, M. T., *Nucl. Instrum. Methods* (1976) **132**, 647.
49. Ishitani, T., Shimizu, R., Murata, K., *Jpn. J. Appl. Phys.* (1972) **11**, 125.
50. Bøttiger, J., Rud, N., "Application of Ion Beams to Materials," G. Carter, J. S. Colligon and W. A. Grant, Eds., Conference Series Nr. 28, p. 224, The Institute of Physics, London, Bristol, 1976.
51. McCracken, G. M., *Rep. Prog. Phys.* (1975) **38**, 241.
52. Bohdansky, J., Roth, J., Sinha, M. K., Ottenberger, W., *J. Nucl. Mater.* (1976) **63**.
53. Anderson, H. H., Lenskjaer, T., Sidenius, G., Sørensen, H., *J. Appl. Phys.* (1975) **47**, 13.

RECEIVED January 5, 1976. Research sponsored by the U.S. Energy Research and Development Administration under contract with the Union Carbide Corp.

# 11

# Depth Distribution and Migration of Low $Z$ Elements in Solids Using Proton Elastic Scattering

ROBERT S. BLEWER

Sandia Laboratories, Albuquerque, N. M. 87115

*Rutherford ion backscattering spectrometry (RIBS) is a powerful analytical technique but is insensitive to light atoms in higher Z hosts. This can be overcome by using nuclear elastic scattering cross section anomalies which exist for ~ 2.5-MeV protons incident on such low Z elements as D, $^3$He, $^4$He, $^9$Be, $^{12}$C, and $^{16}$O. Optimum detection sensitivity (0.5 at. % He in Cu) is achieved using foil targets mounted on an efficient transmitted-beam trap. Since cross section enhancements for D and $^4$He are more than 100 times greater than Rutherford values, thick as well as foil targets can be analyzed. The depth distribution of helium and hydrogen isotopes can be observed nondestructively within the first 10 μm of a solid surface with a resolution of better than 750 Å. Tenets of the technique and results from its application to new energy research materials problems are described.*

Since the introduction of commercially available solid state particle detectors and the publication of "Ion Implantation in Semiconductors" (*1*) in 1970, surface and near-surface elemental analysis by Rutherford Ion Backscattering Spectrometry (RIBS) has become widely used and has joined a host of other techniques which use particles or electromagnetic radiation to probe the character of surfaces. Using incident helium ions of 2–3 MeV energy, several investigators (*2, 3, 4*) have demonstrated the fractional monolayer detection sensitivity and ~ 100-Å depth resolution that are possible when analyzing high Z impurities in low Z lattices. For many applications, however, the need arises to test for the presence and to observe the depth distribution of low Z atoms in higher Z substances, e.g., hydrogen isotopes in metals.

Though conventional helium ion backscattering has become a powerful analytical tool in revealing the concentration and depth profiles of impurities in the near-surface region of solids, it cannot detect helium and hydrogen isotopes in materials for kinematic reasons. Proton backscattering could remove the kinematic limitations. However, for a given energy, the Rutherford elastic scattering cross section of elements decreases as the square of decreasing atomic number, and unfavorable cross section considerations have discouraged experimenters from working along these lines. This paper reports on investigations using a modified form of proton backscattering which have been able to detect less than 7 at. % concentrations of such low Z elements as D, $^3$He, and $^4$He with depth resolutions below 750 Å. Mervine et al. (5) have observed helium in an implanted Pd foil using proton backscattering. Helium has also been detected in metals by forward scattering (6) and nuclear reaction techniques (7), but the experiments reported by the author two years ago (8) are believed to be the first description in the open literature of the use of an ion backscattering technique to determine helium depth distributions. This capability is important in unravelling the physics of the behavior of gases in metals, an area which is becoming particularly important with respect to emerging energy technologies. In fact, largely because of these technologies, there has been a surge of interest in the field. Within the last two years, several other ion beam techniques have been developed to detect and profile low Z elements in materials. Reviews of nuclear reaction methods by F. L. Vook and by J. C. Overly and of elastic recoil techniques by B. Terreault appear in this volume. Backscattering phenomena which occur at lower energies are reviewed by H. Verbeek.

## Helium Ion Backscattering

If a monoenergetic, mass-analyzed beam of ions impinges normally on a solid surface, most particles penetrate into the surface to depths of the order 1–10 $\mu$m for incident 2.5-MeV $^4$He ions or protons and ultimately come to rest within the target. However, many of the incident particles elastically scatter from target atoms, and a small fraction ($\sim 10^{-4}$) undergo large angle collisions ($> 90°$), are scattered in the backward direction, and escape from the target. The energy $E$ of each such ion after scattering is related to its energy before scattering $E_0$ by the formula:

$$E = kE_0 \tag{1}$$

where $k$ is the kinematic recoil factor and is determined from the laws of conservation of energy and momentum. Specifically:

$$k = \left[ \frac{\cos \theta + [(M/m)^2 - \sin^2 \theta]^{1/2}}{1 + M/m} \right]^2 \qquad (2)$$

where $\theta$ represents the scattering angle (laboratory system), and $m$ and $M$ denote the mass of the incident ion and struck target atom, respectively. $k$ equals the fraction of the incident particle energy which is retained after backscattering from a target atom and therefore varies from 0 (for a head-on collision with a target atom of equal mass) to 1.0 (for a head-on collision with a target atom of infinite mass). For a fixed type of incident ion at a given incident energy and backscattering angle, there is a unique value of retained energy as a result of a collision between that ion and a given target atom. Thus each backscattered ion which escapes from the target carries information through its remaining energy about the mass of the target atom with which it collided. Moreover, since the probability that a certain incident ion will scatter through a given angle is known (Rutherford scattering cross section), the concentration of each different species of atoms in the surface of the target can be deduced by counting the fraction of incident ions which are backscattered at the energy which is appropriate for that element.

In addition to the discrete kinematic energy loss, incident ions also lose energy continuously by electronic excitation (ionization) while traveling through the solid to and from the point of collision. The energy loss can be characterized by the relation:

$$\frac{dE}{dx} = -N\epsilon(E) \qquad (3)$$

where $N$ is the target atom density, and $\epsilon$ is called the stopping cross section. $\epsilon$ is a function of ion energy $E$ and, in the energy range of interest in RIBS (1.0–2.5 MeV), decreases with increasing energy. The value of $\epsilon$ has been measured for many materials, and formulas providing good estimates for $\epsilon$ are available for those elements for which experimental data do not exist (9, 10, 11, 12, 13). Because of this additional "drag" energy loss, ions backscattered from target atoms which are beneath the target surface will emerge from the target with a smaller fraction of their original energy than those which rebound from like atoms in the first monolayer. This energy dispersion of backscattered ions from any given type of target atom is a distinct analysis advantage. Since $\epsilon$ can be experimentally measured, it is possible to correlate the energy of the backscattered ion with the depth in the target at which it was scattered through Equations 1 and 3. For a typical experimental run, an incident dose of $\sim 10^{16}$ ions/cm$^2$ (10 monolayers) provides data with adequate statistical accuracy for composition and depth profile analysis of the constituents of the target. Thus the RIBS method is essentially nondestructive. Because of the ability to "see" the depth profiles and identify the mass of both host and all "impurity" atoms in the target, RIBS has been called "mass sensitive depth microscopy" (3).

Several factors influence the choice of incident beam. The variation of $k$ with target atom mass is shown in Figure 1 for four types of incident ion. It is necessary to use an ion beam composed of atoms which have lower mass than the lightest target atom species one wishes to detect. It is also useful for detection and analysis purposes to select a probe beam for which the percentage change in $k$ value between adjacent elements is greatest among the target species of primary interest. For instance, protons are optimum for analyzing targets with low Z constituents ($M = 2$–$20$) while $^{12}C^+$ or $^{16}O^+$ ions would provide greater energy separation of backscattered ions for heavy elements ($M > 50$).

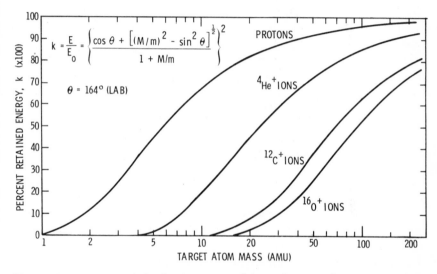

*Figure 1.   Variation of the kinematic recoil factor* k *for each of four incident ions as a function of target atom mass*

Range and stopping cross section characteristics of the incident ion beam also influence the choice of analysis beam. For example, protons penetrate, and therefore sample, much greater depths of a given target than oxygen ions, as shown in Figure 2 for gold, but because the stopping cross section for oxygen in solids is much greater than that for protons, the depth resolution for a gold sample will be considerably better for an incident oxygen beam than for a proton beam. The inherent energy resolution (and thus mass and depth resolution) of the silicon surface barrier detector (energy analyzer) is also a factor, decreasing with increasing mass of the incident beam ions (14). $^4He^+$ ion analysis beams are most

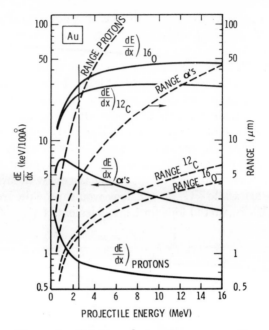

*Figure 2. Range and stopping cross section variation as a function of incident ion energy in gold for $^{16}O$, $^{12}C$, $^{4}He$, and $^{1}H$ (9). The vertical line at 2.5 MeV represents a commonly used ion backscattering analysis energy.*

commonly used because they represent the best trade-off in factors discussed above for analyzing targets containing medium Z materials.

A schematic of the elastic scattering processes which occur in a thin film containing a subsurface impurity layer deposited on a thick substrate is shown in Figure 3a. Typical spectra for an incident $^{4}He^{+}$ ion beam are illustrated at the base of Figure 3b. Details of the accelerator layout and of electronics needed to count detector pulses, are given in Refs. *15* and *16*. For a backscattering event which occurs at the surface at an angle $\phi$ ($= \pi - \theta$), the energy measured by the detector is given by $E_1 = k_{film}E_0$ where the value for $k_{film}$ is determined by Equation 2 or may be taken from Figure 1 for a scattering angle of 164°. Conversely, those incident ions which penetrate to the film substrate interface before scattering suffer (in addition to the recoil energy loss) ionization energy loss before collision, as well as on retraversing the film. These ions will reach the detector with energy $E_3$. The stopping cross section of metals for He ions varies sufficiently slowly with energy that, when analyzing thin surface layers at energies above 1 MeV, an average value of $\epsilon$ may be used for the inbound path and another average value for the outbound path. The measured energy difference $\Delta E_{film} = E_1 - E_3$ is thus a direct measure of the film atomic areal density (atoms/cm$^2$) using the equations in Figure 3. This energy difference is easily converted to film thickness using the appropriate value of the film density.

Likewise the depth of the impurity layer can be deduced by measuring $E_2$ and subtracting its value from the energy $E'_2$ of an incident ion assumed to strike an identical impurity atom on the surface: $\Delta E_{impur} = k_{impur}E_0 - E_2$. Ions which penetrate the thin film and are backscattered in the substrate experience stopping (energy loss) from $\epsilon_{sub}$ which is different from $\epsilon_{film}$ in magnitude although similar in energy dependence and must be taken into account as indicated in Figure 3 in the expression for $E_5$.

Strictly speaking, impurities, such as an implanted helium layer, alter the value of $\epsilon_{film}$, but most impurities are present at low concentrations and, for these cases, the effects are negligible. For compound or alloy

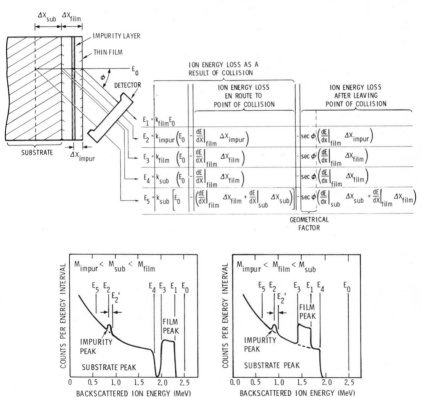

*Figure 3. Representation of characteristic energies at which incident ions are backscattered from various elements of a thin film sample. The resultant spectra generated by $^4He^+$ ions counted per energy interval (channel) on a multichannel analyzer are shown below for a high Z film on a given lower Z substrate (left) and for a lower Z film on the same substrate (right).*

targets, stopping cross sections can be added in proportion to relative atom concentrations, although this procedure (Bragg's Rule) introduces significant errors at energies below ∼ 2 MeV in certain systems (*17, 18, 19, 20*).

An actual data plot of a spectrum for incident He ions is shown in Figure 4. In this case, a scandium deuteride thin film deposited on a ∼ 4 μm molybdenum foil substrate has been analyzed. The sample contains a buried helium layer. The energy at which the $^4$He$^+$ analysis beam ions are backscattered from surface scandium atoms is indicated by the

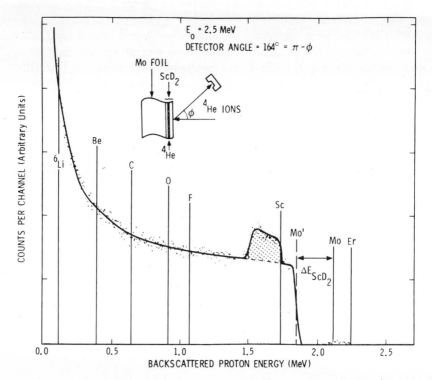

*Figure 4. $^4$He backscattering spectrum of a scandium deuteride thin film deposited on a molybdenum foil. The film has been implanted prior to analysis with ∼ 3 × 10$^{17}$ He$^+$/cm$^2$ at 160 keV.*

vertical line so marked in the figure. In this example the kinematic energy loss from the film atoms is greater than that from the substrate atoms (i.e., the surface layer is lower Z than the substrate), so the film peak appears to "ride on top" of the substrate peak, and it may thus appear that the scandium is not the surface constituent. The leading edge of the molybdenum peak is displaced to an energy lower (by $\Delta E_{ScD_2}$) than it would have appeared had molybdenum been present at the sample surface. $\Delta E_{\mathrm{ScD_2}}$ is an energy decrement equal to the energy lost by incident ions in traversing in and out of the scandium deuteride thin surface film.

The shaded area represents $^4$He ions backscattered from the scandium atoms in the film. $^4$He ions which penetrate the surface thin film and scatter against the molybdenum atoms provide most of the remaining counts in the spectrum. There are no $^4$He$^+$ ions backscattered from the pre-implanted $^4$He layer because this is forbidden by kinematics. The energy at which $^4$He ions would be backscattered from several other elements if they were present on the sample surface is also indicated in Figure 4. The ions backscattered from the surface layer of the thin film arrive at the detector with the least energy loss (thus highest remaining energy) relative to those ions backscattered deeper in the film, which appear at lower energies. This reflects the additional energy loss (drag) of those ions which do not scatter at the surface but which first penetrate into the lattice before scattering. For this reason the depth scale for each element present in the sample surface starts at the right side of its peak and proceeds from higher to lower energy with increasing depth.

Two additional observations are of interest. A small peak occurs at an energy $k_{\mathrm{Er}}E_0$, which indicates a small amount (2.7 at. %) of erbium present in the scandium deuteride film. Its width indicates that this impurity is distributed through the $ScD_2$ thin film but does not extend into the molybdenum substrate. Secondly, no rear side is observed on the molybdenum peak, which indicates that the foil is too thick for incident 2.5-MeV ions to backscatter from rear surface atoms and still possess sufficient energy to return through the front surface and reach the detector. More counts accumulate per energy interval at lower back-scattered ion energies (i.e., at deeper foil penetrations) because of the energy dependence of $\epsilon$ and because the Rutherford scattering cross section decreases as $E^{-2}$ with increasing energy. This results in an upward curve of the substrate peak at low energies.

Peak heights depend on the stopping cross section of the incident ion in the sample undergoing analysis. Materials with higher stopping cross sections produce larger energy loss per unit length of ion path travel and thus result in wider but lower peaks. However, an even stronger dependence exists of the peak height on the elastic scattering cross section of the target; indeed, the fraction of incident ions backscattered at a given angle and energy is, in essence, the basis of the definition of the scattering cross section, and for Rutherford (Coulombic) scattering, the scattering cross section increases as the square of the target atomic number. Thus the greatest sensitivity (typically less than one monolayer) is achieved for high Z elements with RIBS analysis. Moreover, because $\sim$ 2-MeV helium ions obey the Rutherford Scattering Law, the absolute concentration of target atom constituents can be determined by RIBS, an asset not common to many surface analytical techniques.

*Proton Backscattering*

As mentioned previously, in order to observe the presence of light atoms such as helium and hydrogen isotopes in a target using ion backscattering techniques, it is kinematically necessary to use a beam of incident ions of lower mass than the lightest impurity to be observed. Although this necessary condition is fulfilled for incident protons for all elements of atomic mass two or greater, researchers have traditionally been discouraged from developing the proton backscattering technique to observe low Z elements because the rapid decrease with decreasing atomic number of the Rutherford scattering cross section seemed to leave little hope of observing these elements against the host background yield. For example, in the case of equal atom concentrations of helium in copper, the calculated Rutherford backscattering cross section for incident protons at a given energy is 210 times less for collisions with target atoms of helium than for collisions with target atoms of copper. Counting statistics considerably better than 0.5% would be required even to detect a sharply defined helium distribution of 50 at. % above the copper "thick target" yield.

Thus a desirable condition to observe low Z elements is the elimination of backscattering counts from host atoms in the energy region where protons backscattered from D, T, $^3$He, and $^4$He would fall (i.e., $k = 0 \rightarrow$ $\sim 0.4$ or $kE_0 = 0 \rightarrow 1$ MeV, for $E_0 = 2.5$ MeV). Since there is no solid material with an atomic number as small as that of these elements, except at very low temperatures, the thick target spectrum cannot be eliminated from this energy region simply by judicious selection of a very low Z host. However, the thick target background can be reduced to near zero if a sufficiently thin foil (several $\mu$m) is used as a host lattice and if incident beam ions which fail to backscatter in the foil target itself are efficiently trapped after they penetrate it and thus are prevented from reaching the detector. This is achieved by placing a "beam trapping cell" behind the foil and its holder as shown in Figure 5. The trap is designed so that only those incident ions which backscatter in the foil itself are in the proper geometrical position to reach the collimated silicon surface barrier detector. The resultant effect on the spectrum is shown in the lower part of Figure 6. The rear side of the molybdenum peak is now "observable" because of the lower stopping cross section of protons in Mo vs. $^4$He$^+$ ions in Mo, so that the substrate no longer produces backscattered ions below $\sim 1.7$ MeV. Thus, counts registered in this energy range have been reduced from the thick target case by more than 95%.

Other differences appear in the proton spectrum as compared with the $^4$He$^+$ spectrum repeated from Figure 4 in the upper part of this figure. The position of the peaks of medium and high Z elements are shifted upward in energy according to the kinematic relationships expressed in

Equation 2, leaving more than 50% of the spectrum available for elements lower in mass than $^7Li$. In fact, the dispersion in $k$ values is so great for proton backscattering in this mass range that peaks of each of the low Z elements and even isotopes can easily be resolved from each other. This means the depth distribution of each of the low Z elements which may be contained in the target film may be observed simultaneously without deconvolution, if present in sufficient concentrations to be observed.

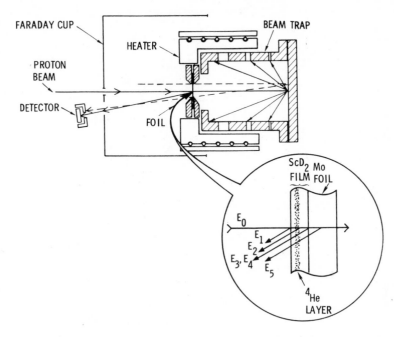

*Figure 5.   Beam trap and its relation to a He-implanted $ScD_2$ film–Mo foil sample and the solid state detector. Even if the incident beam strikes the foil target near its top (worst case), protons which penetrate the foil will not reach the active area of the detector.*

To this point, the kinematic requirement to observe backscattering from low Z elements has been satisfied, and the thick target yield has been reduced to allow a clear field for their observation. However, the calculated Rutherford scattering cross section for helium indicates that the yield from a 50 at. % helium distribution in scandium deuteride should still be unobservable in Figure 6 above the remaining background. As can be seen, the actual experimental yield is many times greater than predicted, especially considering that the $^4He$ concentration in this sample is significantly less than 50 at. %.

The explanation of this phenomenon lies in the fact that for protons of 2.5-MeV energy, a strong nuclear (not Coulombic) elastic interaction

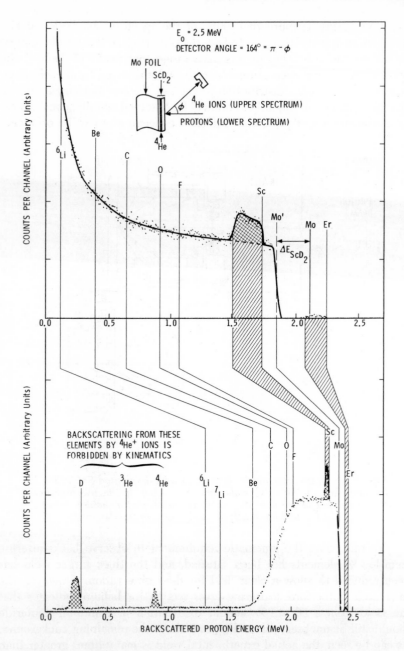

*Figure 6. Comparison of backscattering spectrum of helium-implanted scandium deuteride on a molybdenum foil substrate using a $^4He^+$ ion analysis beam (upper spectrum) as opposed to using a proton analysis beam (lower spectrum). Incident beam energy equals 2.5 MeV in both cases.*

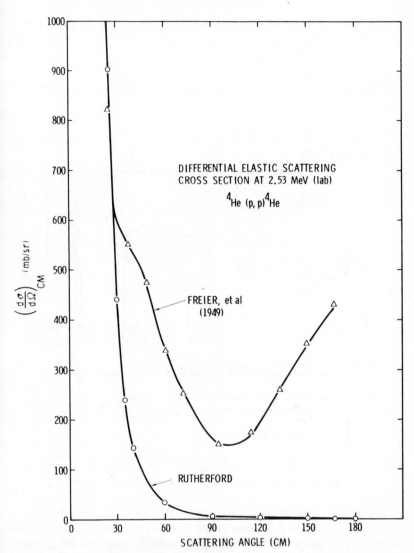

*Figure 7. Differential elastic scattering cross section in center-of-mass units as a function of scattering angle in center-of-mass units (22)*

exists with the He atoms which for scattering angles ≳ 25° results in an elastic scattering cross section much larger than the calculated Rutherford cross section. This fact was first observed by Rutherford's student, James Chadwick, in 1921 (*21*) and thoroughly documented by Freier et al. in 1949 (*22*). The variation of the measured and the calculated (Rutherford) cross sections with scattering angle in the center of mass system is shown in Figure 7. For $\theta = 168°$ (CM) the enhancement is a factor of 214. This enhancement permits detection of concentrations as low as 0.5 at. % He in medium Z (e.g., copper) foils and is sufficiently great even to permit observation of concentrations of a few atomic percent in thick targets (23). A comparison of the spectra obtained from He-implanted foils and thick targets has been published recently (24).

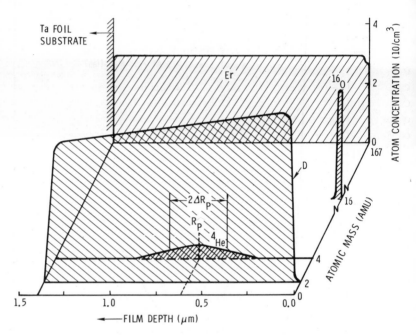

*Figure 8. Atom concentrations and depth distributions of species in a helium-implanted erbium deuteride thin film. The projected range of the helium $R_p$ and its range straggling $\Delta R_p$ are also indicated.*

Similar broad elastic scattering resonances exist for ³He, deuterium, and tritium. In Figure 6 the deuterium peak representing backscattering from deuterium atoms in the scandium deuteride film can be observed at the lowest energies. Its width is greater than that of the scandium peak because protons scattering at 164° (lab) from deuterium atoms give up ~ 90% of their energy. For protons, stopping cross sections are generally higher in the sub-MeV range than above 1 MeV (*see* Figure 2). Thus, the energy lost through electronic excitation on the return path of the proton through the film is much greater than on its inbound path, so the D peak is relatively wide. Conversely, protons which backscatter from scandium atoms retain most of their initial energy so $E_3 \sim E_o$, $\epsilon$ is relatively small, and $\epsilon_{out} \cong \epsilon_{in}$. The detection sensitivity for D and ³He is within the same range as that for ⁴He; concentrations of ~ 7 at. % and ~ 2 at. %, respectively, can be observed, with a sensitivity many times that expected from Rutherford theory. The elastic scattering cross section of the other low Z elements also displays broad as well as sharp-peaked resonances in the low-MeV range of energy. These include $^7$Li, $^9$Be, $^{11}$B, $^{12}$C, $^{14}$N, $^{16}$O, Al, and Si. A comparison of the magnitude of enhancements over Rutherford values can be found in Ref. *24*.

## Applications

RIBS is an appropriate analysis technique for samples whose elemental composition is desired. Its element-by-element depth distribution capability is most useful for samples whose compositional variation with depth is laterally uniform (i.e., layered structures). This includes a host of single or multilayered thin film structures, samples with various oxide, carbide, nitride surface layers, ion-implanted targets, etc. The power of the technique lies in the fact that while it is highly sensitive, it is also nondestructive so that it is unnecessary to erode away surface layers to reveal what lies beneath. For this reason RIBS is particularly well suited for sequential treatment effect studies such as migration of gases in metals, corrosion effect studies, interfacial diffusion, and semiconductor device-aging investigations. Those studies involving low Z elements in higher Z hosts should be conducted using the proton (rather than $^4$He ion) backscattering technique described in the previous section. Several examples of the use of the proton backscattering technique for studying behavior of gases in metals are discussed below.

**Erbium Deuteride Study.** The behavior of inert gas atoms in metals has been of interest for several years (*15, 16, 25, 26, 27*). Activation energies for diffusion of these gases, particularly helium, have been sought by thermally annealing metal samples injected to various concentrations. Because of the low solubility for helium in metals, microscopic bubbles are precipitated in the lattice of both heated and unheated samples at high concentrations (several atomic percent), but the degree of ordinary helium atomic diffusion, as opposed to helium migration via bubble movement, has been uncertain.

The author has injected $^4$He atoms into erbium deuteride thin films at an energy of 160 keV to a fluence of $1.9 \times 10^{17}$ He$^+$/cm$^2$ (He$_{\text{peak}}$/Er = 0.15). The sample films were deposited on a 1.5-$\mu$m Ta foil as a substrate to permit use of the proton backscattering technique to observe the $^4$He and D depth distributions. The foils were mounted as shown in Figure 5 and analyzed with 2.5-MeV protons. The resultant spectra are similar to that in the lower part of Figure 6. The elemental concentration and depth profile information from the spectra can be represented as shown in Figure 8. The depth scale is established by measuring the energy width of the Er peak in the backscattering spectrum and, by use of the equations in Figure 3, solving $\Delta E_{ErD_{1.8}} = E_1 - E_3$ for $\Delta x_{ErD_{1.8}}$. $(dE/dx)_{film}$ must include the electronic stopping contribution of the D atoms in the film as well as that of the Er atoms. The atom concentration of the film can be deduced from the average height of the Er and D peaks and converted into concentrations through a knowledge of $\epsilon(E)$ and the scattering yield. The mean depth $R_p$ of the helium layer may be calculated as described in the previous section and is found to be 6150 Å, agreeing within 6% of a theoretical estimate by D. K. Brice (29, 30). There is a depletion of deuterium within the first few hundred angstroms of the surface, where

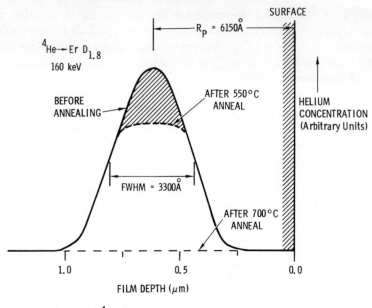

Figure 9. Changes in the helium depth distribution with increasing vacuum anneal temperature. The full width of half maximum (FWHM) of the distribution is unchanged at elevated temperatures.

a thin ($\sim 700$ Å) oxide layer exists. No surface carbon is observable nor are any other low Z impurities. The deuterium concentration appears to decrease slightly with increasing film depth, an effect confirmed by ion microprobe analysis of the same sample. Logan and Davis have observed a similar effect in titanium tritide film using a neutron time-of-flight technique (31, 32). Stopping cross section variations with energy are negligible in sufficiently thin films for all elements except, perhaps, those with $Z \leq 2$, and even for these elements the correction is relatively small and has not been applied here.

It is of interest to determine the migration behavior of the D and [4]He in the Er lattice as the temperature of the sample is increased. By using a heater attached to the sample holder in Figure 5, in situ sequential studies have been performed. The sample whose composition with depth is shown in Figure 8 was heated at temperatures of 150°, 260°, 400°, 550°, and 700°C in the target chamber for 30 min at each temperature. After each anneal a new backscattering spectrum was taken. The sequential behavior of the helium depth distribution could thus be observed.

The [4]He depth profiles after high temperature anneals are exhibited in Figure 9. No change in either the mean depth or mean width (spread) was observed during anneals up to 400°C. However, after the 550°C anneal, the apex of the distribution had been released, although the remainder was unchanged in mean depth or width. This implies that helium did not escape by simple Fickian diffusion processes since the width of the distribution is unchanged even though some gas has been released. This suggests that trapping of the implanted helium has occurred, possibly at damage sites. After the 700°C anneal, all the helium had been released. SEM studies of this foil and similar samples of helium-implanted copper, titanium, and titanium deuteride have revealed that helium-filled bubbles formed, which ruptured the surface and released a portion of the implanted helium directly to the vacuum. It is not clear whether all the helium escaped by this process. Typical SEM micrographs of the sample discussed above are shown in Figure 10.

The release behavior of the implanted helium and the deuterium is summarized in Figure 11. The deuterium in the sample maintained its original concentration and depth distribution with increasing temperature through 550°C and then dissociated and was entirely released after the 700°C anneal. After the 550°C anneal, ~ 20% of the implanted helium was released, the remainder escaping as a result of the 700°C anneal.

A portion of this deuteride sample was shielded during the original [4]He implantation, and analysis of the D-release behavior of that portion of the film is identical with the implanted section. Thus, there was no measurable effect either on the initial D distribution because of the [4]He implant (with its attendant lattice damage) or on the deuterium release behavior as a result of the implantation.

**Fusion Reactor First Wall Investigations.** Magnetic confinement schemes for fusion reactors require vacuum walls for the plasma container which can withstand high particle fluences of neutrons, D, T, and [4]He. In addition, tritium implanted in the first wall will generate [3]He in the lattice of the first wall material by radioactive decay. $(n,\alpha)$ reactions will further contribute to the concentration of gases in this wall so that structural degradation, blistering, and helium and hydrogen embrittle-

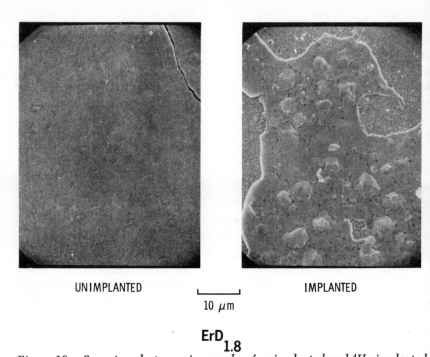

UNIMPLANTED            └─────────┘            IMPLANTED
                           10 μm

ErD$_{1.8}$

*Figure 10.   Scanning electron micrographs of unimplanted and ⁴He-implanted*
*erbium deuteride after a 700°C vacuum anneal*

ment can be expected to result in significant materials problems. Since the walls of such a reactor will reach temperatures of several hundred degrees centigrade, the migrational behavior of these gas atoms in their host material is of particular interest. The proton elastic scattering technique seems ideally suited to study problems of this kind since the depth distribution of each of the low Z isotopes can be observed simultaneously (except ³He and T which are both mass 3). Other ion beam techniques which are isotope specific can be used to measure the behavior of ³He separately (*33*). The author and R. A. Langley (*15, 16, 34, 35, 36*) are conducting studies in this field using the proton backscattering technique.

**"Hydrogen Economy" Energy Concepts.** As natural gas and petroleum sources dwindle, increased consideration is being given to the "hydrogen economy" concept (*37*), particularly in view of its low pollution potential. Transport of gaseous or liquid hydrogen in pipelines or tankers revives concern about hydrogen embrittlement of metals. Protective coatings and the basic nature of hydrogen permeation and diffusion in metals is an area that needs additional research. The use of metal hydrides as a thermally stable, rechargeable storage medium for hydrogen isotope fuels has also been proposed. Proton backscattering has much to offer as an investigative tool in this area as well.

**Intense Neutron Sources.** Interest in neutron irradiation as a therapy mode for cancer patients is growing rapidly. Likewise, an intense large volume source of neutrons for testing materials properties of structural components in fusion reactors is needed. Presently, the most intense 14-MeV neutron generator available (the Lawrence Livermore Laboratory Rotating Target Neutron Source) uses an accelerated deuterium ion beam incident on a titanium tritide target ($T(d,n)^4$He reaction). Tritium in parts of the target beyond the range of the incident D ions apparently does not diffuse toward the surface to replenish it as the near-surface tritium concentration is depleted (*31, 32*). This ultimately limits the life of the targets and may tend to reduce the percentage of time the facility is operational. This effect has more serious consequences in clinical therapy applications because strict maintenance of dose fractionation is important, and tritium handling (target exchange) in a hospital environment necessarily must be minimized. Thus there is great interest in diffusion behavior of hydrogen isotopes in targets under intense particle bombardment (*38, 39*). The proton elastic scattering technique is directly applicable to this problem area and is being used to develop an intense neutron source based on the $T(d, n)^4$He reaction (*40*).

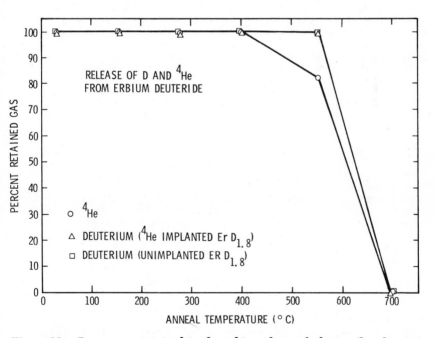

*Figure 11. Percent gas retained in the erbium deuteride lattice (by element) as a function of anneal temperature for both $^4$He-implanted and unimplanted $ErD_{1.8}$*

*Summary*

Analysis of the composition and depth distribution of high Z elements in low Z hosts on substrates has been achievable for several years by standard ion backscattering techniques. However, there is a specialized form of proton elastic scattering which has only recently been shown to be able to detect the presence of small quantities of low Z elements in higher Z substrates, thereby removing what had been generally regarded as an inherent limitation to the range of elemental systems which could be studied by backscattering methods. Examples of studies being conducted in the fields of energy research and medical therapy demonstrate the applicability of the technique to problems of particular current interest.

*Literature Cited*

1. Mayer, J. W., Eriksson, L., Davies, J. A., "Ion Implantation in Semiconductors," Academic, New York, 1970.
2. *Appl. Ion Beams Met. Int. Conf.* Plenum, New York (1974).
3. Meyer, O., Gyulai, J., Mayer, J. W., *Surf. Sci.* (1970) **22**, 263.
4. "Ion Beam Surface Layer Analysis," J. W. Mayer and J. F. Zeigler, Eds., Elsevier Sequoia, S.A., Lausanne, 1974.
5. Mervine, L. R., Der, R. C., Fortner, R. J., Kavanagh, T. M., Khan, J. M., Livermore Lawrence Laboratories, Report **UCRL-73087**, 1971.
6. Pieper, A. G., Theus, R. B., NRL Memo Report **2394**, Naval Research Laboratory, Washington, D.C., 1972.
7. Picraux, S. T., Vook, F. L., *Appl. Ion Beams Met. Int. Conf.* (1974) 407.
8. Blewer, R. S., *Appl. Phys. Lett.* (1973) **23**, 593.
9. Northcliffe, L. L., Schilling, R. F., *Nucl. Data Tables* (1970) **A7**, 233.
10. Whaling, W., "Handbuch der Physik," S. Flugge, Ed., Vol. **34**, p. 193, Springer, Berlin, 1958.
11. Ziegler, J. F., Chu, W. K., *At. Data Nucl. Data Tables* (1974) **13**, 463.
12. Andersen, H. H., Ziegler, J. F., *Nucl. Data Tables*, Pergammon (1977) in press.
13. Turos, A., Wielunski, L., *Nucl. Instrum. Methods* (1972) **104**, 117.
14. Petersson, S. S., Tove, P. A., Meyer, O., Sundqvist, B., Johansson, A., *Thin Solid Films* (1973) **19**, 157.
15. Blewer, R. S., *Appl. Ion Beams Met. Int. Conf.* (1974) 557.
16. Blewer, R. S., *J. Nucl. Mater.* (1974) **53**, 268.
17. Langley, R. A., Blewer, R. S., Nucl. Instrum. Methods (1976) **132**, 109.
18. Thompson, D. A., Mackintosh, W. D., *J. Appl. Phys.* (1971) **42**, 3969.
19. Baglin, J. E. E., Ziegler, J. F., *J. Appl. Phys.* (1974) **45**, 1413.
20. Feng, J. S.-Y., Chu, W. K., Nicolet, M-A., *Phys. Rev.* (1974) **10B**, 3781.
21. Chadwick, J., Bieler, W. S., *Philos. Mag.* (1921) **42**, 923.
22. Freier, G., Lampi, E., Sleator, W., Williams, J. H., *Phys. Rev.* (1949) **75**, 1345.
23. Langley, R. A., *Bull. Am. Phys. Soc. II* (1975) **20**, 499.
24. Blewer, R. S., "Ion Beam Surface Layer Analysis," O. Meyer, G. Linker, and F. Kappeler, Eds., Vol. **1**, p. 185, Plenum, 1976.
25. Barnes, R. S., Mazey, D. J., *Proc. Roy. Soc.* (1963) **A275**, 47.
26. Bauer, W., Wilson, W. D., "Proc. Confr. Radiation Induced Voids in Metals," J. W. Corbett and L. C. Ianiello, Ed., **CONF-710601**, p. 230, U.S. Dept. Commerce, Wash., D.C., 1972.
27. Blewer, R. S., *J. Nucl. Mater.* (1972) **44**, 260.

28. Blewer, R. S., *Radiat. Eff.* (1973) **19**, 243.
29. Brice, D. K., *Radiat. Eff.* (1970) **6**, 77.
30. Brice, D. K., *J. Appl. Phys.* (1975) **46**, 3385.
31. Davis, J. C., Anderson, J. D., *J. Vac. Sci. Technol.* (1975) **12**, 358.
32. Davis, J. C.,Anderson, J. D., private communication, Nov. 1974.
33. Picraux, S. T., Vook, F. L., *J. Nucl. Mater.* (1974) **53**, 246.
34. Blewer, R. S., "Technology of Controlled Nuclear Fusion," G. R. Hopkins, Ed., **USAEC-CONF-740402**, p. 525, U. S. Dept. Commerce, Wash., D.C., 1974.
35. Langley, R. A., Picraux, S. T., Vook, F. L., *J. Nucl. Mater.* (1974) **53**, 257.
36. Blewer, R. S., Langley, R. A., *J. Nucl. Mater.*, in press.
37. "Proceedings Hydrogen Economy Miami Energy Conference," T. N. Veziroglu, Ed., Plenum, New York, 1974.
38. Ormrod, J. H., *Can J. Phys.* (1974) **52**, 1971.
39. Cowgill, D. F., *Trans. Am. Nucl. Soc.* (1975) **22**, 168.
40. Crawford, J. C., Bauer, W., "Radiation Test Facilities for the CTR Surface and Materials Program," **ANL/CTR-75-4**, p. 227, National Technical Information Service, 5285 Port Royal Road, Springfield, Va., 1975.

RECEIVED January 5, 1976. This work was supported by the U.S. Energy Research and Development Administration.

# 12

# Concentration Profiling of Lithium in Solids by Neutron Time-of-Flight

J. C. OVERLEY and H. W. LEFEVRE

Physics Department, University of Oregon, Eugene, Or 97403

*Lithium concentration as a function of depth below the surface of a solid of otherwise-known composition is determined by measuring neutron spectra from the $^7Li(p,n)^7Be$ reaction. Depths of the order of 50 μm are probed nondestructively in a single measurement with a depth resolution of a few microns or better depending on host materials and operating parameters. Sensitivity in the absence of background-producing materials can reach several atomic parts per million at 10% accuracy. Since at high concentrations measurements may be made in a few minutes, changes in lithium concentration caused by beam heating may be followed in real time.*

Diffusion rates and equilibrium concentrations of lithium in solids under various physical conditions are of both basic and applied interest. For example, most fusion reactor designs use $^6Li$ as a coolant and for tritium breeding. The vessel which contains the molten lithium should maintain integrity at high temperatures in intense radiation fields for long times. Lithium which diffuses into the walls of this vessel may affect its strength by its presence and by internal generation of tritium and helium through the $^6Li(n,t)^4He$ reaction. These difficulties may be aggravated if diffusion is enhanced by high radiation fields.

A variety of methods using nuclear reactions to determine depth profiles of selected materials have been described (1, 2, 3, 4). Here we describe a related technique for measuring lithium concentrations in host materials of otherwise-known composition. It involves time-of-flight measurements of neutron spectra from the $^7Li(p,n)^7Be$ reaction within the sample. A range of depths may be probed nondestructively in one measurement. At high concentrations measurements may be made in minutes. They may be made under varying temperature conditions and in the presence of some types of radiation. Similar techniques have been applied elsewhere (5) to deuterium and tritium concentration profiling.

## Method

**Qualitative.** When protons with energy greater than 1.88 MeV are incident on a target containing lithium, neutrons are produced via the $^7Li(p,n)^7Be$ reaction. If the target is thick, a neutron energy continuum results because the protons continuously lose energy as they penetrate the target. Such continua have been studied (6, 7) in detail by time-of-flight techniques for other purposes in our laboratory. The experimentally determined quantities are the number of neutrons detected per flight time interval as a function of flight time.

Figure 1 shows two examples of flight time spectra, obtained at 0° with respect to the incident proton beam. Flight time increases to the left, and therefore neutron energy increases to the right. The sharp peak at the extreme right is caused by γ-rays produced by protons interacting in the target. The remaining structure is attributed to neutrons. The upper curve was obtained with a pure lithium metal target. The kinematic edge at channel number 700 corresponds to neutrons of maximum energy which are produced at the front surface of the target. The decrease in

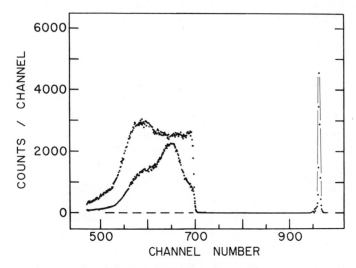

*Figure 1.    Time-of-flight spectra for a pure lithium target (upper curve) and a gold target containing diffused lithium. Flight time is inversely proportional to channel number, and channel width is 0.46 nsec/channel. Data for these spectra were obtained at a proton energy of 2.85 MeV and a 2 m flight path. Integrated beam current was 500 μC for the lower curve and 10 μC for the upper curve.*

the number of detected neutrons near channel 550 is mainly caused by a decrease in reaction cross section which occurs for proton energies below 2.25 MeV. The incident beam energy was 2.85 MeV for these spectra. The lower curve was obtained with a gold target containing diffused lithium. The integrated beam current for this spectrum was 50 times greater than that for the pure lithium target.

The number of neutrons detected at a given flight time is related to the lithium concentration at a certain depth in the target. This depth may be deduced by calculating how much energy the incident protons must have lost to produce neutrons with the given flight time and then applying known proton energy loss relations. The lithium concentration as a function of depth for the gold target shown in Figure 1 is intimately related to the ratio of the two spectra.

Neutrons from the $^7Li(p,n)^7Be^*$ reaction which appear for proton energies above 2.38 MeV complicate the quantitative analysis of lithium concentration, but not intractably so. Since they constitute only about 10% of the total neutron yield at these energies, we shall neglect them here for simplicity.

**Quantitative.** The neutron energy spectrum can be expressed in absolute terms as the number of neutrons per proton produced with energy $E_n$ at angle $\theta$ per unit energy and solid angle interval. For protons slowing down in a thick target and for population of a single final state, the absolute energy spectrum is (7):

$$S_T(E_n) \equiv \frac{N_n(E_n,\theta)}{N_p} = \frac{\sigma(\theta)}{\epsilon} \frac{dE_p}{dE_n} \frac{\text{neutrons}}{\text{proton} \cdot \text{MeV} \cdot \text{sr}} \tag{1}$$

where $\sigma(\theta)$ is the differential reaction cross section for protons of energy $E_p$, and $\epsilon$ is the compound stopping cross section per reactive atom in the target. All quantities except $\epsilon$ are characteristic of the nuclear reaction while $\epsilon$ is characteristic of the atomic composition of the target.

The compound stopping cross section per lithium atom is given by:

$$\epsilon = \frac{n_L\epsilon_L + n_H\epsilon_H}{n_L} \tag{2}$$

where $n_L$ and $n_H$ are number densities of lithium atoms and host atoms, respectively, and $\epsilon_L$ and $\epsilon_H$ are the corresponding atomic stopping cross sections. The latter are known to several percent for most elements. If the host substance consists of several different types of atoms, $n_H\epsilon_H = \Sigma_i n_i\epsilon_i$ where the summation extends over the constituents of the host material.

To obtain the lithium content in a sample, one takes the ratio of the neutron spectrum from the unknown sample (U) to that from a pure lithium sample. From Equations 1 and 2 one obtains:

$$\frac{S_U(E_n)}{S_L(E_n)} = \frac{n_L\epsilon_L}{n_L\epsilon_L + n_H\epsilon_H} \tag{3}$$

This expression contains the ratio of two absolute neutron energy spectra $S(E_n)$. This ratio is obtained from measured time-of-flight spectra which we denote by $N(t)$. If the two time-of-flight spectra are obtained under identical circumstances, $N(t)$ may be substituted directly for $S(E_n)$ since experimental parameters cancel. The atomic percentage of lithium is then:

$$100 \times \frac{n_L(x)}{n_L(x) + n_H(x)} = 100 \times \left[ 1 + \left( \frac{N_L(t)}{N_U(t)} - 1 \right) \frac{\epsilon_L}{\epsilon_H} \right]^{-1} \tag{4}$$

This expression does not depend on experimental parameters, such as the neutron detection efficiency, which are difficult to determine absolutely. Neither does it depend explicitly on nuclear reaction parameters.

When Equation 4 is evaluated at a neutron flight time $t$, the concentration is determined at a specific depth $x$ in the sample. This depth is:

$$x = -\int \frac{1}{n_L\epsilon_L + n_H\epsilon_H} \frac{dE_p}{dE_n} \frac{dE_n}{dt} \, dt \tag{5}$$

We do not evaluate Equations 4 and 5 directly, but we approximate them numerically. A depth increment $\Delta x$ is chosen arbitrarily. The proton energy at depth $\Delta x$ in the target is calculated using the atomic stopping cross section for pure host material for protons at the known incident energy. Nuclear reaction kinematics are then used in a calculation of flight times $t$ and $t + \Delta t$ for neutrons produced from lithium by protons at these energies. The average lithium content in the spatial interval $\Delta x$ is calculated with Eqaution 4 with the ratio $N_L(t)/N_u(t)$ determined by summing counts in the flight time spectra between $t$ and $t + \Delta t$. This first order lithium content is used in Equation 2, and the stopping cross section obtained is used to calculate a revised proton energy at depth $\Delta x$. Revised neutron flight times and lithium content are calculated, and the entire process is repeated until self-consistency is achieved. The lithium concentration in the next depth interval is similarly calculated, using the now-known proton energy at depth $\Delta x$. We step through the target in this manner. The number density of host atoms is held constant for the calculations presented here. One could easily use different assumptions in the iteration, however, such as keeping the total number density of atoms constant or maintaining constant total atomic volume.

## Experimental

A pulsed proton beam is produced by the University of Oregon's 5-MV Van de Graaff accelerator. Beam pulse repetition rates are variable by factors of two from 0.0625 MHz to 2 MHz. The beam current is correspondingly variable without changing other operating parameters up to a maximum of about 5 $\mu$A. This is helpful when a wide range of neutron yields is encountered, as occurs when yields from pure lithium targets are compared with those from targets with dilute lithium concentrations. A Klystron beam buncher compresses the time width of the beam pulses at the target to 1.2 nsec, full width at half maximum (FWHM) by impressing a 5–8 keV time-correlated energy modulation on the beam. The beam diameter at the target is determined by collimation and is 0.15–0.8 cm.

Two types of target chambers have been used in this work. One is equipped with a lithium furnace which enables thin lithium layers to be evaporated in situ. Since secondary electron emission during bombardment depends strongly on surface constitution, secondary electrons are

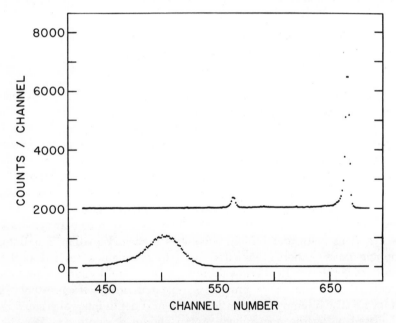

*Figure 2.   Partial time-of-flight spectra for a thin lithium layer evaporated on one side of an aluminum foil. The curve displaced upward by 2000 counts was obtained with the beam incident on lithium, the lower curve is for the beam incident on aluminum. Beam current was 0.2 μA at a 0.125 MHz repetition rate at a beam energy of 2.815 MeV. Integrated charge was 100 μC. Flight path was 2 m, and the channel width 0.45 nsec/channel. The γ-ray peak is off scale at channel 959.*

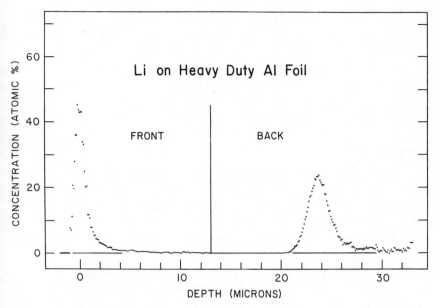

*Figure 3. The time-of-flight spectra of Figure 2 transformed as if they repre-*
*sented concentration profiles. The depth increment is 0.1 μm.*

suppressed by biasing the target at + 300 V with respect to its surround-ings. The other target assembly contains a 1.3-μm thick nickel foil through which the beam passes from the accelerator vacuum into air. Target samples are placed about 1 cm beyond the exit foil where they may be quickly positioned without venting the beam tube vacuum.

The neutron detector is a proton-recoil scintillation counter consisting of a 1.27-cm thick plastic scintillator optically coupled to a high gain photomultiplier tube. A flight path of 2 m was used for most of this work. Biases are set to detect neutrons with energies greater than about 50 keV.

A fast timing signal from the photomultiplier is used to start a time-to-amplitude converter (TAC). The stop signal is derived from a ferrite core beam pulse pickoff. The TAC output pulses are analyzed by an analog-to-digital converter and stored by an on-line PDP-7 computer. Overall time resolution may be gauged by the γ-ray peak in Figure 1. Its full width at half maximum is about 1.5 nsec. Beam currents are inte-grated with an accuracy of 0.1% to a preset level. The current integration is corrected by a live charge measurement technique for large electronic dead times which are encountered with pure lithium targets.

Figure 2 shows two more examples of flight time spectra. The target for these spectra was a thin lithium layer evaporated on heavy duty aluminum kitchen foil. The upper curve was obtained with a 2.8-MeV proton beam incident on the lithium layer. The two peaks are caused by neutron groups populating the ground state and first excited state of $^7$Be. The lower curve was obtained with the target reversed so the beam traversed the aluminum before encountering lithium. The neutron peak is displaced to longer flight time (lower energy) because the proton beam lost energy in the aluminum. The peak is wider both because of energy loss straggling and because of effective stretching of the energy scale for long flight times.

*Resolution*

To illustrate depth resolution, the flight-time spectra of Figure 2 have been transformed as though they represented concentration profiles. Results of the transformations are combined in Figure 3. The left half shows the profile when the beam is incident on the lithium layer; the right half is for the beam incident on the aluminum side of the target. The widths of the peaks—1.1 μm (FWHM) at the front surface and 2.4 μm at a depth of 24 μm—quantitatively portray depth resolutions.

Figure 4 summarizes expected depth resolutions (FWHM) as a function of depth for protons with energies of 4.8, 4.0, and 2.8 MeV incident on aluminum, copper, and gold. The curves terminate at a proton energy of 1.8 MeV and indicate the maximum depth which may be probed for each incident proton energy. The resolutions consist of the quadratic combinations of spreads from straggling, flight time spreads of 1.5 nsec, and a 5-keV incident beam energy inhomogeneity. Stopping cross sections and resolutions caused by straggling were calculated from the tables by Janni (8). The effects of flight time resolution were calculated for a neutron flight path of 2 m. For these parameters, depth resolution near the front surface of the target is limited by flight time resolution. Depth resolution may be improved by increasing neutron flight path, but ultimately incident beam energy spreads will dominate. Where the curves rise with penetration depth, straggling is the main limitation on depth resolution. The calculations are in excellent accord with depth resolutions observed, as in Figure 3, for example.

The diameter of the beam determines areal resolution. In principle, the beam diameter may be made as small as desired by appropriate collimation (4). However, as beams are made smaller, sensitivity suffers because beam intensities also decrease. In practice, beam diameters of about 1 mm are not difficult to obtain without appreciable loss.

*Sensitivity*

We define the sensitivity of the technique in terms of the minimum concentration of lithium atoms which can be conveniently measured. At low concentrations Equation 4 may be written as:

$$\frac{n_\text{L}}{n_\text{L} + n_\text{H}} \cong \frac{N_\text{U}(t)}{N_\text{L}(t)} \frac{\epsilon_\text{H}}{\epsilon_\text{L}} \tag{6}$$

Uncertainty in the measurement is governed mainly by counting statistics in the number of neutrons detected from the sample. For 10% uncertainty, $N_\text{U}(t) = 100$ counts. We assume these counts are contained in a 100-keV neutron energy interval which is fairly typical of one depth resolution interval. We assume further that the measurement requires 1 hr at a 1-$\mu$A beam current, that the neutron detection efficiency is 30%, and that the detector has a diameter of 10 cm and is placed 1 m from the target.

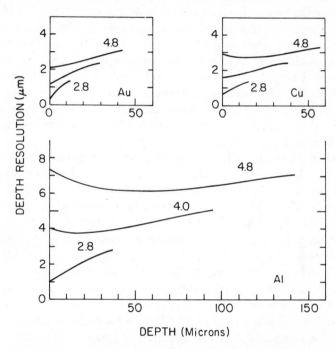

Figure 4.  *Depth resolution (FWHM) as a function of depth for targets of Al, Cu, and Au at incident beam energies of 2.8, 4.0, and 4.8 MeV. Effects included are: a beam energy spread of 5 keV, time resolution of 1.5 nsec, a flight path of 2 m, and energy loss straggling.*

The number of neutrons detected from pure lithium under these experimental circumstances is calculated to be about $2.1 \times 10^8$. Combining these factors with typical values of stopping cross sections yields measureable concentrations of about 2 ppm of lithium in aluminum, 3.5 ppm in copper, and 5 ppm in gold.

These figures are somewhat misleading because the effects of background subtraction have not been included in the uncertainty in $N_\text{U}(t)$.

One type of background is uncorrelated with time. This arises in part from room-thermalized neutrons inducing $(n, \gamma)$ reactions in the target room. This type of background may be substantially reduced by appropriate shielding and through rejection of $\gamma$-ray events by pulse shape discrimination. This type of background may be reduced to less than 100 counts/hr per 100 keV neutron energy interval. Since this background is small and because it can be estimated from the interval between the $\gamma$-ray peak and the neutron kinematic edge (*see* Figure 1), it does not importantly affect sensitivity.

A more important source of background arises from neutrons scattered from the target chamber and other nearby objects into the detector. This background is time correlated and appears as a long flight time tail on monoenergetic neutron peaks (*see* Figure 2). It may often be measured by shadow bar techniques, but since it may exceed the signal of interest by a large factor, it may severely limit sensitivity. One trick for minimizing the effect of this background is to have the incident proton beam travel from low to high lithium concentrations. Under these circumstances a dynamic range of $10^5$ in detected lithium concentration is possible. This is illustrated in Figure 3. With the beam incident on the lithium surface layer, a lithium concentration of several percent is inferred within the aluminum. With the beam incident on the other side, backgrounds are not as large, and the lithium concentration is inferred to be less than 0.02%. This run has not been fully corrected for background, and the data were obtained in a few minutes instead of 1 hr.

A more important source of background stems from $(p,n)$ reactions on particular constituents in the host material. Fortunately, the $^7Li(p,n)^7Be$ threshold energy is uncommonly low so that for most substances there is at least a small proton energy interval where contaminant reactions cannot occur. Cross sections for the $^7Li(p,n)^7Be$ reaction are also uncommonly high, with a value in excess of 40 mb/sr for a range of energies. Cross sections for competing reactions generally decrease with atomic number. For example, $^{37}Cl(p,n)^{37}Ar$ has a differential cross section at 2 MeV of about 2 mb/sr while that for $^{59}Co(p,n)^{59}Ni$ is about 0.05 mb/sr (9).

To check the influence of contaminant reactions in a practical case, we measured the neutron time-of-flight spectrum resulting from bombardment of type 301 stainless steel by a 3.8 MeV proton beam. The spectrum was transformed as though it represented a lithium concentration profile without subtracting backgrounds of any kind. Type 301 stainless steel contains about 16–18% chromium and 6–8% nickel. Target isotopes which can produce contaminant reactions and the proton energy at which they become possible are $^{53}Cr$ (1.41 MeV), $^{57}Fe$ (1.64 MeV), $^{54}Cr$ (2.2 MeV), $^{64}Ni$ (2.5 MeV), $^{58}Fe$ (3.15 MeV), $^{61}Ni$ (3.06 MeV). All in all, about 5% of the atoms in the target can contribute to the neutron yield.

If these neutrons were mistakenly ascribed to lithium, a lithium concentration of 0.4% would be inferred. With a separate background measurement and assuming that a 10-to-1 background-to-signal ratio is tractable, sensitivities of 400 ppm of lithium should be attainable in type 301 stainless steel. Higher sensitivities would be achieved if pulse shape discrimination were used.

### Examples

Several years ago, while investigating suitable substrates for evaporated lithium targets, we evaporated some lithium on gold. It was apparent after washing off the lithium with water that the surface had

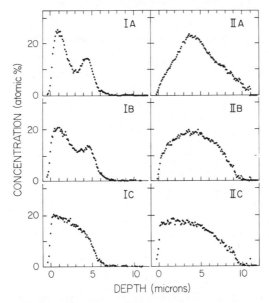

*Figure 5.   Lithium concentration profiles for two Li–Au targets (I and II). For insets A, running parameters were the same as for Figure 2. Insets B are after about 5 min of heating with a beam current of about 1.8 μA. Insets C are after 15 min of heating with a beam current of about 3.3 μA.*

been altered. When we started these studies it was natural that this sample was one of the first candidates for investigation despite its indefinite history. Figure 5, IA and IIA, show the lithium profiles obtained from two targets (I and II) cut from the original piece. Results were puzzling, particularly since the concentration deduced at any depth depended on incident proton energy. Either the assumption that the lithium and gold were intimately mixed was faulty, perhaps because of grain structure, or there were unknown, irregularly distributed contaminations in the gold sample.

The results of Figure 5 IA and IIA, were obtained at low beam currents and a proton energy of 2.8 MeV. After obtaining them, beam current was increased, and the targets were heated with a beam power of about 5 W. After about 5 min of heating at an indefinite temperature, probably of the order of 100°C, the results shown in IB and IIB were obtained. Adequate counting statistics could be compiled in less than a minute, and changes were observed in successive flight time spectra.

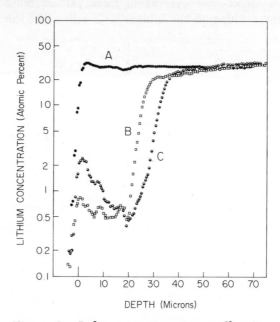

*Figure 6. Lithium concentration profiles for two Li–Al samples. Curve A is from a Li–Al melt in vacuum. Curves B and C are from a melt done in an argon atmosphere. Data for B and C were obtained at positions separated by 2.5 mm. For these data the 4 MeV beam was brought into air, and the flight path was 10 m.*

After an additional 15 min of heating at a beam power of about 10 W, the lithium distributions stabilized at the values shown in IC and IIC. These deduced concentrations were independent of incident proton energy.

Figure 6 shows an example of areal resolution. We have made a few attempts at producing a Li–Al intermetallic compound for use as a stable neutron source target of standard strength. In one such attempt, a lithium and aluminum mixture in a BN crucible was heated in vacuum. In another attempt, the mixture was contained in a stainless steel boat and heated in a tube furnace in an argon atmosphere. Curve A shows the lithium profile from the first melt. Curves B and C are for the second melt and were obtained at points separated by 2.5 mm. In each case, the sample surface which had been next to the crucible wall was bombarded. For the second sample the lithium concentrations are low near the surface, and significant changes occur over distances of less than 2.5 mm. These data were obtained with a proton bombarding energy of 4.0 MeV, the samples were located in air, and the neutron flight path was 10 m. No corrections for excited state neutrons have been made, and we have assumed that the number density of aluminum atoms is the same as that for pure aluminum. Deviations from the latter assumption will primarily alter the depth scale.

### Conclusions

There are some advantages in using neutron time-of-flight techniques in depth profiling analyses. When probing with charged particles at depths of several microns or more, resolution is usually limited by energy straggling. Since neutrons are not susceptible to atomic energy loss mechanisms, greater depths and better resolutions may be achieved than with charged particle reactions at comparable energies. Resonance techniques involving $\gamma$-rays also have this advantage, but concentrations must be determined point-by-point. For near-surface concentration analysis, heavy ion reactions and backscattering measurements can provide superior depth resolution. However, resolution in these cases is often limited by energy uncertainties introduced by common charged particle detectors. At sufficiently low energies, time-of-flight techniques are limited by incident beam energy spreads which are often much smaller.

There are, to be sure, disadvantages inherent in neutron time-of-flight methods. The apparatus required is more complex and not as generally available as that used for other nuclear techniques. Low cross sections and inefficient neutron detection may require large beam currents which might prove destructive to some samples. Low cross sections and high backgrounds may also severely limit sensitivity. The situation for lithium profiling by neutron time-of-flight is particularly felicitous because of the large $(p,n)$ cross section and low threshold energy.

### Acknowledgments

We wish to acknowledge the assistance of J. D. MacDonald in maintaining the accelerator and computer systems.

*Literature Cited*

1. Leich, D. A., Tombrello, T. A., *Nucl. Instrum. Methods* (1973) **108**, 67.
2. Pronko, P. P., Pronko, J. G., *Phys. Rev.* (1974) **89**, 2870.
3. Möller, E., Nilsson, L., Starfelt, N., *Nucl. Instrum. Methods* (1967) **50**, 270.
4. Pierce, T. B., in "Characterization of Solid Surfaces", P. F. Kane and C. B. Larrabee, Eds., p. 419, Plenum New York, 1974.
5. Davis, J. D., Anderson, J. C., *J. Vac. Sci. Technol.* (1975) **12**, 358.
6. Burke, C. A., Lunnon, M. T., Lefevre, H. W., *Phys. Rev.* (1974) **C10**, 1299.
7. Wylie, W. R., Bahnsen, R. M., Lefevre, H. W., *Nucl. Instrum. Methods* (1970) **79**, 245.
8. Janni, J. F., unpublished data.
9. Johnson, C. H., Galonsky, A., Ulrich, J. P., *Phys. Rev.* (1958) **109**, 1255.

RECEIVED January 5, 1976. Work supported in part by the National Science Foundation.

# 13

# Two Methods Using Ion Beams for Detecting and Depth Profiling Light Impurities in Materials

B. TERREAULT, M. LEROUX,[1] J. G. MARTEL, and R. ST-JACQUES

Centre de l'Energie, INRS, Université du Québec, Varennes, Québec, Canada

C. BRASSARD, C. CARDINAL, J. CHABBAL, L. DESCHÊNES, J. P. LABRIE, and J. L'ECUYER

Laboratoire de Physique Nucléaire, Université de Montréal, Montréal, Québec, Canada

*We describe two methods of detecting and depth profiling light elements (H to O) implanted into higher Z materials without sample erosion. The first method, to detect simultaneously C, N, and O in films a few micrometers thick uses Li-induced nuclear reactions producing fast forward alphas and deuterons. The detectability limits are about $10^{-7}$ $g/cm^2$, and the resolution is 0.15 $\mu m$. The second method can profile all elements lighter than carbon in thin or thick samples. The atoms knocked out of the targets are detected by 10–40 MeV heavy ions. Isotopes are distinguished; a detectability limit of $10^{-9}$ $g/cm^2$ and depth resolution of 300 Å were obtained, but $10^{-10}$ $g/cm^2$ and 100 Å are possible.*

This paper deals with methods of detecting, in a nondestructive way, light impurity elements in the first few micrometers or less of the surface of materials and of measuring their depth distributions. This is accomplished by using a collimated beam of mono-energetic particles as a probe. One then detects and energy-analyzes particles emerging from the interaction of this beam with impurity atoms which may be elastically scattered beam particles, recoiling impurity atoms, or the products of a nuclear reaction. A general microanalysis apparatus is shown in

---

[1] Present address: Alcan Aluminium, Laboratoire Analytique, Bâtiment 109, Arvida, Québec, Canada.

Figure 1, which is referred to throughout for the definition of kinematical and geometrical quantities. Any particular method is characterized by (1) specificity and mass resolution, (2) sensitivity, and (3) depth resolution.

A nuclear reaction is highly specific; it picks out one isotope of one element. With the help of the kinematics one can often weed out all but the interesting process and obtain very clean signals. Conversely, to detect a wider range of impurities simultaneously, one may use elastic scattering. Impurities are distinguished by the atomic mass dependence of the energies and angles of the scattered beam and recoil.

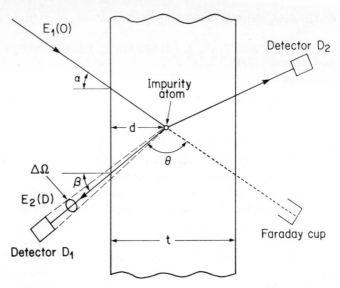

*Figure 1. A typical microanalysis apparatus (not to scale)*

A high sensitivity is important not only because of time and money considerations but also because beam intensity and/or exposure time cannot be increased arbitrarily without excessively heating the sample and/or inducing structural changes, migration of impurities, or radiation damage. In this chapter we will use the following conventional definitions:

Sensitivity, $S$, is the mass of contaminant per unit area which can be measured in 1000 sec with maximum practical beam current, with a 10% statistical uncertainty after subtraction of background (if any).

Detectability limit is a quantity 1/10 as large.

If the beam and/or detected particles are charged, they lose energy continuously on their way into or out of the target so that the depth of implantation of the impurity can be deduced from the energy $E_2(D)$ of the particle as it reaches the detector. From Figure 1, one obtains:

$$E_2(D) = f[E_1(0) - \sec \alpha d(dE_1/dx)] - \sec \beta \left\{ \begin{matrix} d \\ (t-d) \end{matrix} \right\} (dE_2/dx) \ (1)$$

where $f$ is the fraction of the beam energy transferred to the detected particle in the collision and depends on $E_1$, $\theta$, and the masses, and $(dE_{1,2}/dx)$ are the average energy losses per unit distance of the beam and detected particle, respectively. In the curly bracket of Equation 1, $d$ applies in the "backward" geometry (detector $D_1$) and $(t-d)$ in the "forward" or "transmission" geometry (detector $D_2$). Hence, while mass resolution depends on the energy resolution $\delta E_2$ and the kinematical factor $f$, depth resolution depends on $\delta E_2$, kinematics, $dE/dx$, and the geometry.

## Previous Work

The Rutherford backscattering (RBS) of light ions (*1, 2, 3*), so successfully used in detecting heavy impurities, is hardly applicable to light elements because of the $Z^2$ dependence of the cross section. Only at high concentrations and with suitable modifications has it been applied in this case. Blewer (*4*) devised a method applicable to thin targets (ca. 1 $\mu$m) and impurity concentration of the order of 1 at. % or more. The method of Roth et al. (*5*) is practical at concentrations of the order of 100 at. %, is rather indirect, and cannot separate the effects of two or more different impurities.

Other authors (*6, 7, 8*) have detected in coincidence the elastically scattered beam and the recoil, in the forward direction, through thin samples. Of these methods, that of Moore et al. (*8*), thanks to the use of heavy ion beams and hence high Z and large $(dE/dx)$, is capable of the best sensitivity and depth resolution, and they obtained $S \simeq 10^{-8}$ g/cm$^2$ and $\delta d \simeq 1000$ Å with a mass resolution $\delta M/M \simeq 0.1$. The limitations are the necessity of very thin samples ($\simeq 0.2 \mu$m) and complicated electronics.

The most-used methods of microanalysis of light elements have been nuclear reaction methods. The extensive literature on the subject has been well reviewed (*1, 2, 9, 10, 11*). Sensitivities of $10^{-8}$g/cm$^2$ and depth resolutions down to a few hundred angstroms have been obtained in some cases. One may note that the best results for a particular chemical impurity have sometimes been obtained for a rare isotope rather than the one most likely to be encountered in practical problems.

Recently we performed some analyses (*12, 13*) with two new methods we had devised. In the descriptions which follow we will stress the principles, some aspects of which were not treated in any detail in these short communications, and we will report some refinements of the analyses.

*Detection of Carbon, Nitrogen, and Oxygen by*
*Lithium-Induced Nuclear Reactions*

**Principle of the Method.** The method is distinguished by the
following:

(a) It uses nuclear reactions induced by a $^6$Li beam which allows
simultaneous detection of the three common contaminants $^{12}$C, $^{14}$N, and
$^{16}$O.

(b) These reactions give in the forward direction fast deuterons
and alphas which are easily separated by a $\Delta E - E$ telescope.

(c) These reactions have high-to-very high cross sections, but Li
reactions on heavier elements (background) fall rapidly with mass.

(d) The detection around $\theta = 0°$ allows a very large counter solid
angle without loss of energy resolution because $dE_2/E_2 \approx \sin \theta d\theta$, so the
kinematical spread is only a second-order effect.

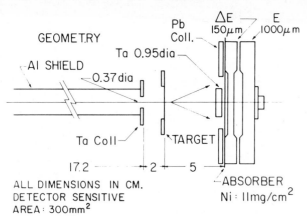

*Figure 2.   Target and detection apparatus used in C,*
*N, and O detection by Li-induced nuclear reactions*

The detecting apparatus is shown in Figure 2. The 7-MeV $^6$Li beam
was provided by the Université de Montréal tandem accelerator. The
tantalum disc protects the detector from the direct beam and leaves an
effective counting solid angle of $5 \times 10^{-2}$ sr. The nickel foils stop elastically scattered Li atoms but let through fast alphas, deuterons, and
protons which are separated digitally on-line by comparing their energy
losses in the $\Delta E$ and $E$ counters. A detector at 45° monitors the yield of
Rutherford scattering, giving the product of target thickness by beam
current.

The cross sections of the reactions observed are shown in Table I
with the expected sensitivities. The cross sections of the three reactions
actually used in the analysis were measured in $CO_2$ and $N_2$ gas targets as
a function of energy. The results, shown in Figure 3, illustrate one

**Table I. Forward Differential Cross Section at 6.7 MeV and Corresponding Sensitivities with a 100-nA $^6$Li$^{+3}$ Beam**

| Reaction | $d\sigma/d\Omega$ ($\mu b/st, \pm 8\%$) | S ($\mu g/cm^2$) |
|---|---|---|
| $^{12}$C$(^6Li,d)^{16}$O | 200 | 1.2 |
| $^{14}$N$(^6Li,d)^{18}$O | 200 | 1.3 |
| $^{14}$N$(^6Li,\alpha)^{16}$O | 65 | 4.0 |
| $^{16}$O$(^6Li,d)^{20}$Ne | 480 | 0.6 |

difficulty with nuclear reactions—the unpredictable cross sections complicate the analysis and require tedious measurements. The depth resolution is limited by straggling in the nickel foils and is calculated to be 0.15 $\mu$m (FWHM).

**Results with Niobium Thin Films.** Thin targets of niobium to be used in a study of sputtering (*14, 15*), ca. 1 $\mu$m thick, were prepared by sputter deposition on copper backings which were then dissolved in nitric acid. X-ray diffraction and electrical resistivity measurements showed that the films prepared under some well defined conditions of deposition had the structure and atomic spacing of bcc niobium and contained little impurity.

The capabilities of the method are demonstrated in Figure 4. The effectiveness of particle separation is shown in 4a. In 4b the energy

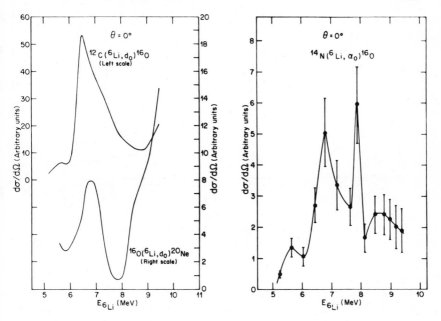

*Figure 3. Energy dependence of the forward differential cross sections of the reactions used in the analysis of C, N, and O in niobium thin films*

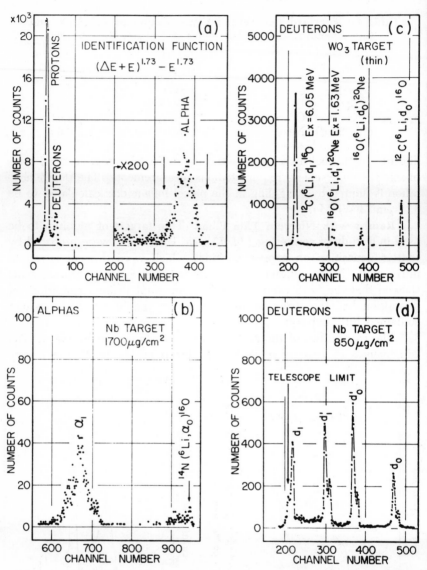

*Figure 4.   (a) Particle identification spectrum obtained in the analysis of a Nb thin film by a ⁶Li beam. (b) Energy spectrum of the alphas from the above run. (c) Deuteron energy spectrum obtained with a target of tungsten oxide on carbon. (d) Deuteron energy spectrum obtained with a niobium thin film, showing carbon and oxygen contaminants on the surfaces.*

resolution is indicated. The upper and lower peaks are caused, respectively, by the reactions where the residual nucleus $^{16}O$ is left in its ground and first excited states. In 4c we show a spectrum obtained with a very thin layer of $WO_3$ on a carbon backing. The carbon and oxygen signals

**Table II. Concentrations by Weight of Contaminants Measured in or on Four Sputter-Deposited Niobium Thin Films**

| Film Thickness ($\mu g/cm^2$) | $^{12}C$ (ppm) | $^{14}N$ (ppm) | $^{16}O$ (ppm) |
|---|---|---|---|
| 850 | 1040 | 21 | 650 |
| 900 | 500 | 12 | 300 |
| 1250 | 340 | 7 | 280 |
| 1600 | 175 | 39 | 150 |

are well separated. Finally in 4d we see the same peaks as in 4c, but with a niobium film. The splitting of each of the peaks shows that most of the contaminants are on the surfaces of the films; the width of the sub-peaks indicates the depth resolution. The concentrations of contaminants found in or on four films analyzed are given in Table II. The nitrogen amounts detected correspond to less than one monolayer of $N_2$.

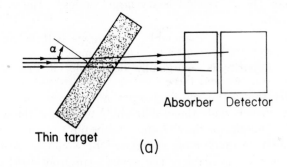

(a)

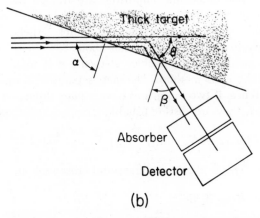

(b)

*Figure 5. A schematic ERD apparatus in two different geometries. Particles stopping in the target or in the absorber are distinguished from those stopping in the detector (not to scale).*

### Depth Profiling of Light Impurities by Elastic Recoil Detection (ERD)

**Apparatus.** In the ERD method, two detecting setups, both very simple, are possible. They are shown schematically in Figure 5. The first variant (a) works with targets up to 5 $\mu$m thick. A beam of heavy ions ejects forward the target atoms by Rutherford collisions. The (target + absorber) thickness is designed to stop the beam and the ejected heavy target atoms, but not the light impurities. A large detection solid angle ($\approx 10^{-2}$ sr) is allowed thanks to the kinematical singularity referred to earlier in the section on the Li-beam method. The second variant of the method (b), with thick targets, works on the same principle except that the detecting angle $\theta$ cannot be zero, and consequently the counting solid angle must be closed down to about $10^{-4}$ sr.

**Detectability and Mass Resolution.** To obtain good energy and mass resolution, the ejected impurity atom must lose not too large a fraction of its energy in the (target + absorber) which must stop "background" atoms (beam, ejected heavy target atoms). Hence follow the following criteria:

(1) The impurity is "detectable" if its range is three times that of the longest-ranged background atom (at the angle $\theta$).

(2) The impurity is "marginally detectable" if the ratio of the ranges is between two and three.

Table III shows the recoil energies and ranges for various contaminants in nickel, with three different beams and three detection angles $\theta$. The energies of the $^{16}$O and $^{35}$Cl beams optimize roughly the depth resolution. A much more energetic $^{79}$Br would be desirable, but it is not commonly available. The ranges are from Northcliffe and Schilling (16). The detectable impurities are framed by solid lines in the table, those marginally detectable by dashed lines. The Cl beam is clearly preferable. It is also clear that with thick targets, the detection angle $\theta$ will have to be kept smaller than about 30°. With "fine tuning," carbon should be detectable; heavier elements will require substantial modification of the ERD method. The mass separation is ca. 2 MeV/amu (Table III). Hence, if two impurities have a mass difference $\delta M$ (amu), from Equation 1, they are distinguishable if:

$$d(\mu\text{m}) \lesssim \delta M \ (2 \text{ MeV/amu})/(10 \text{ MeV}/\mu\text{m}) = 0.2 \ \delta M \text{ (amu)}$$

**Sensitivity.** The Rutherford differential cross section:

$$d\sigma/d\Omega = Z_1^2 Z_2^2 \ e^4 (M_1 + M_2)^2 / 4 M_2^2 E_1^2 \cos^3 \theta_r$$

simplifies for $M_1 >> M_2$ and $M \simeq 2\,Z$ to:

$$d\sigma/d\Omega \simeq Z_1^4 \ e^4 / 4 E_1^2 \cos^3 \theta_r, \ (\theta_r = \text{recoil angle})$$

i.e., it is nearly independent of $Z_2$ contrary to the case of RBS of light ions on heavy impurities. With a thin target and 1 W/cm² beam, the Cl and O beams give $S \simeq 10^{-8}$ g/cm², and the Br beam gives $S \simeq 10^{-9}$ g/cm². With thick targets, the sensitivity would be worse by about one order of magnitude.

**Table III.   Ranges and Recoil Energies (in parentheses) of Various Atoms Embedded in Nickel under the Impact of Various Projectiles and at Various Angles of Detection $\theta$**

| | | Range (mg/cm²) and Energy (MeV) | | | | | | |
|---|---|---|---|---|---|---|---|---|
| Beam | $\theta$ | Beam | $^{59}Ni$ | $^1H$ | $^3H$ | $^4He$ | $^7Li$ | $^{12}C$ |
| $^{16}O$ | 0° | 2.8 (10) | 1.2 (6.7) | 19 (2.2) | 14 (5.3) | 12.3 (6.4) | 7.7 (8.5) | 3.6 (9.8) |
| | 30° | 2.6 (9.3) | 1.0 (5.0) | 12 (1.7) | 10 (4.0) | 8.5 (4.8) | 5.6 (6.4) | 2.9 (7.4) |
| 10 MeV | 60° | 2.2 (7.6) | 0.4 (1.7) | 2.8 (0.55) | 2.2 (1.3) | 2.5 (1.6) | 2.0 (2.1) | 1.3 (2.5) |
| $^{35}Cl$ | 0° | 3.9 (30) | 3.0 (28.1) | 33 (3.2) | 27 (8.7) | 26 (11) | 18.5 (16.7) | 8.0 (22.8) |
| | 30° | 3.5 (25.5) | 2.5 (21.1) | 22 (2.4) | 18 (6.5) | 17 (8.3) | 12.5 (12.5) | 5.9 (17.1) |
| 30 MeV | 60° | 2.6 (15.7) | 1.3 (7.0) | 4.5 (0.81) | 3.9 (2.2) | 4.3 (2.8) | 3.7 (4.2) | 2.3 (5.7) |
| $^{79}Br$ | 0° | 3.7 (40) | 3.7 (39.2) | 16 (2.0) | 14 (5.6) | 14.5 (7.2) | 7.3 (9.0) | 6.4 (18.4) |
| | 30° | 2.9 (26.5) | 3.1 (29.4) | 11 (1.5) | 9.5 (4.2) | 10 (5.4) | 6.0 (6.8) | 4.9 (13.8) |
| 40 MeV | 60° | none | 1.6 (9.8) | 2.5 (0.5) | 2.3 (1.4) | 2.8 (1.8) | 2.2 (2.3) | 1.9 (4.6) |

**Depth Resolution.** The depth resolution is dominated by the energy loss of the heavy ion and the straggling of the light atom. Also as a result of the dependence on $M_1$ and $M_2$ of the energy transfer factor $f$, the $Z_1$ dependence of the energy loss, and the $Z_2$ dependence of the straggling, it turns out that the depth resolution is, very roughly, independent of beam and impurity, except if $M_2$ is very different from $2 Z_2$. With a combined (target traversed + absorber) thickness of 5 mg/cm² as required to stop background atoms, one gets for the straggling $\delta E_{(s)}$ 45 $Z_2$ (keV) and with a detector intrinsic resolution of 30 keV, then:

with $M_2 = 2 Z_2$ ,    $\delta d \simeq 300 \cos \alpha / \cos^2 \theta$   (FWHM in Å)
$\quad M_2 = Z_2$ (protons) ,   $\delta d \simeq 500 \cos \alpha / \cos^2 \theta$
$\quad M_2 = 3 Z_2$ (tritons),   $\delta d \simeq 200 \cos \alpha / \cos^2 \theta$

**Results with Hydrogen and Lithium in Copper.** We fabricated by evaporation on carbon or copper backings sandwich targets with two minute layers ($\approx 1 \, \mu g/cm^2$) of "impurity" (isotopic $^7LiF$ or natural LiOH) separated by $25\text{–}135 \, \mu g/cm^2$ of copper. Beams of 25–40 MeV and about 10 nA of $^{34}Cl^{+5}$ were used. Absorbers were nickel foils of 5–7 $\mu$m. Only the thin target geometry ($\theta = 0°$, $\Delta\Omega = 10^{-2}$ sr) has been tested successfully because the standard target holder shadows the sample for $\alpha > 65°$.

Figure 6a shows the mass discrimination capability. The ratios of the heights of the $^6Li$ and $^7Li$ peaks agree with the natural abundances.

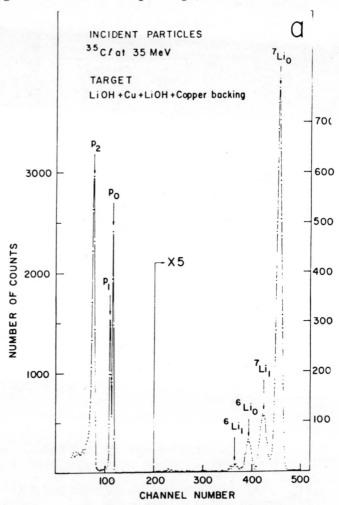

Figure 6. (a) Energy spectrum of particles ejected from two layers of natural LiOH separated by 1000 Å of copper. Particles coming from the first and second layer are labeled 0 and 1. Peak $P_2$ is hydrogen on the back surface of the copper backing, probably caused by accumulation of water vapor.

The H-to-Li ratios however are different in the two layers of LiOH, giving evidence for water vapor introduced at the interface between successive evaporations.

Figure 6b–d shows the depth resolution in more detail than in 6a. The target thicknesses are nominal, based on evaporation time. The peaks are somewhat wider than calculated, so it is possible that the energy resolution was underestimated. However the distances between the peaks are also larger than expected by factors of 1.5–1.9, indicating that either the ($dE/dx$) of the heavy ions or the copper thicknesses are larger than we think. These thicknesses have been concurrently or subsequently measured by other methods. Alpha particle energy loss gave values consistent with the peak separations above while both the yield of Rutherford scattering of the beam on the copper and the Cu x-ray fluorescence yield are consistent with the nominal values. The sensitivity

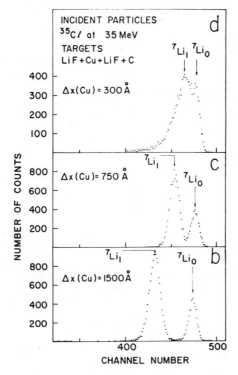

Figure 6. (b–d) Energy spectra of particles from two layers of isotopic $^7LiF$ separated by various amounts of copper. In (b) and (c) the target surface was normal to the beam; in (d) it was tilted at 30° to the beam.

estimated by using the nominal amounts of lithium compounds is $S = 10^{-8}$ g/cm² which agrees with calculations.

More carefully prepared and characterized targets will have to be used to check the details of the method. However a mass separation of 1 amu, a depth resolution of a few hundred angstroms, and a detection limit of about $10^{-9}$ g/cm² have been achieved by ERD without special care.

**Future Prospects.** We attempted in thick target geometry with $\alpha = \beta = 65°$ and $\theta = 50°$ to detect helium implanted in niobium. The signal as expected was a broad shoulder on the upper edge of the background from the elastic scattering of the beam. It will be a relatively simple matter to build a target holder allowing incidence and detection at grazing angles. Clearly this thick target method is the one which must be developed for many applications.

Going a step further one realizes that the depth resolution and the possibility of analyzing heavier impurities by ERD are both limited by the need for an absorber to eliminate the backgorund. The obvious solution is to use magnetic separation. One needs a magnet of $10^{-4}$ sr acceptance (with thick targets), and one has then in reality a high energy secondary ion mass spectrometer ("HESIMS") differing from the standard SIMS in that atoms are extracted from depths down to thousands of angstroms instead of the first few monolayers, with a depth resolution better than 100 Å and without significant erosion of the surface.

*Acknowledgments*

The authors at the Centre de l'Energie are grateful to M. Kaminsky and J. A. Davies for comments. The outstanding work of the Université de Montréal tandem operators made the runs rapid successes.

*Literature Cited*

1. Chu, W. K., Mayer, J. W., Nicolet, M. A., Buck, T. M., Amsel, G., Eisen, F., *Thin Solid Films* (1973) **19**, 423.
2. "New Uses of Ion Accelerators," J. F. Ziegler, Ed., Plenum, New York, 1975.
3. Verbeek, H., Adv. Chem. Ser. (1976) **158**, 245.
4. Blewer, R. S., Adv. Chem. Ser. (1976) **158**, 262.
5. Roth, J., Behrisch, R., Sherzer, B. M. U., *Appl. Phys. Lett.* (1974) **25**, 643.
6. Cohen, B. L., Fink, C. L., Degnan, J. H., *J. Appl. Phys.* (1972) **43**, 19.
7. Smidt, Jr., F. A., Pieper, A. G., *J. Nucl. Mater.* (1974) **51**, 361.
8. Moore, J. A., Mitchell, I. V., Hollis, M. J., Davies, J. A., Howe, L. M., *J. Appl. Phys.* (1975) **46**, 52.
9. Amsel, G., Nadai, J. P., D'Artemare, E., David, D., Girard, E., Moulin, J., *Nucl. Instrum. Methods* (1971) **92**, 481.
10. Turos, A., Wielunski, L., Barcz, A., *Nucl. Instrum. Methods* (1973) **111**, 605.

11. Bottiger, J., Picraux, S. T., Rud, N., "Ion Beam Surface Layer Analysis," O. Meyer, G. Zinker, and F. Käppeler, Eds., p. 811, Plenum, New York, 1976.
12. Cardinal, C., Brassard, C., Chabbal, J., Deschênes, L., Labrie, J. J., L'Ecuyer, J., Leroux, M., *Appl. Phys. Lett.* (1975) **26**, 543.
13. L'Ecuyer, J., Brassard, C., Cardinal, C., Chabbal, J., Deschênes, L., Labrie, J. P., Terreault, B., Martel, J. G., St-Jacques, R., *J. Appl. Phys.* (1976) **47**, 381.
14. Leroux, M., Master's Thesis, Université du Québec, 1975.
15. Leroux, M., Martel, J. G., St.-Jacques, R., Terreault, B., unpublished data.
16. Northcliffe, L. C., Schilling, R. F., *Nucl. Data* (1970) **A7**, 233.

RECEIVED January 5, 1976. Work supported by the National Research Council of Canada and the Quebec Ministry of Education.

# 14

# Lattice Location Studies of Gases in Metals

F. L. VOOK and S. T. PICRAUX

Sandia Laboratories, Albuquerque, New Mex. 87115

*Ion channeling and nuclear microanalysis techniques can determine the crystallographic lattice locations of implanted or dissolved gases in single crystal metals. Ion-induced nuclear reaction, x-ray, or backscattering measurements of channeling angular yield distributions along specific axes and planes clearly indicate lattice locations of gases for high symmetry sites. Interstitial and substitutional sites can be clearly distinguished. Light gas atoms such as hydrogen and oxygen occupy well defined interstitial positions, whereas heavier noble gas atoms do not occupy well defined interstitial or substitutional sites. In bcc metals deuterium is located in tetrahedral sites in W, Cr, and Mo at low temperature and in distorted octahedral sites in Cr and Mo at higher temperatures. For implanted He in W, multiple He atoms are trapped by lattice vacancies.*

$\text{M}$any of the physical properties of metals are modified by the presence and lattice location of gases. In particular, the nature of hydrogen and helium in metals has applications to reactor technology, aerospace propulsion, the hydrogen economy, and hydrogen-doped superconductors. The interaction of hydrogen with structural materials has had renewed interest with the increased use of ordinary [1]H as an aerospace and terrestrial chemical fuel and with the growing interest in the heavier isotopes, deuterium and tritium, as nuclear fusion fuels. The technological importance of hydrogen and helium in metals includes the effects of embrittlement at low concentrations and hydrogen storage in hydrides at high concentrations. In addition, the first wall structural material for controlled thermonuclear reactors and the fuel cladding for fast breeder reactors will include metals that will be subject to large fluences of hydrogen and helium.

The microscopic lattice location of hydrogen and helium in solids bears on the solubility and migration of these gases, which in turn is important for understanding such effects as damage trapping, bubble nucleation, embrittlement, and blistering in reactor environments. Most of these physical processes involve the near-surface regions of solids and can be examined by newly developed experimental techniques using ion channeling. Reference 1 presents a recent review of the lattice location of impurities by channeling.

## Lattice Location by Channeling

The channeling technique is one of the few techniques that gives direct lattice location information. For interstitials it has been applied only in a few cases to give detailed results. In this section we use lattice location channeling measurements of implanted deuterium in W, Cr, and Mo to illustrate the technique.

The controlled introduction of hydrogen isotopes into the near-surface region by ion implantation allows the study of metal systems with low hydrogen solubility. Such systems are difficult to investigate by traditional techniques such as neutron diffraction or NMR which normally require high concentrations throughout the crystal. In contrast, the ion channeling technique requires moderately high concentrations ($\sim 0.1$ at. % hydrogen) only near the surface, and this can easily be achieved using ion implantation. Ion channeling also has been used to investigate the lattice location of dissolved gases in metals. For example, deuterium in Nb (2, 3), a metal with high hydrogen solubility, has been investigated. Other location experiments are summarized below.

Results have been reported for the lattice location of deuterium implanted in the VIB transition metals W, Cr, and Mo (4, 5, 6). The early experiments were made at room temperature since the hydrogen diffusivity in these metals is relatively low. More recently the measurements have been carried out at 90 K as well as room temperature for Cr, Mo, and W (6). No other information is available for the lattice location of deuterium in these metals because of their low hydrogen solubility. Results of the studies show that the implanted deuterium occupies well defined lattice sites in these materials and that the location depends on implantation and anneal temperatures, implanted concentration, and state of radiation damage.

**Measurements.** In the studies of W, Cr, and Mo the D or $^3$He atoms were introduced by ion implantation into high-purity $<100>$ single-crystal samples, and the nuclear reaction, $D(^3He,p)^4He$, was used to detect implanted D or implanted $^3$He. Initial experiments were conducted by implanting the appropriate ions along non-channeling direction $7°$ from the $<100>$ axes. Fifteen- and 30-keV D were introduced into Cr and W, respectively, by accelerating the molecular species $D_3^+$ to 45 and 90 keV; $^3He^+$ implants were done at 60 and 200 keV. Implantation fluences were approximately $3 \times 10^{15}$ D atoms/cm² and $1 \times 10^{15}$ He atoms/cm². The projected range for 15-keV D in Cr is 1140Å, for 30-keV D in W is 1270Å, and for 60- and 200-keV ³He⁺ into W, it is 1140 and 3310Å (7, 8), respectively. Analysis of the lattice location of the implanted D(³He) was performed with incident 750-keV ³He (500-keV D) ions by monitoring the ion backscattering from the samples and by using the high energy protons from the nuclear reaction to monitor the implanted ions. This nuclear reaction has a Q value of 18.3 MeV and a peak cross section of ∼ 70 mb/sr at a ³He beam energy of ∼ 650 keV. A 300 mm² surface barrier detector with 500-μm depletion depth was located at a scattering angle of ∼ 135° with a solid angle ≈ 0.1 sr. Angular collimation of the incident beam resulted in an angular divergence (full angle) ≤ 0.06°, and the angular resolution of the goniometer was 0.01°. The elastic backscattering signal from the samples was accepted within an energy region corresponding to a depth region ∼ 1000Å centered at the projected range of the implanted impurity.

The lattice location results for implanted deuterium in W and Cr at room temperature follow. Channeling angular distributions for <100> scans of D in Cr and W are shown in Figure 1 (5). The backscattering and nuclear reaction yields have been normalized by the random (non-channeled) yields. The random levels indicated by the dashed lines at 1.0 correspond, respectively, to ∼ 4000 and 14,500 counts for scattering from Cr and W and 610 and 180 counts for the reaction yield from D in Cr and W. The low backscattering minimum yields of 1.7 and 1.1% for the <100> axes in Cr and W, respectively, observed pior to implantation indicate good crystalline quality. For W the strong enhancement in the D yield (flux peak) is seen with a full angular width of 0.2° and a maximum relative yield of 1.8, whereas the W channeling critical angle is 2.01°. In contrast to W(where the D yield does not drop below the random level), the <100> D data for Cr show a sharp flux peak with maximum relative D-yield of ∼ 1.3 and width ∼ 0.3° superimposed on a broad dip with minimum relative D-yield of ∼ 0.73. The angular width of the D-dip of ∼ 1.4° is similar to that observed for Cr of 1.38°. These data as discussed below suggest that D is localized near the tetrahedral interstitial site in W and the octahedral interstitial site in Cr. In addition to these axial data, corresponding data were taken of the {100} plane for W and Cr. These planar angular scans also confirm the location of the D site near the tetrahedral interstitial position in W and the octahedral position in Cr.

As seen from the example in Figure 1, the lattice location information is contained in the angular scans of the yield of close impact parameter events. In Figure 1 these events are elastic backscattering for the host lattice and a nuclear reaction for the impurity gas atoms. The technique is not limited to these processes, and for heavy mass impurities in light mass hosts, backscattering alone may be used, whereas ion-induced x-rays may be used for intermediate mass impurities in heavy hosts.

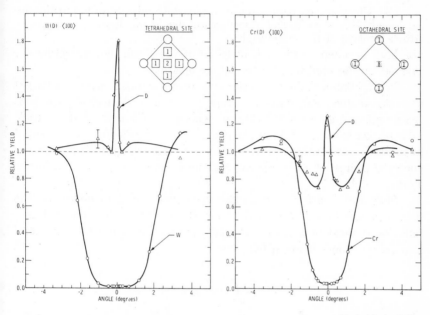

Physical Review Letters

*Figure 1.   Angular scans through the $\langle 100 \rangle$ axis for (a) W and (b) Cr for 3 × $10^{15}/cm^2$ 30- and 15-keV D implants, respectively, at 296 K. A 750-keV $^3$He analysis beam was used; $\bigcirc$, backscattered yields for W and Cr; $\triangle$, proton nuclear reaction yields from D (5).*

**Analysis.** The analysis of the angular scans requires a knowledge of the channeling effect which has been reviewed by Gemmell (9) and others (10). Channeling of a collimated beam of ions incident on a single crystal along an axial or planar direction occurs by a series of correlated small angle-screened Coulombic collisions. Using a continuum potential averaged along atomic rows or planes the critical angle of an incident beam relative to a row with which axial channeling will occur is $\psi_{\frac{1}{2}} \approx (2Z_1Z_2e^2/Ed)^{\frac{1}{2}}$ as long as $E \geq (2Z_1Z_2e^2d/a^2)$, where $Z_1e$ and $E$ are the atomic charge and the energy of the incident ion $Z_2e$, $d$ is the average atomic charge and spacing along the row, and $a$ is the Thomas–Fermi screening distance. Typical critical angles as seen in Figure 1 are $\leq 2°$.

The most important feature of channeling is the spatial distribution of impact parameters with respect to the crystal rows or planes imposed on the incident beam of particles. Figure 2 is a schematic of channeling in a crystal with a plot of the average potential and particle density across the channel. Two features are evident for the channeled particles:

(1) Close impact parameter processes such as elastic backscattering and nuclear reactions are forbidden for substitutional impurities or host lattice atoms

(2) Close impact parameter processes with interstitial impurities are enhanced with a yield that depends on the beam density and precise lattice location of the interstitials.

For high symmetry sites, a knowledge of the beam density across the channel can be used to infer the lattice location.

The spatial probability density of particles across a channel is largest near the center. The transverse energy $E_\perp$ of a channeled particle is conserved, and $E_\perp = E\psi^2 + U(x,y)$ where $\psi$ is the angle of the particle with respect to the crystal rows or planes, $E\psi^2$ is the kinetic energy caused by particle momentum transverse to the channel, and $U(x,y)$ is the potential energy of the screened coulombic interaction with the lattice atoms.

Multi-row continuum potentials have been used to determine the spatial distribution of the incident ion flux density in the channel as a function of incident angle $\psi$. The flux density $f$ as a function of lateral channel position $\vec{r}$ is given by:

$$f(\vec{r},\psi) = \int dA(\vec{r}_{in})/A(E_\perp)$$

$$E_\perp(\vec{r}_{in},\psi) \geq U(\vec{r})$$

where $A$ is the transverse area inside the equipotential contour corresponding to $E_\perp = E\psi^2 + U(\vec{r}_{in})$. This analysis assumes statistical equilibrium and thus does not take into account oscillations in the beam

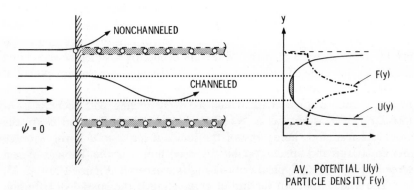

Figure 2.   Schematic of channeling in a crystal with plot of representative average potential and particle density across the channel

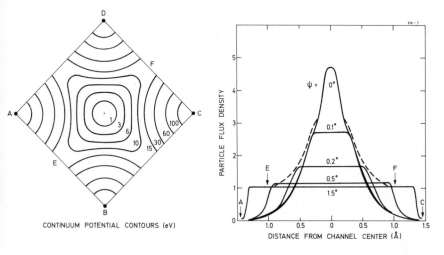

CONTINUUM POTENTIAL CONTOURS (eV)

DISTANCE FROM CHANNEL CENTER (Å)

Proceedings of International Conference
on Ion Beam Surface Layer Analysis

*Figure 3. Calculated ⟨100⟩ continuum potential contours and flux density curves for 0.7-MeV He in bcc Cr with lattice rows at ABCD and distance AC = 2.88Å. The flux density along direction AC is given as a function of incident angle ψ relative to ⟨100⟩ direction (——) and along direction EF is given for*
$$\psi = 0 \; (\text{---}) \; (6).$$

probability density with depth as discussed in Refs. 9 and 10. However these effects, which are most pronounced for planar channeling, usually do not preclude location analysis by statistical equilibrium models. Often the impurities are distributed in depth over several oscillations. This equation has been solved numerically, and Figure 3 gives an example of the continuum potential contours for He incident along the <100> direction in bcc Cr and the flux density along the line EF for $\psi = 0$ and along AC as a function of $\psi$. The calculation used a minimum value for $A = \pi a^2$ at the center, the Moliere approximation to the Thomas–Fermi potential and a static lattice.

Figure 4 gives a <100> scan for $1 \times 10^{15}$ D/cm² in W, implanted at 15 keV at room temperature and measured at room temperature with a 750-keV ³He beam (*11*). The upper solid line through the data indicates that the resolution was high enough to resolve the side band flux peaks not seen in Figure 1a. The dashed curve labeled "T site" gives the calculated angular distribution for the D localized in the tetrahedral interstitial position including a rms transverse vibrational amplitude of $\rho = 0.2$Å. The good agreement in the maximum may be fortuitous, but the agreement is within the localization distance of 0.3Å obtainable from the calculations for this site and confirms the lattice site. Similar calculations confirm the nearly octahedral site for D in Cr in Figure 1b.

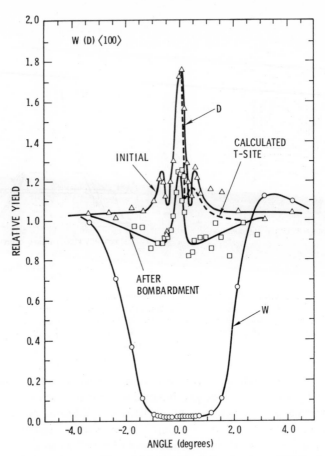

*Figure 4. Angular scan through ⟨100⟩ for 1 × 10¹⁵/cm²
15-KeV D implanted W before and after 71-μC irradiation
with 750-keV ³He⁺ (11). (- - -), calculations for T-site
occupancy (6).*

The reason different D locations exist for similar transition metals is
not known, but recent low temperature implantation and annealing
results have given added insight (6). For low fluence implantations at
90°K into both Cr and W, the D is located predominately near the
tetrahedral interstitial position. However for Cr the D moves into octahedral interstitial sites upon annealing to room temperature. Analogous
behavior is observed for D implanted into Mo. For low fluence implantation at 90°K the D is located predominately near tetrahedral sites,
whereas upon annealing to room temperature, it moves to a position near
the octahedral site. Detailed analysis of the D sites after annealing
suggests that the D equilibrium position is displaced slightly (0.2–0.3Å)
from the octahedral sites, being along the axis of the two lattice nearest

neighbors for Cr and in the plane perpendicular to this axis for Mo. These positions are shown in Figure 5 and labeled I and II for the Cr and Mo cases, respectively. Two possible interpretations of this change in lattice site for D in Cr and Mo upon annealing are that the new near-octahedral positions are stabilized by D clustering or D–defect interactions. Both hydrogen–hydrogen and hydrogen–defect interactions are significant in the bcc transition metals, and for these VIB metals with low hydrogen solubility, the binding energies could be sufficient for such centers to be stable at room temperature.

## Complications

**Distorted Sites.** Complications to the interpretation of the lattice location of impurities can occur if the impurity gases are in slightly distorted positions near high symmetry sites. Figure 5 shows interstitial sites in the bcc lattice with two possible distortions. Also shown is a comparison of measured <100> angular distributions using a 0.7-MeV

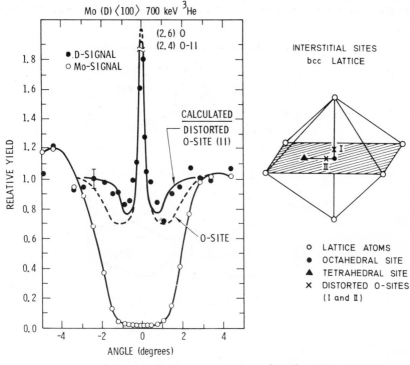

Proceedings of International Conference on Ion Beam Surface Layer Analysis

*Figure 5. Comparison of measured ⟨100⟩ angular distributions using 0.7-MeV ³He for D in Mo with calculated O-site and distorted O-sites I and II where δ = 0.36 and 0.2Å, respectively, along the ⟨100⟩ directions shown (6)*

³He analysis beam for D in Mo with calculations for the O-site and a distorted O-site II where the distortion is 0.2Å along the <100> direction shown. The narrowing of the dip caused by the 1/3-O sites along the <100> direction is consistent with the distortion postulated. As discussed above, the detailed physical reason for these distortions is not known, but these results illustrate how the channeling technique can give rather detailed information in favorable cases.

**Bulk-Doped Crystals.** Channeling analysis can be used for bulk-doped host lattices. Carstanjen and Sizmann (2, 3) have used channeling in single crystal niobium together with the $D(d,p)T$ nuclear reaction to study the lattice location of D in Nb. An incident 300-keV D beam penetrating through Nb loaded with 2 atomic % D was used. The Nb was loaded in a 50-Torr $D_2$ atmosphere at 600°C, cooled to room temperature, and subsequently cooled to 150 K in vacuum for measurement.

Figure 6 gives the angular dependence of the measured yields (proton/nC) of the $D(d,p)T$ reaction (□). Each experimental point required $\approx 1\ \mu C$ of incident D. The solid lines give the theoretical yield profiles assuming tetrahedral, octahedral, and hexahedral deuterium sites

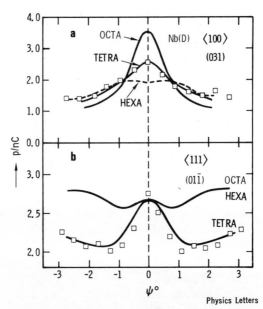

*Figure 6. Measured yields (protons/nC) of the D(d,p)T reaction (□) and calculated yield profiles assuming tetrahedral, octahedral, and hexahedral deuterium sites in niobium for (a) ⟨100⟩ with angular variation ψ in the (031) plane and (b) ⟨111⟩ with angular variation ψ in the (011) plane (2, 3)*

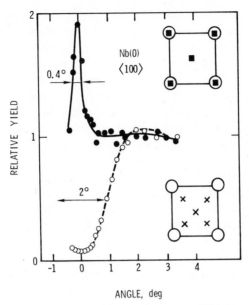

*Figure 7.   Angular scan along $\langle 100 \rangle$ axis of Nb in which $^{18}O$ was dissolved to 0.1–1.0 at %. Elastically scattered protons ($\bigcirc$) incident at 700 keV and emitted $\alpha$ particles ($\bullet$) from the $^{18}O(p,\alpha)^{15}N$ nuclear reaction are shown. Insets give projected positions for octahedral site location ($\blacksquare$) and tetrahedral site location ($+$) (12).*

in niobium. For the theoretical curves, a rather large value of 0.4° standard width for beam divergence and crystal mosaic spread was assumed to achieve good agreement with the data. In contrast to the flux peak angular full widths of $\sim 0.2°$ shown in Figures 1 and 4, the angular widths of the flux peaks in Nb are unusually broad, $\sim 1.5°$, and nearly of the order of the channeling critical angle. This complicates the analysis and the identification of tetrahedral site occupation by channeling measurements.

Figure 7 shows proton channeling studies of the lattice location of $\sim 0.1$ at % oxygen dissolved in niobium (*12*). The $^{18}O(p,\alpha)^{15}n$ nuclear reaction was used. The figure shows an angular scan through the <100> axis. The flux peaks observed for the <100> and several other directions clearly indicated that the oxygen occupies a well defined interstitial location. Based primarily on their {100} planar data, the authors concluded that the oxygen occupies the octahedral interstitial position. However, the data do not indicate the 1/3 dip expected for the 1/3 octahedral sites along the <100> axis. The reason for a lack of a 1/3 dip along the $<100>$ axis is not known but could possibly be explained by the broad enhancement in yield along channeling directions in bulk-doped crystals. This arises since a channeled beam will lose energy with depth more slowly than a randomly aligned beam, and in a bulk-doped crystal, will therefore encounter more target atoms while still able to induce a nuclear reaction. Thus, this increase in yield along a channeling direction with an angular width of the order of the channeling host lattice full width could cancel the 1/3 dip. This broad enhancement effect can complicate lattice location measurements of interstitial impurities in bulk-doped crystals.

**Interaction with Defects.** The lattice location of defects in metals can be influenced strongly by defects. In several cases the influence of radiation induced defects can be observed by their direct influence on impurities in changing the lattice site. Such radiation damage was commented upon by Carstanjen and Sizmann (2, 3) in the location studies of D in Nb.

A detailed experimental observation of a radiation-induced change in the lattice location of implanted D in W was reported by Picraux and Vook (11). Figure 4 shows $<100>$ angular scans of the D lattice location in W before and after 750-keV $^3$He fluences of $\approx 5 \times 10^{16}/cm^2$. The $^3$He beam was the same beam used to determine the D lattice location by means of the nuclear reaction, and $5 \times 10^{16}$ He/cm$^3$ corresponded to a total of only 74 $\mu$C of $^3$He incident on $\approx 1$ mm$^2$. Thus, in such cases where gas-atom defect interactions may be significant, it is important that the lattice location measurements use small measuring fluences. Typically $\approx 2$ $\mu$C are required to obtain one data point at these gas concentrations of a few tenths of an atomic percent.

The rate at which the change in lattice site is induced by the $^3$He beam for a channeled and a random (non-channeled) irradiation was also measured. The flux peak decreased fairly rapidly with irradiation and reached a relatively stable level, $\approx 1.2$–1.3 for both aligned and random irradiation, even though only for random irradiation did the W minimum yield increase slightly. The envelope of a shallow dip as well as the stability of the flux peak after the initial $\sim 40$ $\mu$C of $^3$He$^+$ irradiation would appear to confirm a change in the D location and cannot be explained simply by damage modification of the channeling. The data suggest that the change in location is induced fairly rapidly by a $^3$He fluence of only 20–30 $\mu$C ($\approx 1$–2 $\times 10^{16}/cm^2$) and that the new position remains relatively stable.

The angular scan after bombardment in Figure 4 gives some indication of a small dip ($\sim 0.9$) along the $<100>$ direction as well as the central flux peak of somewhat lower magnitude. Additional angular distributions are needed to determine the new D location. However, the

<100> scan suggests that the D is distorted appreciably from the tetrahedral interstitial site.

Another example of the interaction of a gas in a metal with defects is the lattice location studies of implanted He in W. Experiments of Kornelson (*13*) and calculations by Wilson and Bisson (*14*) indicate that a purely interstitial He atom would migrate rapidly through W, whereas vacancy trapping of single and multiple He atoms results in highly stable structures.

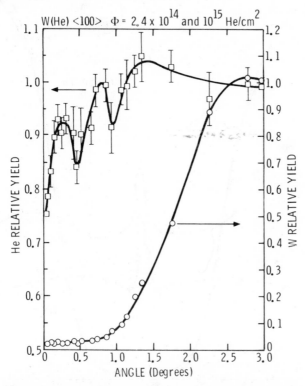

*Figure 8.   Angular scans through the ⟨100⟩ axis of W for an implant of 1 × 10¹⁵ (60-keV) ³He/cm² and for a multiple implant of 1.5 × 10¹⁴ (200-keV) ³He/cm² plus 0.9 × 10¹⁴ (60-keV) ³He/cm². Analysis is done using a 500-keV beam, and results are plotted on a single curve of relative yield vs. absolute angle. Data from Ref. 4.*

Figure 8 shows a new plot of combined angular scans (*4*) through the <100> axis of W for axial channeling measurements for 60-keV ³He implanted to a fluence of 1 × 10¹⁵/cm² at room temperature and similar results for a multi-energy implant consisting of 1.5 × 10¹⁴ 200-keV ³He/cm² + 0.9 × 10¹⁴ 60-keV ³He/cm². For the nuclear reaction yield from the implanted ³He, the scale has a factor of two magnification compared with the data for the elastically backscattered D ions. The flux peaks superimposed on a narrow dip indicate that the He location is neither a simple interstitial nor a purely substitutional location. The

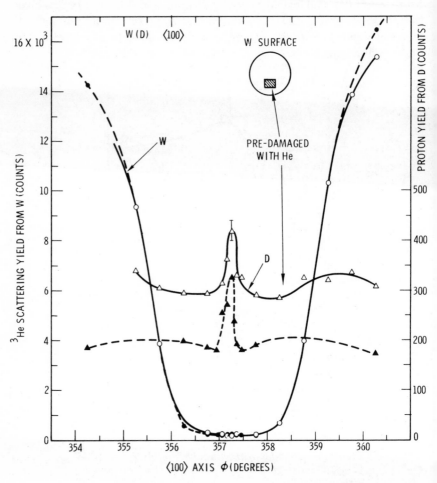

*Figure 9.   Effects on ⟨100⟩ angular scan by 750-keV ³He bombardment of W to a fluence < 5 × 10¹⁵/cm² prior to 30-keV D implantation (16)*

central narrow dip and the wider envelope of the dip indicate that some component of the He is located along or near the ⟨100⟩ rows. The flux peaks on either side of the narrow central axial direction indicate that some of the He atoms are located at positions displaced from the centers of the channel.

Comparisons of the experimental angular scans with site location predictions for two and for three He atoms trapped in a W lattice vacancy (*15*) agree well with most of the implanted He atoms being trapped as (3He + vacancy) defects with a smaller component as (2He + vacancy) defects. Some fraction of (6He + vacancy) centers may also be present since these also may have relative He locations similar to the (3He + vacancy) center.

Another damage-related complication is the enhanced trapping of hydrogen by prior radiation damage (*16, 17*). Figure 9 shows the angular scans of 30-keV D implanted into W with and without a prior 750-keV $^3$He bombardment to a fluence of $< 5 \times 10^{15}$/cm$^2$. The prior $^3$He implantation enhanced the retention of D in the near-surface regions by 50–200%. The effect shown was initially observed for D introduction by implantation in pre-damaged W and Au and by soaking pre-damaged Pd in D$_2$ gas (*16*). In the case of the W crystal shown, the amount of D retained in the pre-damaged region corresponds approximately to that introduced by implantation.

In addition to monitoring the total D retained in the solid, the channeling lattice location was measured. The open triangles correspond to a pre-damaged region on an adjacent area of the surface. As can be seen, for the pre-damaged region an almost 50% increase in the total D retention is observed, but no increase is observed in the magnitude of the central flux peak. This indicates that the additional deuterium which has been trapped is not located simply on tetrahedral interstitial sites and may be randomly located with respect to the host lattice.

Recently damage trapping of He has been observed in Mo following implantations of He, O, Ne, or Bi (*17*). Large enhancements in the amount of hydrogen retained in pre-implanted samples were observed over samples without prior implantation. Depth profiles of the implanted hydrogen indicated that the hydrogen is trapped at the damage created by the initial ion implantation. Lattice location measurements of the trapped H were not made, however.

## Other Location Measurements

A summary of all the published lattice location measurements of gases in metals known to the authors is given in Table I, including introduction and detection methods of the impurity and the observed predominant lattice location. For substitutional and nearly substitutional impurities having aligned channeling fractions $> 50\%$, the symbol S is used. For aligned fractions between 25% and 50%, S–R is used. In some cases the assigned location appears to be tentative. Also the effects of experimental conditions such as temperature and radiation damage may not be completely determined.

Most of the lattice location experiments have been made for impurities introduced by ion implantation. In some cases impurities have been introduced by diffusion or growth from the melt. Where both techniques of impurity introduction have been used, the agreement in site location has been good.

**Table I.   Summary of Location Studies of Gases in Metals**

| Host | Impurity | Method Intro.[a] | Method Detect.[b] | Location[c] | References |
|------|----------|---------|----------|----------|------------|
| Ti | O | I,D,M | $(p,\alpha)$ $(d,p)$ | O | 18, 19, 20 |
| Cr | D | I | $(^3\text{He},p)$ | T,O | 5, 6 |
| V | Kr | I | BS | R | 21 |
| Fe | F | I | $(p,\alpha\gamma)$ | R | 22 |
| | Ar | I | x-ray | R | 23, 24 |
| | Xe | I | BS | R-S | 25, 26 |
| Ni | F | I | $(p,\alpha\gamma)$ | R | 22 |
| | Ar | I | x-ray | R | 24 |
| Cu | D | I | $(d,n)$ | I | 27 |
| | Xe | I | BS | R | 26 |
| Nb | D | D | $(d,p)$ | T | 23, 29, 30 |
| | O | D | $(p,\alpha)$ | O | 12 |
| Mo | D | I | $(^3\text{He},p)$ | T,O | 6 |
| Ta | Xe | I | $\beta^-$ emit. | — | 31 |
| W | D | I | $(^3\text{He},p)$ | T | 4, 5, 6 |
| | $^3\text{He}$ | I | $(d,p)$ | D | 14 |
| | Rn | I | $\alpha$ emit. | S | 31 |

[a] Method of impurity introduction: I = ion implantation, D = diffusion, M = grown from the melt.

[b] Method of impurity detection: BS = ion backscattering, (a,b) = notation for ion-induced nuclear reaction with the incident particle and b the emitted particle, x-ray = ion-induced x-rays, $\alpha$ emit. and $\beta^\pm$ emit. correspond to detecting radioactive $\alpha$, $\beta^+$, or $\beta^-$ decay.

[c] Categories of lattice site location: S = ≥ 50% of the impurity on or near substitutional sites, I = interstitial with T = tetrahedral and O = octahedral interstitial sites, R = random sites, i.e., no strong orientation dependence, D = defect association.

In general the lighter gas atoms such as hydrogen and oxygen occupy well defined interstitial positions, whereas the heavier noble gases Ar, Kr, Xe do not occupy simple interstitial or substitutional sites. This could indicate a tendency for these heavier gas atoms to occupy low symmetry sites or a distribution of sites. Often only small channeling dips have been observed in these cases. Based on a series of channeling and hyperfine measurements of Sb, Te, I, and Xe implanted in Fe, de Ward and Feldman (25) have suggested that lattice vacancies may be associated with noble gas impurities.

*Summary*

The channeling lattice location technique can give detailed information on the site location of gases in metals in favorable cases. Interstitial impurities can be localized to $\lesssim 0.4$Å from the structure of flux peaks and to $\lesssim 0.2$Å from the structure of dips. In some cases, such as D in Cr and Mo, detailed information can be obtained on distorted interstitial sites. Site locations can generally be obtained for sharp flux peaks by comparing angular scans for axial directions with continuum potential model calculations in combination with planar angular scans. Relatively few lattice location measurements have been made for gases in metals; the most extensive studies have been carried out for hydrogen in the transition metals. In general the lighter gas atoms such as hydrogen and oxygen occupy well defined interstitial positions. In contrast, the heavier noble gases do not occupy simple interstitial or substitutional sites.

Results indicate D is located in tetrahedral positions in W, Mo, and Cr at low temperature. At higher temperatures hydrogen migration can result in new lattice locations because of defect and/or D–D interactions (e.g. D moves to distorted octahedral sites in Cr and Mo). In W, implanted He is trapped in vacancies and has been observed as multiple He atom configurations. Interaction of defects with interstitial gases can cause a change in site, and this has been observed for D in W. In bulk D-doped Nb, the D site exhibits wider than expected flux peaks that complicate exact site determination. Hydrogen can be trapped at ion damage giving a randomly appearing lattice location. Further work with careful attention to experimental details is required to unravel the sites of gases in metals.

*Literature Cited*

1. Picraux, S. T., "New Uses of Low Energy Accelerators," J. F. Ziegler, Ed., p. 229, Plenum, New York, 1975.
2. Carstanjen, H. D., Sizmann, R., *Phys. Lett.* (1972) **40A**, 93.
3. Carstanjen, H. D., Sizmann, R., *Ber. Bunsenges Phys. Chem.* (1972) **76**, 1223.
4. Picraux, S. T., Vook, F. L., *Appl. Ion Beams Met. Int. Conf.* (1974), 407.
5. Picraux, S. T., Vook, F. L., *Phys. Rev. Lett.* (1974) **33**, 1216.
6. Picraux, S. T., "Proceedings of International Conference on Ion Beam Surface Layer Analysis," O. Meyer, Ed., Vol. 2, p. 527, Plenum, New York, 1976.
7. Brice, D. K., private communication.
8. Brice, D. K., *Radiat. Eff.* (1970) **6**, 77.
9. Gemmell, D. S., *Rev. Mod. Phys.* (1974) **46**, 129.
10. "Channeling: Theory, Observation and Applications," D. V. Morgan, Ed., John Wiley, 1973.
11. Picraux, S. T., Vook, F. L., *Ion Implantation Semiconduct. Sci. Technol., Proc. Int. Conf.* (1974) 355.
12. Matyash, P., Skakun, N., Dikii, N., *JETP Lett.* (1974) **19**, 18.

13. Kornelson, E. V., *Radiat. Eff.* (1972) **13**, 227.
14. Wilson, W. D., Bisson, C. L., *Radiat. Eff.* (1974) **22**, 63.
15. Bisson, C. L., Wilson, W. D., *Appl. Ion Beams Met., Int. Conf.* (1974) 423..
16. Picraux, S. T., Vook, F. L., *J. Nucl. Mater.* (1974) **53**, 246.
17. Picraux, S. T., Bottiger, J., Rud, N., *Appl. Phys. Lett.* (1976) **28**, 179.
18. Della Mea, G., Drigo, A. V., Russo, S. Lo., Mazzoldi, P., Yamaguchi, S., Bentini, G. G., Desalvo, A., Rosa, R., *Phys. Rev.* (1974) **B10**, 1836.
19. Alexander, R. B., Petty, R. J., *At. Collisions Solids Proc. Int. Conf. 5th* 1974.
20. Della Mea, G., Drigo, A. V., Russo, S. Lo., Mazzoldi, P., Yamaguchi, S., Bentini, G. G., Desalvo, A., Rosa, R., *At. Collisions Solids, Proc. Int. Conf. 5th* (1974) 791.
21. Linker, G., Gettings, M., Meyer, O., *Ion Implantation Semiconduct. Other Mat., Proc. Int. Conf. 3rd* (1973) 465.
22. MacDonald, J. R., Kaufmann, E. N., Darcey, W., Hensler, R., *Radiat. Eff.*, in press.
23. MacDonald, J. R., Boie, R. A., Darcey, W., Hensler, R., *Phys. Rev.* (1975) **B12**, 1633.
24. Kool, W. H., Wiggers, L. W., Viehbock, F. P., Saris, F. W., *Radiat. Eff.* (1975) **27**, 43.
25. deWaard, H., Feldman, L. C., *Appl. Ion Beams Met. Int. Conf.* (1974) 317.
26. Feldman, L. C., Murnick, D., *Phys. Rev.* (1972) **B5**, 1.
27. Fischer, H., Sizmann, R., Bell, F., *Z. Phys.* (1969) **224**, 135.
28. Borders, J. A., Poate, J. M., *Phys. Rev.* (1976) **B13**, 969.
29. Iferov, G. A., Pokhil, G. P., Tulinov, A. F., *JETP Lett.* (1967) **5**, 250.
30. Skakun, N. A., Matyash, P. P., Dikii, N. P., Svetashov, P. A., *Sov. Phys. Tech. Phys.* (1975) **20**, 432.
31. Domeij, B., *Nucl. Instrum. Methods* (1965) **38**, 207.

RECEIVED January 5, 1976. This work was supported by the U.S. Energy Research and Development Administration under contract No. AT(29-1)-789.

# Surface Chemical Analysis with a Combined ESCA/Auger Apparatus

PAUL W. PALMBERG

6509 Flying Cloud Dr., Eden Prairie, Minn. 55343

*There are several unique advantages in combining x-ray photoelectron spectroscopy (ESCA) and Auger electron spectroscopy (AES) into a single instrument. The occurrence of surface segregation can often be detected without sputtering by measuring the magnitude of photoelectron and/or Auger peaks at two substantially different kinetic energies. ESCA contributes in-depth chemical bonding information which can be correlated with composition profiles obtained by AES.*

A number of analytical techniques capable of determining the chemical composition of the outermost layers of solids have been developed within the past 10 years. These include Auger electron spectroscopy (AES) (*1, 2*); photoelectron spectroscopy (*3*), which is commonly termed ESCA (electron spectroscopy for chemical analysis); ion scattering spectroscopy (ISS) (*4*); and secondary ion mass spectroscopy (SIMS) (*5*). While each of these techniques can provide elemental analysis of a surface, important differences exist with respect to sensitivity, depth of analysis, quantitative capability, spatial resolution, radiation damage, data acquisition speed, and chemical bonding information. Because of these differences, it can be very advantageous to combine two or more of the techniques into a single instrument.

There is an instrument which has the high sensitivity and excellent two-dimensional spatial resolution capability of the AES technique combined with ESCA, which is the least damaging of all surface analytical techniques and which provides direct information on chemical binding. Experimentally, this combination is simplified because both techniques require energy analysis of emitted electrons. In switching from one technique to the other, it is only necessary to change the operational mode of the electron spectrometer and the excitation source from an electron beam to a monochromatic x-ray beam.

*Experimental Apparatus*

A detailed description of the double-pass cylindrical mirror analyzer (CMA) for operation as a combined ESCA/Auger spectrometer has been reported elsewhere (6) and will be discussed only briefly here. This apparatus, shown schematically in Figure 1, includes spherical grids for pre-retardation followed by a double pass cylindrical mirror analyzer. A coaxial electron gun is used to excite the target for AES measurements, and an x-ray source mounted external to the CMA is used for photoelectron excitation.

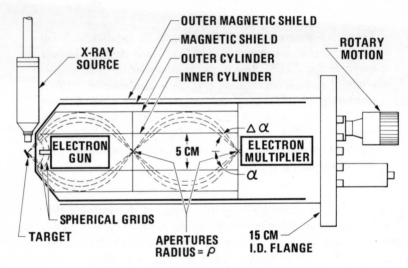

*Figure 1.   Schematic of double pass CMA*

In the Auger mode of operation, the inner cylinder, which is connected directly to the outermost spherical grid, is placed at ground potential so that electrons emitted from the target enter the CMA with their initial energy, $E$. In this mode the double pass CMA operates identically to conventional, single stage CMA's. The resolution, $R = \Delta E/E$, is 0.6% in this non-retarding mode which is used for AES measurements to optimize the sensitivity. Because photoelectron peaks are generally narrower in energy than Auger peaks and because chemical bonding information in ESCA requires precise energy definition, higher energy resolution is required. To meet the energy resolution requirements for ESCA, the spherically shaped retarding grids are used to slow the electrons from their initial kinetic energy to a lower pass energy. Using a CMA with a relative resolution of 1%, for example, 1000 eV electrons can be analyzed with an absolute resolution $\Delta E$ of 0.5 eV by first reducing their energy to 50 eV. To improve the sensitivity for ESCA measurements relative to that obtained with aperture sizes appropriate for AES, the apertures between the two CMA stages and those at the exit of the second stage can be enlarged by a rotary motion feedthrough.

The electron multiplier is operated in a pulse counting mode to detect the ESCA signals and in the analog mode to detect the Auger signals. Because the photoelectron current exiting from the analyzer is typically in the range $10^{-14}$–$10^{-15}$ amp, the pulse counting mode is required for optimum signal-to-noise performance. In the AES mode of operation, however, the current exiting from the analyzer is in the range of $10^{-10}$ amp, which is much too large for pulse-counting detection. The voltage across the multiplier is therefore reduced to limit the electron multiplier output current, and the signal is processed by standard lock-in techniques.

In addition to electron beam and x-ray excitation, the sample can also be sputter etched with an inert gas ion beam without repositioning. Rapid in-depth composition profiles can be made by simultaneous sputter etching and Auger measurements (7). In-depth information can also be obtained with ESCA by turning the ion beam off prior to ESCA analysis. The secondary electron signal generated by the incident ion beam is small compared with the Auger signal and therefore does not cause an interference problem. In ESCA, however, where the signal levels are much lower, the secondary electrons generated by the ion beam prevent simultaneous ESCA analysis. To obtain in-depth chemical bonding information with ESCA with good in-depth resolution, it is necessary to defocus or raster the ion beam to obtain a uniform etch rate over the area sampled in the ESCA mode. The area sampled in ESCA with the double pass CMA is circular with a diameter of 1.5–3 mm, depending on the operating parameters.

*Applications*

In addition to combining the best features of each instrument, the combined ESCA/Auger apparatus provides more peaks from each element, resulting in better element discrimination and some information related to the depth distribution of surface elements without sputtering. It is also possible to obtain in-depth chemical bonding information by obtaining photoelectron spectra after sputtering for various time intervals.

**Element Discrimination.** In the AES and ESCA techniques, the sensitivity to a particular element can sometimes be reduced when a second element is present which produces peaks in the same energy range. As seen in Figure 2a, Mn is difficult to detect with high sensitivity in an Fe matrix because of peak overlap between the major LMM Auger transitions. In the ESCA spectrum of Figure 2b, however, the Mn peaks are completely removed from the iron spectrum, and hence the detection

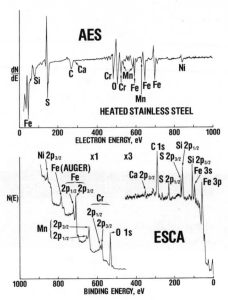

*Figure 2. Element identification. Auger and ESCA spectra of heated stainless steel.* $E_p = 5\,kV$, $Al\,k_\alpha$.

limit and quantitative capability are greatly improved. The reverse can also occur where a peak overlap problem occurs in the photoelectron spectrum but not in the Auger spectrum.

**Nondestructive, In-depth Composition Information.** In-depth composition analysis can be readily obtained by simultaneous AES and inert gas sputtering (7), but can sometimes lead to erroneous conclusions because of sputtering effects such as preferential removal of certain elements or knock-in collisions. Hence it may be desirable to determine the occurrence of surface segregation, for example, without sputtering. When peaks of significantly different energy are generated from the element of interest by ESCA and/or AES, the degree of surface segregation on a specimen can be determined by comparing ESCA and AES spectra with similar data from a specimen in which the element of interest is distributed uniformly in the escape depth region.

An illustration of this method for sulfur segregated on vanadium is presented in Figures 3 and 4. Sputtered cadmium sulfide is used as a reference in which the S is distributed uniformly throughout the escape depth region. In Figure 3, within the escape depth of 150 eV electrons, the concentration of S in V is about 44% of that in CdS. Within an escape depth of 2120 eV, however, the concentration of S on V is only 13% of that in CdS. In CdS the larger escape depth at 2120 eV contributes proportionately to the Auger signal. In the heated V specimen,

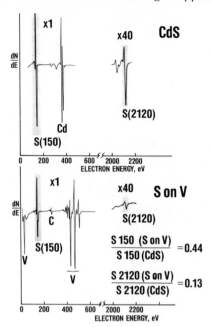

Figure 3. Surface sensitivity—AES. Auger spectra from CdS and heated vanadium. $E_p = 5$ kV.

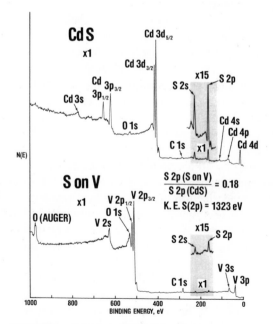

Figure 4. Surface sensitivity—ESCA. ESCA spectra from CdS and heated vanadium. Al $k_\alpha$.

however, where the S is likely segregated into a layer of thickness less than either escape depth, the variable escape depth has little effect on the observed ratio of high and low energy Auger signals. The ratio of escape depths can then be derived (8) from the formula:

$$\lambda(E_2)/\lambda(E_1) = I_V(E_1)I_{CdS}(E_2)/I_V(E_2)I_{CdS}(E_1) \tag{1}$$

where $\lambda(E)$ is the escape depth for an Auger or photoelectron of kinetic energy $E$, and $I$ is the measured Auger or ESCA signal. In the derivation of Equation 1, it is assumed that the backscattering contribution to the Auger signal is identical in CdS and V. From Figure 3 it can be determined that $\lambda(2120)/\lambda(150) = 3.34$ which compares favorably with a ratio of 3.75 expected (9) from a square root dependence of escape depth on electron energy. If sulfur were segregated uniformly over a thickness of $\lambda(2120)$ the experimental value for $\lambda(2120)/\lambda(150)$ (from Equation 1) would be near unity. Since the measured value is substantially different from unity and near that expected for a $\lambda \propto E^{1/2}$ dependence, it can be concluded that S is segregated almost entirely within a depth of $\lambda(150)$.

It is also possible to use a combination of photoelectron and Auger peaks to determine the extent of surface segregation. The intensity of the S-$2p$ photoelectron peak from CdS and V may be determined from the ESCA data in Figure 4. When the ESCA signals are inserted into Equation 1 along with the Auger signals from the 150-eV S peak, a value of 2.44 for $\lambda(1323)/\lambda(150)$ is obtained, which again reasonably agrees with a value of 2.96 expected from an $E^{1/2}$ dependence.

**Radiation Damage.** In general, the radiation damage caused to the sample surface during AES analysis is several orders of magnitude greater than that during ESCA analysis. This difference can be explained in terms of the relative ionization cross sections between core and valence band levels in the two techniques. In the AES technique, where the solid is excited by an electron beam, the probability for ionization of a lattice electron decreases rapidly with increasing binding energy. Hence, valence band electron excitation is much more probable than core level ionization. Since Auger transitions require initial ionization of a core level, only the less probable core level ionization events contribute to the Auger signal, and the more probable valence band excitations only contribute to lattice damage. In the ESCA technique, where the sample is irradiated with an x-ray beam, the photoelectron ionization cross sections for core levels are considerably more probable than direct photoionization of the valence band. Hence, most of the x-ray energy is lost through production of photoelectrons which contribute to the ESCA signal.

The observed damage in the AES technique is also increased because the electron beam is normally focused to a small spot whereas the x-ray beam floods the entire area accepted by the energy analyzer. This effect could, of course, be eliminated by either defocusing the electron beam or rastering the beam over the acceptance area of the analyzer.

Radiation damage effects are generally most severe for organic materials and physisorbed surface layers. In many organic materials the electron beam damages the surface so rapidly that the Auger analysis is meaningless. Also, the electron beam can remove physisorbed layers so rapidly that they do not contribute to the Auger spectrum. In such systems, it is therefore necessary to use the ESCA technique where the radiation damage is usually sufficiently low to allow meaningful data to be taken. Radiation damage can be ascertained by observing the time dependence of the spectrum.

Although the radiation damage effects are most severe in organic systems, they can also occur in many inorganic compounds such as alkali halides and some metal oxide materials. An example of electron decomposition in metal oxides is shown in Figure 5 where the oxygen and

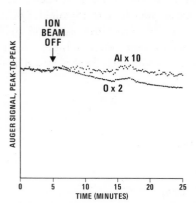

Figure 5. Electron beam fragmentation—sapphire $(Al_2O_3)$. Auger peak-to-peak amplitudes of O and Al in $Al_2O_3$ vs. time.

aluminum Auger signals are monitored as a function of time. During the initial 5 min, the sample was simultaneously sputtered to eliminate the effect of electron beam fragmentation. After the ion beam was turned off, the oxygen signal decreases slowly while the aluminum signal remains at a nearly constant level. This effect occurs because the electron beam preferentially desorbs oxygen from the surface. When the same sample is observed with the ESCA technique, no radiation damage is observed.

**In-Depth Profiling.** The combination of ion-etching and the AES analysis has been used extensively to profile the depth distribution of elements in thin film structures. As mentioned previously the secondary electron signal generated by the ion beam is sufficiently small in com-

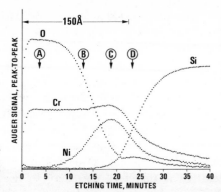

*Figure 6. In-depth composition profile of Nichrome film on silicon. Nichrome heated in air at 450°C for 30 sec.*

parison with the Auger signal to permit simultaneous Auger analysis and sputter etching. The rapid data acquisition capability of the Auger technique permits several elements to be monitored at frequent intervals while the surface is continuously eroded with the ion beam.

In a combined ESCA/Auger apparatus the Auger in-depth composition profile can be used as a convenient bench mark for determining the level within the thin film structure at which more detailed ESCA information is obtained. An example of this procedure is given in Figures 6 and 7. The Auger in-depth composition profile (Figure 6) is for a Nichrome film on a silicon substrate which was heated in air at 450°C.

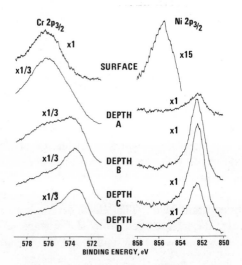

*Figure 7. Variation of chromium and nickel $2P_{3/2}$ photoelectron spectra as a function of depth in Nichrome film. Nichrome heated in air. Depth indications are in Figure 6.*

As indicated, the effect of heating in air is to form a chrome oxide layer on the surface. To determine the chemical state of chrome and nickel throughout the thin film structure, a second identical film was examined, and the sputtering was interrupted at points A, B, C, and D for detailed chemical analysis with the ESCA. As indicated in Figure 7, the energy position of the chrome peak is consistent with the formation of chrome oxide throughout the oxygen-rich region. The chrome peak shifts to a lower binding energy corresponding to the metallic state as the oxide layer is sputter removed. The ESCA data indicates that the nickel is oxidized in the outer surface layers but that it is in the metallic state in the interior of the chrome oxide film. The complete compositional profile could have been obtained with the ESCA technique alone but the time required to generate the profile of Figure 6 would be much longer because of the necessity to terminate sputtering during the ESCA measurement and because of the lower data acquisition speed of the ESCA technique.

## Summary

The combination of AES and ESCA in a single instrument not only combines the strongest capabilities of each individual method but also improves the capability for element discrimination and nondestructive in-depth composition information by comparing the magnitude of Auger and photoelectron peaks. The power of the in-depth profiling technique is also considerably enhanced by using the ESCA technique to evaluate the chemical state of thin film constituents at various depths.

The range of materials and problems which can be examined with the combined ESCA/Auger system is considerably increased over that for either single purpose instrument. For materials such as organics, which are easily damaged by an electron beam, ESCA is a more appropriate tool. For inorganic materials which generally resist radiation, AES offers the advantages of high spatial resolution and speed.

## Literature Cited

1. Palmberg, P. W., "Electron Spectroscopy," D. A. Shirley, Ed., p. 835, North Holland, Amsterdam, 1972.
2. Riviere, J. C., *Contemp. Phys.* (1973) **14**, 513.
3. Siegbahn, K., *J. Electron Spectrosc. Relat. Phenom.* (1974) **5**, 3.
4. Goff, R. F., Smith, D. P., *J. Vac. Sci. Technol.* (1970) **7**, 1.
5. Benninghoven, A., Loebach, E., *Rev. Sci. Instrum.* (1971) **42**, 49.
6. Palmberg, P. W., *J. Vac. Sci. Technol.* (1975) **12**, 379.
7. *Ibid.* (1971) **9**, 160.
8. Palmberg, P. W., *Anal. Chem.* (1973) **45**, 549A.
9. Chang, C. C., "Characterizatiaon of Solid Surfaces," P. F. Kane, G. B. Larrabee, Eds., p. 509, Plenum, New York, 1974.

RECEIVED January 5, 1976.

# 16

# Interface Phenomena in Solar Cells and Solar Cell Development Processes

WIGBERT SIEKHAUS

Materials and Molecular Research Division and Energy and Environment Division, Lawrence Berkeley Laboratory, University of California, Berkeley, Calif. 94720

*Processes on interfaces critically influence the formation and performance of thin film as well as single crystal silicon solar cells. The present limitations of the single crystal ribbon growth process are resolvable through surface reactions at the gas/liquid silicon interface. The diffusion of carbon substrate material into thin film silicon can be measured as a function of substrate crystallinity by Auger depth profiling. An ultrathin layer of aluminum affects thin film crystallinity. Impurities are distributed between grain boundaries and crystallites in thin films and may affect the chemical vapor deposition mechanism. Electronic states measured by electron loss spectroscopy strongly depend on surface structure and impurities.*

$\mathbf{P}$hotovoltaic electric power generation will become economical if the cost of the generating equipment can be brought to $\sim$ \$2.40/m$^2$ times efficiency (%) (1). There are two options available if one wants to achieve that goal (Figure 1): to develop inexpensive concentrators in conjunction with high cost, high efficiency cells or to develop inexpensive cells.

Both options have a multitude of interface problems. For the first option, inexpensive optical coatings or lenses must be developed with high reflectivity or transmissivity throughout the solar spectrum. These properties must be preserved for 30 years in an atmosphere made corrosive by air pollution. That is a formidable interface problem.

High concentration factors imply collectors tracking the sun. These cannot be built for less than \$100/m$^2$. Consequently, solar cells must be developed with efficiencies higher than 25%. This appears to be achievable only by multilayered cells in which materials are stacked with increasing band gap energy. At the layer interfaces of such cells, problems of interdiffusion and mismatch of optical properties must be controlled.

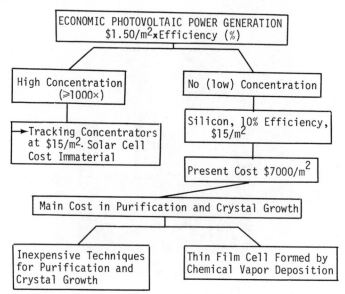

*Figure 1.   Cost reduction options*

The second option, development of inexpensive solar cells, is our concern. Examples will be restricted to silicon, the material presently receiving the most attention. The present cost of silicon solar at 12% efficiency is $7000/m², and radical changes in the techniques of cell formation are necessary before the cost can be substantially reduced. Two techniques are presently under development—growth of thin single crystal ribbons and growth of thin films on various substrates.

### The EFG Process

In the edge fed growth (EFG) ribbon process (Figure 2), liquid silicon rises in a size-defining capillary slit and forms—with the top of the capillary at the bottom and a solid silicon ribbon at the top—a meniscus from which the ribbon is withdrawn at 2.5 cm/min. The ribbon is ~ 200μ thick and 2.5 cm wide. There are three interfaces involved in this process—the liquid silicon–solid silicon interface, the liquid silicon–gas interface, and the capillary–liquid silicon interface. The capillary is usually made of carbon, which goes into solution at the bottom of the capillary and segregates at the meniscus, forming SiC crystallites at the

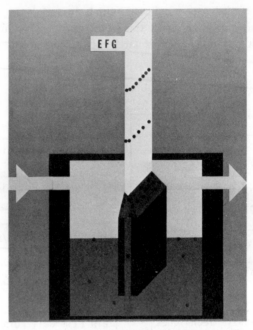

*Figure 2.   EFG single crystal ribbon growth*

top of the capillary which change its and the ribbon's shapes. The crystallites occasionally detach themselves and are included into the ribbon. This is at the moment a serious, unsolved interface problem which must be controlled. At the liquid silicon–solid silicon interface, solidification proceeds rapidly, $\sim 10^{-6}$ sec/monolayer, which leaves little time to

| Impurity | $\dfrac{D \times 10^{5}}{cm^{2}/sec}$ |
|:---:|:---:|
| B | 33±4 |
| P | 27±3 |
| Sb | 14±5 |
| Ga | 6.6±0.5 |
| In | 1.7±0.3 |
| Ca | 2.3 |

Diffusion data, Vol. 3, p. 86
Vol. 4, p. 144

*Figure 3.   Impurity diffusion coefficients in
liquid silicon at 140°C*

correct stacking faults, but which is slow enough for impurities to segregate and diffuse into the meniscus (Figure 3). In contrast to the Czochralsky process, these impurities cannot return to the melt, but rather are accumulated in the meniscus. They influence meniscus surface tension, which changes the shape of the meniscus, and apparently introduce periodic fluctuations in ribbon shape, ribbon purity, and crystallinity. To eliminate this, and possibly also the SiC carbide formation problem, we propose to use the liquid silicon–gas interface at the meniscus to remove impurities by gas–surface reactions. Figure 4 shows the process, in which impurities concentrated in the meniscus are segregated to the

*Figure 4.    Edge fed growth surface segregation purification process*

meniscus surface and removed by reactions with a gas. At a meniscus height of 0.5 cm, a ribbon thickness of 200$\mu$, and a gas–surface reaction time of 1 msec, an impurity concentration of $\sim 1\%$ could potentially be removed through the meniscus interface. This surface segregation purification process may allow crystal growth from low grade silicon, significantly reducing cost. Alternately, it may permit growth of extremely pure, single crystal ribbons with a very low density of crystal imperfections. There are three problem areas in which research must be carried out before the EFG surface segregation purification process (EFG–SSPP)

An economically competetive solar cell consists of a substrate on which a 100 $\mu$ thick silicon layer is deposited, covered by a transparent conductor containing a metallic grid.
The problems to which research is directed are identified in the cell by the letters a, b, b', c

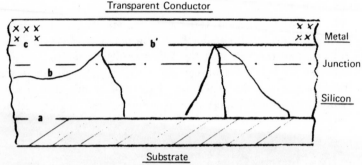

Research Areas:

a) Low temperature catalysis of the decomposition of $SiH_4$, $SiCl_4$ on a low melting point substrate material (e.g. Al) to form large area silicon sheets.
   Research tool: molecular beam deposition.
b) Catalysis of large grain crystallization by use of suitable impurities.
   Segregation of impurities onto grain boundaries.
   Research tool: low energy electron diffraction, Auger spectroscopy.
   a,b,b',c. Dependence of electronic surface states on impurities.
   Research tool: photoelectron spectroscopy, energy loss spectroscopy.

*Figure 5.   Thin silicon film solar cell design and related research areas*

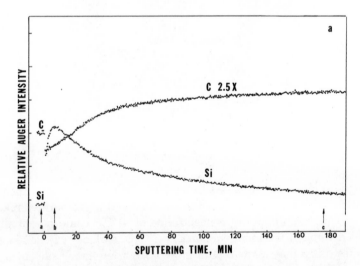

*Figure 6.   Interdiffusion at the carbon–silicon interface. Auger depth profiles of high temperature deposited (1150–1200°C) silicon films on (above) (a) edge (prism) plane of pyrolytic graphite, (right) (b) basal plane of pyrolytic graphite, (c) extruded graphite, and (d) glassy carbon.*

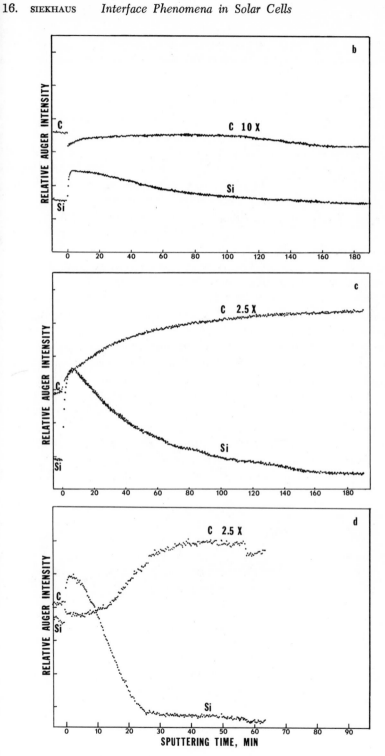

can be perfected—surface segregation of impurities, effect of impurities on surface tension, and gas/liquid metal impurity surface reactions.

## Thin Film Cells

The other option to reduce silicon solar cell cost is to develop thin film solar cells. There are interface problems encountered there also. Figure 5 shows a thin film cell which may soon become economical and lists the interface problem areas to which research must be directed. The cell consists of a conducting substrate upon which silicon is deposited, preferably by chemical vapor deposition, forming crystallites comparable in size with the film thickness ($100\mu$). The $p$–$n$ junction is formed during deposition by changing the dopant in the carrier gas. A transparent (semi) conducting layer is deposited on top of the silicon film. This layer contacts each grain so that current transport across grain boundaries is minimized. A metallic grid is imbedded in the transparent layer to reduce its sheet resistance. In an alternate cell design, the transparent conductor is one part of a silicon-transparent conductor heterojunction.

**Interface Diffusion.** A serious difficulty in thin film development is the diffusion of substrate material into the silicon layer. A high temperature material suitable as a substrate is carbon. Figure 6 summarizes the silicon–carbon interdiffusion for various allotropic forms of carbon (2); carbon diffuses deeply into silicon, and for all subtrates except glassy carbon, silicon also diffuses into carbon. In both cases, a high resistivity SiC layer is formed, which reduces cell efficiency.

Since many attempts to build diffusion barriers have failed, it is highly desirable to investigate whether chemical vapor deposition and growth of large crystal grains can be accomplished at a lower tempera-

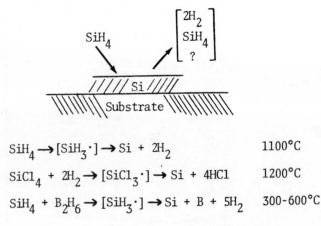

$$SiH_4 \rightarrow [SiH_3 \cdot] \rightarrow Si + 2H_2 \qquad\qquad 1100°C$$

$$SiCl_4 + 2H_2 \rightarrow [SiCl_3 \cdot] \rightarrow Si + 4HCl \qquad 1200°C$$

$$SiH_4 + B_2H_6 \rightarrow [SiH_3 \cdot] \rightarrow Si + B + 5H_2 \qquad 300\text{-}600°C$$

*Figure 7.   Chemical vapor deposition reactions*

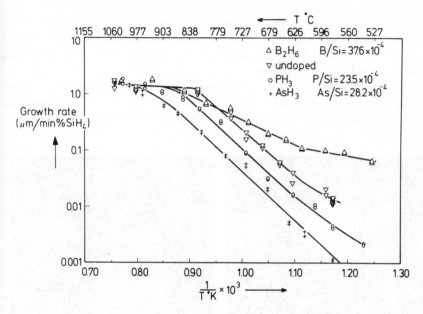

*Figure 8. Influence of deposition temperature on the growth rate of poly-erystalline silicon films without and with additional doping of AsH$_3$, PH$_3$, and B$_2$H$_6$*

ture. Figure 7 shows the reaction equations of two reactants, SiH$_4$ and SiCl$_4$, and points out that addition of diborane (B$_2$H$_6$) to the chemical vapor allows deposition at a substantially lower temperature. The actual reaction mechanism is now being investigated (3), but much more research using modulated molecular beam spectroscopy is needed before it can be determined how the deposition mechanism is influenced by impurities in the gas stream. Figure 8 shows that a *p*-type impurity increases the deposition rate while a *n*-type dopant decreases the deposition rate. Since both SiH$_4$ and SiCl$_4$ are tetrahedral molecules carrying a slight negative excess charge on the hydrogen atoms, it is tempting to propose (Table I) (5) that the *p*-dopant forms a charge layer at the surface which attracts such molecules, whereas *n*-dopants form a repulsive layer. Table I shows that a prediction of the model, the reduction of the deposition rate in the presence of B$_2$H$_6$ of diamond from CH$_4$ (a molecule with a slight positive excess charge on the hydrogen atoms), is verified by the literature. Experiments to study the influence on the deposition rate of surface charges induced by electric fields are presently carried out in our laboratory. There is, in summary, justifiable optimism that CVD deposition rates may be catalyzed to proceed at lower temperatures.

**Low Temperature Crystal Growth.** Crystal growth at temperatures much below the melting point of silicon can be achieved if the silicon surface is covered during deposition by a thin layer of a low-melting point liquid alloy of silicon (vapor–liquid–solid (VLS) techniques). Experiments have shown that both on $SiO_2$ (6) and pyrolytic graphite (7), crystal grain size and crystal orientation is significantly improved if Si deposition is preceded by the deposition of a thin aluminum layer (~ 500 Å).

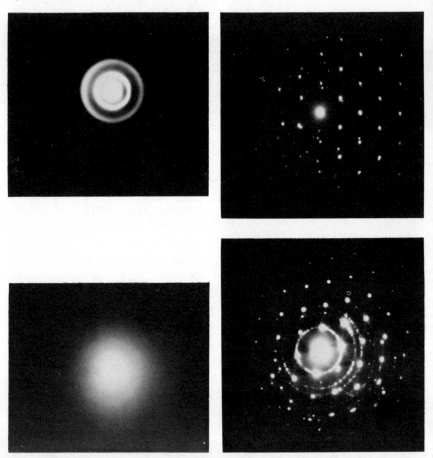

*Figure 9.   Electron diffraction micrographs of (a) silicon film deposited on quartz at 600°C and (b) silicon film deposited on quartz at 600°C with prior coating of a Si–AC–Si layer*

Figures 9a and b compare diffraction electron micrographs of Si films deposited at 600°C without (a) and with (b) the predeposition. Figure 10, an Auger depth profile of such a film on quartz, confirms that

**Table I. Predicted and Observed Dopant Gas Effects on the Deposition Rates of Silicon and Carbon Chemical Vapors[a]**

|  | $SiH_4$ | $SiCl_4$ | $CH_4$ | $CCl_4$ |
|---|---|---|---|---|
| Bonding characters | $Si^+ - H^-$ | $Si^+ - Cl^-$ | $C^- - H^+$ | $C^+ - Cl^-$ |
| $B_2H_6$ | I | I | D | I |
|  | I[b] | I[b] | D[c] |  |
| $PH_3$ | D | D | I | D |
|  | D[b] |  |  |  |
| $AsH_3$ | D | D | I | D |
|  | D[b] |  |  |  |

[a] I and D indicate an increase and decrease, respectively, in deposition rate. The first row for each dopant gas is for the predicted effects, the second row for the observed effects.
[b] Data from Ref. *9*.
[c] Data from Ref. *10*.

the aluminum initially deposited onto the quartz surface is carried along during deposition and concentrated on top of the silicon layer. Figure 11 shows a similar deposition at 600°C on a graphite surface. Aluminum is apparently distributed uniformly throughout the silicon layer and even diffused into the graphite substrate. A clear silicon–graphite interface is formed with little interdiffusion, and no SiC formation is observable in the Auger spectrum. The depositions reported in Refs. *6* and *7* were done using silicon evaporation. Further work will show whether this technique can be used at all during silicon vapor deposition from silicon.

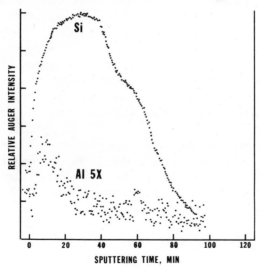

*Figure 10. Auger depth profile of a silicon film deposited at 600°C onto a thin aluminum layer on a quartz substrate*

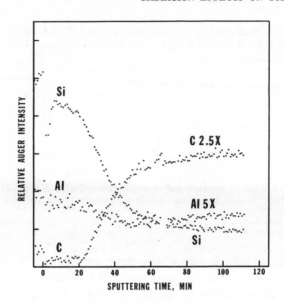

*Figure 11. Auger depth profile of a silicon film deposited at 600°C onto a graphite substrate covered with a thin aluminum layer*

**Grain Boundary Effects.** It is likely that even under the best condition, thin films will contain grain boundaries, although probably substantially fewer than today. To assess their influence on the electrical performance and the time-stability of thin films, one needs to know the distribution of impurities at the grain boundaries. Experiments in our laboratory to determine the impurity distribution by Auger spectroscopy of surfaces created by fracture along grain boundaries have not been successful thus far because silicon cleaves easily. There is, however, indirect evidence that at least doping impurities are to be found in high concentration along grain boundaries of films formed by chemical vapor deposition. Figure 12 (8) shows that while there is an almost linear relationship between the donor concentration in the solid and the dopant concentration during chemical vapor deposition, the donor concentration in polycrystalline films falls abruptly by five orders of magnitude as the dopant concentration is reduced below one part in 10,000. The authors show then (8), that the total concentration in the chemical vapor:

$$N_D^T = \text{phosphorous/silicon } 0.35 \times 10^{22} \qquad (1)$$

and, starting from the assumption that all grain boundaries are covered with dopants before any dopant is incorporated into the grain, that the electrically active dopant concentration:

$$N_D = N_{DT} - N_{GB} \qquad (2)$$

can be fitted nicely to a relationship (Figure 13):

$$N_D = N_{DT} - 4.05 \times 10^{20} S \qquad (3)$$

where $S$ ($S = 0.024$) is the fraction of the surface atoms covered or associated with an electrically inactive dopant atom. An average grain size of cubic shape is assumed to define the number of surface atoms.

**Surface States.** To analyze surface and interface states associated with all boundries in cells, polycrystalline or single crystal, we use electron loss spectroscopy, which, in conjunction with uv spectroscopy, allows definition of empty and filled surface states.

Figure 14 shows the loss spectra of the silicon ($111 - 2 \times 1$) surface as a function of oxygen exposure. On the clean surface, bulk and surface plasmons were observed at loss energies of 16.5 eV and 10.6 eV, respectively. The 5-eV transition is possibly a bulk transition. Three surface

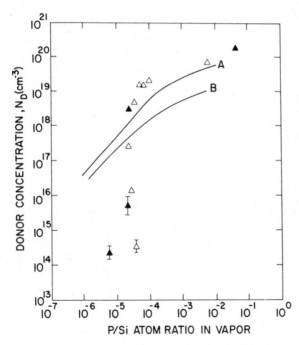

*Figure 12.   Donor concentration, $N_D$ vs. P/Si atom ratio in vapor for poly Si samples deposited on thermal SiO₃ substrates. Deposited at 650°C from SiH₄ ($\triangle$), deposited at 840°C from SiBr₄–H₂ ($\blacktriangle$). Phosphorus-doped, single-crystal silicon.*

state transitions were observed, $S_1$ at 2.5 eV, $S_2$ at 7.4 eV, and $S_3$ at 14.5 eV. The $S_0$ transition at 0.6 eV, as seen in high resolution energy loss spectroscopy, is lost in the elastic background. However, according to recent theoretical calculations, the $S_0$ and $S_1$ transitions have similar initial states and are therefore directly correlated. By exposing the surface to oxygen, the $S_1$ and $S_3$ transitions are effectively removed at a coverage of 0.1 monolayer while $S_2$ increases. At the same time, a loss peak at 3.3 eV appears and becomes well established at monolayer coverage. This transition lies far below the bandgap and is not important to the electrical performance of Si cells.

It is apparent that interfaces introduce a multitude of electronic states and that these can be controlled by adsorption or segregation of suitable impurities onto such interfaces. We are continuing to investigate the effect of impurities likely to be found in practical cells onto electronic

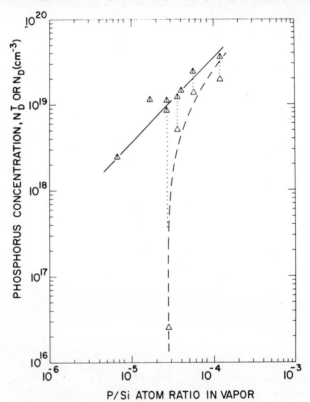

Figure 13. Total phosphorus concentration, $N_D{}^T$, and donor concentration, $N_D$, vs. P/Si atom ratio in vapor for poly Si deposited from $SiH_4$ at 650°C. Total phosphorus concentration (△), donor concentrations (△)— see Equation 1, calculated from Equation 3.

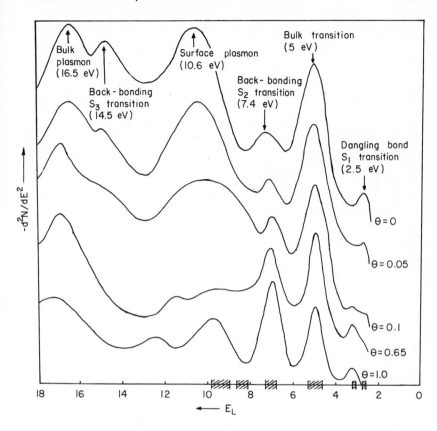

*Figure 14.    Effect of oxygen adsorption on the strength and position of energy loss peaks as a function of θ, oxygen coverage. (θ=1 represents one monolayer). The most effective trapping state $S_1$ (tied to $S_0$) close to the top of the valence band is clearly eliminated by oxygen adsorption. The abscissa gives the loss energy in eV, measured from the elastic peak, and the shaded bands are absorption bands as measured by uv transmission spectroscopy of $S_1O$ molecules.*

interface states. The following interface problems are encountered in thin film cells.

(1) Catalysis of chemical vapor deposition

(2) Catalysis of crystallization

(3) Impurity segregation on surfaces, interfaces, and grain boundaries

(4) Control of surface and interface states

(5) Control of interface diffusion

Research in our and other laboratories is directed toward solving or circumventing them.

## Literature Cited

1. Wolf, Martin, IEEE Photovoltaic Specialist Conf. 9th, 1972.
2. *J. Appl. Phys.* (1975) **46**, 3402.
3. Farrow, R. F. C., *J. Electrochem. Soc., Solid-State Sci. Technol.* (1974) **121** (7), 889.
4. Eversteyn, F. C., Put, B. H., *J. Electrochem. Soc., Solid-State Sci. Technol.* (1973) **120** (1), 107.
5. Chang, Chin-An, *J. Electrochem. Soc.* (1976) **123** (8), 1245.
6. Chang, Chin-An, Siekhaus, W. J., Kaminsky, T., *Appl. Phys. Lett.* (1975) **26** (4), 178.
7. Chang, Chin-An, Siekhaus, W. J., *Appl. Phys. Lett.* (1976) **29** (3), 208.
8. Cowher, M. E., Sedgwick, T. O., *J. Electrochem. Soc., Solid-State Sci. Technol.* (1972) **11**, 1565.
9. Cotton, F. A., Wilkinson, G., "Advanced Inorganic Chemistry," p. 466, John Wiley & Sons, New York, 1966.
10. Poferl, D. J., Carducer, N. C., Augus, J. C., *J. Appl. Phys.* (1973) **44**, 1428.

RECEIVED January 5, 1976. Work supported by U.S. Energy Research and Development Administration.

# Ion Microprobe Studies of Surface Effects of Materials Related to Fission and Fusion Reactors

CARL E. JOHNSON and DAVID V. STEIDL

Chemical Engineering Division, Argonne National Laboratory,
9700 Cass Ave., Argonne, Ill. 60439

*Secondary ion mass spectrometry performed with an ion microprobe mass analyzer has been used to investigate surface chemical problems related to fission and fusion reactors. Ion microprobe analysis of nuclear reactor fuel elements has shown that oxygen and fission products are involved in the corrosive attack of fuel cladding, and these data have helped to establish conditions for optimum fuel performance. For fusion reactors, the containment of tritium is a principal criterion in reactor design. Ion microprobe analysis has contributed to a better understanding of the relation between the permeability of containment materials to hydrogen isotopes and the impurities present on the surface of these materials.*

Growing interest in the detailed characterization of materials and in fundamental surface studies has stimulated the development of a variety of new analytical techniques. Among the most promising of these is secondary ion mass spectrometry (SIMS), which is performed with an ion microprobe mass analyzer (IMMA). The SIMS technique can provide surface (lateral) and in-depth analysis of the concentrations and distributions of elements and/or isotopes in solids with excellent resolution. Analysis is based on the emission and subsequent mass analysis of secondary ions ejected from a sample when it is bombarded with a high energy ionic beam. Most elements can be detected with a sensitivity in the ppm range, and for some elements the sensitivity can extend even to the ppb range. However, the rare gases present a most difficult challenge to the analyst, and high sensitivity for the elements gold and carbon can be achieved only by bombarding the respective samples with cesium and nitrogen ion beams.

Ion bombardment of a solid can produce secondary ions by two different processes—kinetic (1, 2) and chemical (3, 4, 5, 6). When a primary ion of $\sim$ 20 keV energy strikes the surface layers of a material, it transfers kinetic energy to the atoms in the lattice, initiating a collision cascade. This interaction results in the ejection of sample atoms from the surface and their excitation to metastable and ionized states. Any unbound electrons in the excitation volume (for example, from conduction bands) will have a much higher velocity than the solid state ions, causing most of these ions to be neutralized before they can escape into the vacuum of the sample chamber. However, an atom may escape from the sample surface as a neutral particle while maintaining an excited state. Such a metastable atom can eject an Auger electron in the vacuum above the sample and become ionized. The kinetic ionization process predominates in the bombardment of metals and some semiconductors with ions of inert gases.

The chemical ionization process depends on one or more chemically reactive species being present in the sample to reduce the number of conduction band electrons available for neutralization of the ions produced at the surface of the solid. With an increase in the surface work function of the sample, more positive ions are produced at the sample surface and thus become available for mass analysis. It is assumed that the kinetic ionization process continues concurrently with the chemical process. However yields of secondary positive ions are significantly higher when chemical ionization is taking place than when kinetic ionization alone is occurring. In the chemical ionization process, most ions that are found in the vacuum were originally present as atoms in the outer 10–20 Å of the sample.

Two techniques have been developed to take advantage of the higher ion yields produced in the presence of chemically reactive species. In one, Guenot (3) intentionally introduced oxygen into the sample chamber. With this technique, an oxide layer is continually produced on the sample surface, and chemical ionization predominates. The advantages of the other technique—bombarding with a chemically active gas (principally oxygen) to achieve similar results—has been demonstrated by Andersen (4, 5). Bombardment with a chemically active ion controls the sample surface chemistry and forces the predominance of the chemical ionization process. Andersen's experiments with aluminum (4, 5) demonstrate the two phenomena dramatically. When aluminum is bombarded with $Ar^+$ ions, the positive $Al^+$ ion yield decreases exponentially with time (see Figure 1) (4, 5). The ability to extract ions from the sample is progressively lost with time, probably because of the destruction of the oxide layer which exists on the sample surface. The final level of $Al^+$ ion output under $^{90}Ar^+$ bombardment is interpreted to represent a dynamic equilibrium between the arrival rate of the bombarding gas at the sample surface and the arrival rate of reactive gas molecules from the ambient gas.

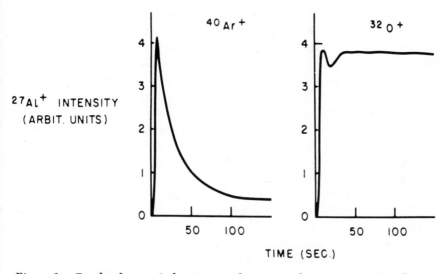

*Figure 1.    Bombardment of aluminum with argon and oxygen ions. Acceleration potential, 11 kV; sample current, $4 \times 10^{-9}$ A; probe size, 20 $\mu^2$.*

As can also be seen in Figure 1, when an aluminum sample is bombarded with $^{32}O_2^+$ ions, a result indicative of the chemical process is obtained. In this case, the preliminary rise in $Al^+$ ion output is interpreted to indicate the removal of an oxygen-rich layer on the surface. Its removal rate directly depends on the current density of the bombarding ion beam. After a slight drop, the $Al^+$ ion signal recovers to maintain a high, stable ion output. Under oxygen ion bombardment, it appears possible to maintain appropriate surface conditions so that positive ion formation is significantly enhanced. As is pointed out in subsequent discussion, this kinetic and chemical relationship can be used to great advantage to examine material that is suspected to have undergone a corrosive oxidative reaction.

The SIMS technique has excellent capability for determining the concentration of a given element of interest as a function of depth in a sample. McHugh (6) has studied in great detail the capability of the ion microprobe to resolve a 50-Å $^{31}P$ layer located some 230 Å below the surface of a 1050-Å thick $Ta_2O_5$ film. The resolution of the experimentally determined depth (concentration) profiles was significantly influenced by the primary ion beam kinetic energy. This readily predicted result is

caused by atomic displacement cascades in the sample under the impact
of the primary ion beam. McHugh showed that the best results were
obtained with a 1.75-keV $^{16}O^+$ ion bombardment ($O_2^+$ at 3.5 keV); the
$^{31}P$ peak centered about 230 Å but had a FWHM (full width half maximum) of 78 Å. The spread in the peak, which exceeded the expected
50 Å, resulted from atomic mixing as the primary ion beam etched away
the sample. Another technique that could be used to improve the peak
resolution would be to increase the angle of incidence that the primary ion
beam makes with the surface of the sample. The oblique angle of incidence would reduce the mean penetration depth of the particle and would
give results equal to reducing the primary ion beam energy.

Using the IMMA, we have recently analyzed the depth profile of
$^{58}Ni$ in a sample of aluminum. In this experiment, the two samples were
bombarded with $^{58}Ni$ accelerated to 4 MeV, one with a beam current
that was three times greater than the other. The predicted penetration
depth (7) was calculated to be 2.08 $\mu$m. Both samples were analyzed
using a 20-keV $^{28}N_2^+$ primary ion beam. Excellent results were obtained
with one sample showing a maximum in the $^{58}Ni^+$ intensity at 2.17 $\mu$m
and the other at 2.12 $\mu$m. The crater depth was generally determined
using interference microscopy techniques. However, when surface roughness of the sample crater became too severe, light microscopy coupled
with direct measurement proved very effective. The FWHM was 0.65
$\mu$m and 0.60 $\mu$m, respectively. Integration of the area under each peak
showed a relationship of 3/1.

### Instrumentation

The general concept of producing a spatially resolved mass analysis
of the surface of a solid by bombarding with a beam of primary ions and
analyzing the secondary ions sputtered from the sample was introduced
by Castaing and Slodzian (8). Their efforts resulted in the production of
an ion emission microscope. More recently, Liebl (8) designed an ion
microprobe that has been more fully developed by Andersen and Hinthorne (10).

Figure 2 is a schematic of the instrument used at Argonne. The ions
for sample bombardment are generated in a hollow cathode duoplasmatron ion source which can produce a variety of gaseous ions. The ions,
which are either positive or negative, are first accelerated to energies of
5–22.5 kV. After being accelerated, the ions are passed through a primary
mass spectrometer, which allows the experimenter to select a specific
ionic species for the bombardment of the sample being analyzed. This
beam is focused to form a small probe-like beam by means of an electrostatic lens column consisting of a condenser and an objective lens. The

ion probe, which can be varied in diameter from about 2 to 300 mμ, is allowed to impinge on the sample and can be viewed during bombardment through an optical microscope.

The sputtered ions from the sample surface are collected and mass analyzed by a double-focusing mass spectrometer that is similar to the Mattauch–Herzog type inasmuch as the velocity dispersions of the mag-

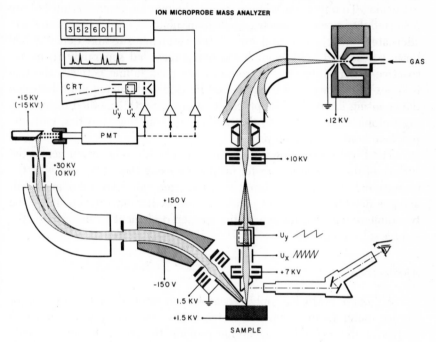

*Figure 2.   Schematic for the Applied Research Laboratories ion microprobe mass analyzer*

netic and electric sectors are matched so that sputtered ions with a wide range of initial energies may be accommodated. No entrance slit is used, and the bombarded area is stigmatically focused directly onto the resolving slit. A retrofocal lens in front of the spectrometer increases the instrument's solid angle of acceptance.

Both positive and negative sputtered ions can be detected by suitable arrangement of potentials on the conversion electrode and scintillator of a high gain device similar to the Daly type (*11*). At the conversion electrode, secondary electrons are ejected by the sputtered ions; these electrons are accelerated toward a scintillator. The light emitted by the subsequent impact is detected by a photomultiplier tube. The resolved signals can be read either as direct current on chart recorders or as count rates in scalers capable of counting in the megacycle range. In scaler counting, mass discrimination effects are minimized, and the counting efficiency is approximately the same for both light and heavy ions.

The detector assembly allows automatic positioning of isotopes that are within a 15% mass range of each other. Our experience has shown that the precision of the data obtained from isotope ratio measurements is higher than that achievable by simple mass scanning and counting of ion peaks. To obtain the isotopic ratios, the two ion beams of interest are deflected by a set of electric deflection plates so that the beams are alternately switched in and out of the resolving slit and detector 500 times per second, with each beam being recorded separately in its respective scaler. Interference from any ion peak falling between the two isotopes being measured is eliminated by using a blanking pulse between the counting intervals. The precision attained by this method of measuring isotope ratios is close to that defined by counting statistics, with the practical limit being about 0.1%.

The IMMA also allows the primary ion beam to be swept across the surface of the sample in a raster fashion by using the sweep plates of the primary column. The sputtered ions thus generated are mass analyzed, and the detected signals are viewed on an oscilloscope, which shows the two-dimensional distribution of the sputtered element in the area of the sample selected for analysis.

### Applications

Although SIMS has only recently been applied to the analysis of surface materials in fission and fusion reactor systems, the unique capabilities of this analytical technique provide the means to solve a wide variety of problems. We have selected only a few applications to illustrate these capabilities. Presented later are detailed discussions of studies devoted to understanding the mechanisms of cladding attack in fission–reactor fuel elements and investigations of the surface characteristics that limit permeation of hydrogen through materials of construction for fusion reactor systems. In addition, highlights of several on-going studies in both of these areas are presented below.

One of the suggested mechanisms of fission–reactor fuel cladding attack emphasizes the strong influence that carbon can exert on the chromium concentration at grain boundaries in the stainless steel cladding (i.e., in changing the surface composition of grains within the bulk stainless steel). It is well established (12) that austenitic stainless steels become more susceptible to oxidation when the chromium concentration drops below 12%. If cladding attack is influenced by such a mechanism, it is expected that carbide precipitates (most likely chromium carbide) will reside at grain boundaries within the cladding. Studies of fractured cladding surfaces are under way to identify any carbide aggregates adhering to grain boundary surfaces. Although these studies will be qualitative, owing to roughness of the fractured surface, the isolation and identification of any such precipitates will significantly help to unravel the complex mechanism of cladding attack.

In studies of fusion reactor systems, assessment of the stability of first-wall candidate materials to the lithium metal–plasma environment is of considerable interest. A few years ago, Brehm et al. (*13*) reported that niobium containing 300–500 ppm oxygen exhibited corrosive attack at the solid–liquid lithium interface and preferentially at the niobium grain boundaries. Because this metal is a prime candidate for the reactor first wall, the conditions under which such a reaction may limit the usefulness of this material must be understood. In the fusion reactor, lithium wets the first wall and may enter the metal by a recoil reaction resulting from impingement of the plasma on the lithium-wetted surface. In our work, the extent of penetration of lithium into niobium by recoil is studied by means of the $^{10}B(n,\alpha)^{7}Li$ reaction. The boron is vacuum sputtered onto a niobium plate and then bombarded by thermal neutrons. The SIMS technique is used to identify the presence of lithium, and depth-profiling techniques are used to determine the penetration depth of the recoil reaction. A knowledge of the kinetics of this reaction allows assessment of the significance of the reaction at reactor operating conditions. With this as background, attention is now focused on the more detailed, complete studies.

**Cladding Attack in Fission Reactor Fuel.** In studies of irradiated mixed oxide fuels, chemical reactions that occur during fissioning of the fuel are being identified by postirradiation examination of the fuel using ion microprobe mass analysis. Many techniques have been used in efforts to elucidate the behavior of fission products in irradiated fuels, but none have been as successful as ion microprobe analysis. The ion microprobe mass analyzer has contributed significantly to our knowledge of the effects of irradiation on nuclear fuel. This is caused mainly by the relatively high sensitivity of the instrument, with detection sensitivity to the ppb level in certain cases. This high detection sensitivity is especially important for the low mass range ($7 \leq 11$), where other standard techniques, such as electron microprobe, either have poor sensitivity or do not work. Further, since the ion microprobe collects isotopic data, it has provided a direct means for differentiating between fission products and structural materials.

In our studies of irradiated fuels (*14, 15*), particular emphasis has been placed on the chemical reactions and redistributions of fuel constituents that could contribute to corrosive attack of the stainless steel cladding. Oxidation of the cladding caused by redistribution of oxygen within the fuel and interaction of the cladding with certain highly reactive fission products such as cesium are examples of these processes. Figure 3 shows the radial distribution of fission-product cesium in a fast reactor fuel.

Postirradiation examination of fast reactor fuel elements has frequently shown extensive intergranular attack of the inner wall of the stainless steel cladding along with less serious uniform oxidation of the inner wall. Since the lifetime of fuel elements may be limited by the intergranular attack, it is important to understand the mechanism and parameters that control the reaction. A section of cladding from a urania-plutonia fuel irradiated to a 5 at. % burn-up at 20 kW/ft was analyzed

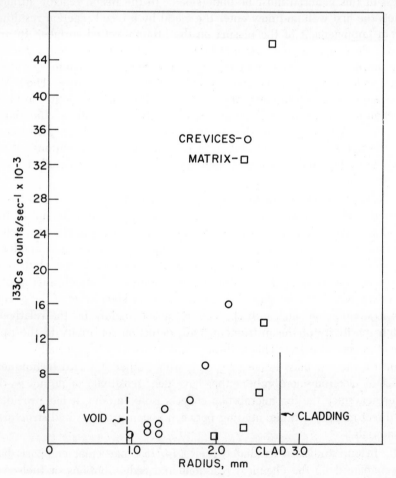

*Figure 3. Radial distribution of $^{133}Cs$ in mixed oxide fast reactor fuels*

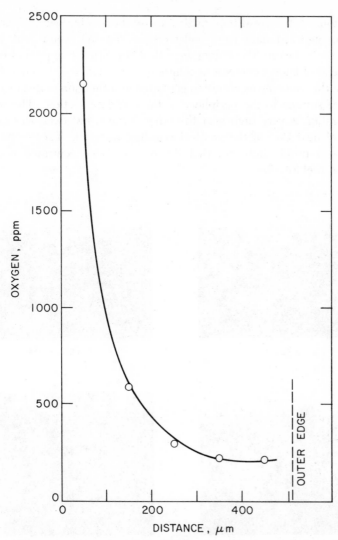

OXYGEN, ppm

DISTANCE, $\mu$m

OUTER EDGE

*Figure 4. Oxygen profile in irradiated stainless steel cladding.*
$^{28}N_2^+$ *at 4.6 × 10⁻⁹ A with 10.5-μm beam diameter.*

for oxygen, using the IMMA. A section of this cladding had been analyzed previously for oxygen by vacuum fusion analysis and had contained
3000 ppm oxygen. Archival samples (unirradiated) of this cladding
contain about 100 ppm oxygen. The ion microprobe analysis for oxygen
was carried out using a $^{28}N_2^+$ ion beam. The beam current was 4.6 ×
10⁻⁹ A, and the beam diameter was 10.5 μm. Each sample location was
first thoroughly cleaned by etching the surface with the nitrogen ion
beam to ensure that the area being analyzed was free of surface con-

tamination. After consistent analyses had been obtained for a given location, digitized data were collected for the $^{16}O^-$ mass peak in the secondary ion beam. These data were then converted into ppm of oxygen using steels of known contents as standards. The data are shown in Figure 4, where the oxygen concentration is plotted as a function of distance from the inner surface to the periphery of the cladding section. The oxygen concentration is very high near the inner surface, but decreases quickly to a level near that of the archival cladding material. The steepness of the oxygen profile indicates that the oxygen is concentrated near the cladding–fuel interface.

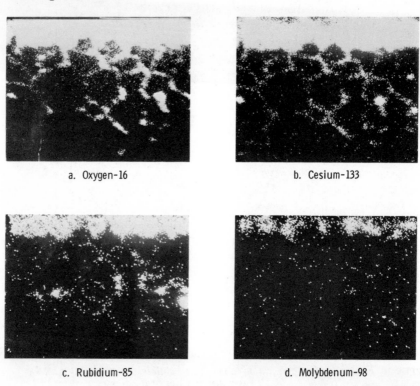

a. Oxygen-16                                          b. Cesium-133

c. Rubidium-85                                        d. Molybdenum-98

*Figure 5. Ion microprobe scanning images of $^{16}$oxygen, $^{133}$cesium, $^{85}$rubidium, and $^{98}$molybdenum at urania–plutonia fuel–stainless steel cladding interface. All images 80 μm × 100 μm.*

Figure 5 shows ion microprobe scanning images of the distribution of $^{16}O$, $^{85}Rb$, $^{133}Cs$, and $^{98}Mo$ in the cladding at the fuel–cladding interface. In each image, the degree of whiteness increases with ion concentration. Examination of the oxygen scanning image shows a very high concentration near the interface. This layer is probably composed of chromium and iron oxides. The center of the first large grain near the interface, shown at the top center of Figure 5a, is the location of the first sample point in Figure 4 (50 $\mu$m from the inner edge of the cladding). Most of the oxygen found by vacuum fusion analysis (3000 ppm) is located in a thin layer on the inner surface of the cladding. Calculations have shown that complete oxidation to chromic oxide of the chromium in a 1-$\mu$m thick layer of cladding would contribute 7800 ppm oxygen to the cladding. Close inspection of the image for oxygen shows that the degree of whiteness decreases with increasing distance from the interface. Thus, this qualitative pictorial information confirms the distribution shown in Figure 4.

As is discussed above, the primary ion beams selected for a particular analysis can contribute significantly to elemental sensitivity. It has been well established that bombarding with reactive gases can provide significant signal enhancement, thereby increasing detection sensitivity. This feature can be used to advantage in studies where oxidation or nitridation may be of concern in assessing material performance.

Ion-scanning images were obtained for a second area that showed no intergranular attack but was suspect on the basis of earlier nondestructive analysis results. Figure 6 shows ion images for $^{12}$C, $^{52}$Cr, $^{55}$Mn, $^{56}$Fe, and $^{58}$Ni. In this figure, the cladding makes up most of the upper portion of the ion images; the outer edge of the cladding is near the bottom of each image. In all the ion-scanning images shown in Figure 6, there is close correspondence in the location of each of the isotopes. The enhancement of the images suggests the presence of a chemically active material. Because the primary ion beam used in these investigations was argon, which is chemically inactive, the observed signal enhancement is believed to be caused by an oxidative reaction that had occurred earlier at this location. Subsequent analysis proved oxygen to be present in these locations, suggesting that the formation of ferrite, a complex iron–chromium oxide containing impurity levels of nickel and manganese could account for the enhancement. Examination of cladding by this technique provides the opportunity to identify areas where attack is just beginning. This is extremely important if the mechanism of this corrosive reaction is to be unraveled.

**Hydrogen Permeation in Fusion Reactor Structural Material.** The distribution and migration characteristics of tritium in a reactor system are of concern to those working on the design of controlled thermonuclear reactors. To ensure containment of tritium (hydrogen) gas, its solubility and diffusion characteristics through structural materials must be determined. Although solution and diffusion are largely bulk phenomena, diffusion often strongly depends on the properties of and interaction at surfaces.

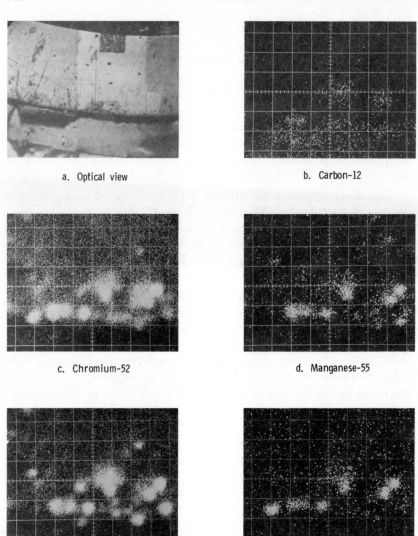

a. Optical view

b. Carbon-12

c. Chromium-52

d. Manganese-55

e. Iron-56

f. Nickel-58

*Figure 6.   Ion microprobe scanning images of outer edge of type 316 stainless steel fuel cladding—Area 1.   All images $100 \times 126$ μm.   Original magnification: a, 110×; b–f, 548×.*

Permeation of a gas through a solid depends on the adsorption of and desorption from the metal surface. The rates of adsorption and release of a species from a surface strongly depend upon the number of active sites available at the surface. Hence, the rate of any process occurring as a result of adsorption or release can be expected to depend on the number of such sites. In terms of the behavior of hydrogen isotopes in fusion devices, it is reasonable to assume that some of the impurity species likely to accompany hydrogen, such as O, N, and C either can permanently lodge on active sites, making these sites unavailable to hydrogen, or can compete with hydrogen for the active sites, reducing the probability of occupation by hydrogen.

A frequently observed consequence of this kind of passivation is the reduced permeability of metals with surface layers of chemically bound impurities. Earlier experimental studies (*16*) showed that to achieve maximum permeability of hydrogen through vanadium, the metal had to be activated before each run. In that work, the vanadium foils were activated by heating to 850°C in vacuum, followed by cooling to 450°C. Permeability measurements were then made at 450°–550°C. A least squares refinement of the data of Heinrich et al. (*14*) is plotted in the upper region of Figure 7.

More recently (*17*), a study was initiated to measure the hydrogen permeability of high purity vanadium that had not been subjected to the vacuum baking procedure. These studies have shown that impurity layers on vanadium can reduce its hydrogen permeability by more than three orders of magnitude. A least squares analysis of the data collected for untreated (unactivated) vanadium samples at 350°–550°C is given in the lower half of Figure 7. With further heating above 600°C, there is a rapid increase in the hydrogen permeability, such that at 780°C the measured rate is more than one order of magnitude greater than the rate predicted by the curve for unactivated vanadium. In all cases, the vanadium samples became deactivated unless they were periodically baked at 850°C. In general, the lower curve in Figure 7 has been found to represent samples that are fully deactivated. This implies that a uniform, relatively thin layer of impurities is probably responsible for the reduced permeability.

Ion microprobe analyses were made of as-received vanadium samples to identify the elemental constituents of the impurity layer. The positive ion spectra shown in the upper scan of Figure 8 were obtained from samples of the high purity, as-received vanadium, and the lower scan in Figure 8 was taken after the same material had been subjected to permeation measurements at high temperatures. Clearly, the surface of the as-received sample had picked up a significant quantity of oxygen, as evidenced by the appearance of the $^{16}O^+$ signal and the very large increase in the $^{67}VO^+$ signal. The change in the surface characteristics undoubtedly arose from impurities on the surface while the sample was in the permeation apparatus. Traces of $^{65}(VN)^+$ and $^{63}(VC)^+$ are also observed in the spectra, but the most prominent new mass peaks are clearly from oxygen-containing species. All samples were mounted on an aluminum substrate;

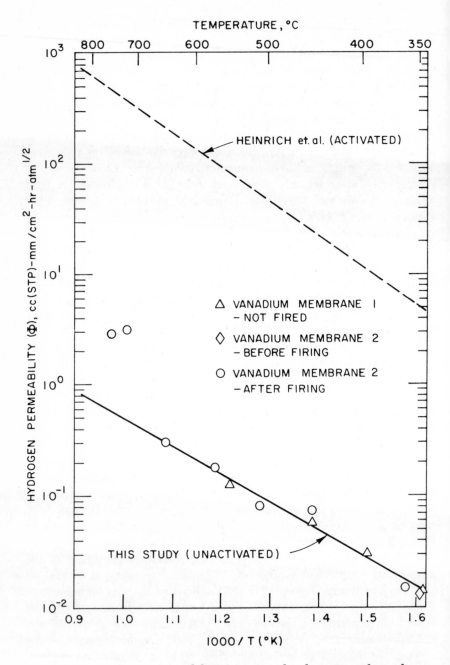

*Figure 7.   Hydrogen permeabilities of activated and unactivated vanadium*

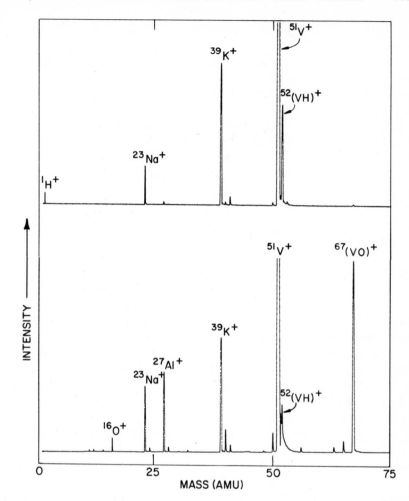

*Figure 8. Ion microprobe spectra of as-received vanadium (upper trace) and a vanadium sample that had been used for permeation measurements (lower trace)*

as a result, a $^{27}Al^+$ signal was frequently observed, the intensity of which was independent of the history of the sample and which is believed to be the result of contamination from the substrate.

Ion microprobe data as a function of depth in the sample for the two vanadium specimens of different histories are given in Table I. For the sputtering parameters used in these studies, the time in seconds approximately corresponds to the number of atom layers removed. The mass peak intensities at each time were normalized to a value of 100 for the $^{50}V^+$ signal (natural isotopic abundance of $^{50}V \simeq 0.25\%$), the assumption being that $^{50}V$ is uniformly distributed throughout the bulk of the vanadium.

For the as-received vanadium (sample I), the concentrations of the oxygen-containing species (O and VO) decrease rapidly in the first several-thousand atom layers, whereas hydrogen appears to be rather uni-

**Table I.  Ion Microprobe Mass Analysis Data for Two Vanadium Samples**

| Sample | Time (sec) | Mass Signal (arbitrary units) | | | | |
|---|---|---|---|---|---|---|
| | | $^1H^+$ | $^{16}O^+$ | $^{28}Si^+$ | $^{50}V^+$ | $^{67}(VO)^+$ |
| I | 0 | 28 | 28 | — | (100) | 101 |
| | 900 | 28 | 7 | — | (100) | 61 |
| | 2100 | 29 | < 2 | — | (100) | 35 |
| II | 0 | 17 | 57 | — | (100) | — |
| | 315 | 16 | 38 | 116 | (100) | — |
| | 1535 | 72 | 32 | 116 | (100) | — |
| | 2030 | 96 | 28 | 124 | (100) | — |

formly distributed. This is to be expected for a high purity material that could have suffered some surface oxidation prior to its use. However, because the material was maintained under ambient conditions, the oxygen did not have a chance to migrate very far into the bulk. Hydrogen, on the other hand, having a much greater mobility in metal lattices under ambient conditions, would be expected to reach a uniform distribution in the bulk material over a long time.

Sample II, which had been subjected to continuous permeation measurements for several weeks at up to 550°C, was cooled overnight to room temperature with continuous pumping at a pressure of $\leq 10^{-4}$ torr prior to probe examination. The ion microprobe data for this sample (given in Table I) also appear to reflect its laboratory history. The oxygen level again decreases with depth in the metal, but not nearly as rapidly as for sample I, probably because the high temperature treatment resulted in greatly accelerated oxygen diffusion. The hydrogen level is depleted at the surface relative to the hydrogen level in the bulk metal, indicating that on cool-down from 550°C under vacuum, a concentration gradient developed between the surface and bulk vanadium. At 550°C, the vanadium lattice is much expanded, thus allowing an increase in the solubility and permeability of hydrogen in vanadium. At ambient temperatures, the loss of hydrogen from the surface, owing to the vacuum, can easily be visualized. However, at ambient temperature, the diffusion of hydrogen through the bulk vanadium would be very slow, and thus a large hydrogen concentration gradient would be established between the surface and bulk vanadium.

Two other features of the data in Table I should be noted. First, silicon is a common impurity in vanadium-rich ores. Therefore, it would be expected to be uniformly distributed throughout refined vanadium metal. The results for sample II and for all other samples where the $^{28}Si^+$ signal was on scale confirm this uniform distribution and give credence to the use of the $^{50}V^+$ signal as an integral intensity standard. Second, the relative intensities of the $^{16}O^+$ and $^{67}(VO)^+$ peaks are observed to vary with penetration depth. This may be a consequence of changes in the chemical state of oxygen as a function of depth in the sample, but it may also result from atomic and molecular recombinations occurring at the metal surface.

## Literature Cited

1. Slodzian, G., Hennequin, J. F., *CR Acad. Sci.* (1966) **263B**, 1246.
2. Castaing, R., Hennequin, J. F., *Adv. Mass Spectrom.* (1971) **5**, 419.
3. Guenot, D., Diplôme d'Etudes Supérieures, Orsay, 1966.
4. Andersen, C. A., *Int. J. Mass Spectrom. Ion Phys.* (1969) **2**, 61.
5. *Ibid.* (1970) **3**, 413.
6. McHugh, J. A., *Radiat. Eff.* (1974) **21**, 209.
7. Lindhard, J., Scharff, M., Schiott, H. E., *K. Dan. Vidensk., Selsk., Mat.-Fys. Medd.* (1963) **33**, 14.
8. Castaing, R., Slodzian, G., *J. Micros. Paris* (1962) **1**, 395.
9. Liebl, H., *J. Appl. Phys.* (1967) **38**, 5277.
10. Andersen, C. A., Hinthorne, J. R., *Science* (1972) **175**, 853.
11. Daly, N. R., *Rev. Sci. Instrum.* (1960) **31**, 264.
12. Seybolt, A. U., *J. Electrochem. Soc.* (1960) **107**, 147.
13. Brehm, Jr., W. F., Gregg, J. L., Li, Che Yu, *Trans. Metall. Soc. AIME* (1968) **242**, 1205.
14. Johnson, C. E., Johnson, I., Blackburn, P. E., Crouthamel, C. E., *React. Technol.* (1972–73) **15**, 303.
15. Johnson, I., Johnson, C. E., Crouthamel, C. E., Seils, C. A., *J. Nucl. Mater.* (1973) **48**, 21.
16. Heinrich, R. R., Johnson, C. E., Crouthamel, C. E., *J. Electrochem. Soc.* (1965) **112**, 1071.
17. Maroni, V. A., *J. Nucl. Mater.* (1974) **53**, 293.

RECEIVED January 5, 1976.

# 18

# Surface Effects on Tritium Diffusion in Materials in a Radiation Environment

G. R. CASKEY, JR.

Savannah River Laboratory, E. I. du Pont de Nemours and Co., Aiken, S.C. 29801

*Tritium transport and distribution in a material are controlled by chemical potential gradients, thermal gradients, and cross-coupling to impurities and defects. Surfaces influence tritium diffusion by acting as sources and sinks for defects and impurities. Surface films restrict tritium transfer between the solid and surrounding fluids. Radiation directly affects boundary processes such as dissociation or adsorption, may erode a surface film or the surface itself, and introduces defects and impurities into the solid by radiation damage, transmutation, or ion implantation, thereby modifying tritium transport within the solid and its transfer across external interfaces. There have been no definitive investigations of these effects, but their practical significance has been demonstrated in tritium release or absorption studies with stainless steel, Zircaloy, niobium, and other materials.*

Current interest in tritium migration in solids arises from technical, economic, and environmental considerations associated with operating fission reactor plants and projected fusion reactor systems. In the first case, the interest is almost entirely caused by the possible environmental contamination by tritium leakage to plant effluents (*1, 2*). Small quantities of tritium are generated by both ternary fission and reaction of neutrons with boron and lithium. There are many ways that this tritium may eventually escape to the environment; for example, through permeation through the steam generator system. Fusion reactors, on the other hand, are designed to process continuously large (kilogram) quantities of tritium both in the feed and exhaust from the deuterium–tritium (D–T) plasma and in the tritium breeding, extraction, and purification systems (*3, 4, 5, 6, 7*). Efficient containment and recovery are essential in this case. Potential problems associated with fusion systems have been described in detail (*4, 5, 6*), and several alternative approaches to minimize tritium release have been advanced (*4*). In all cases, knowledge of the diffusion and permeation characteristics of tritium in structural materials of principal interest is helpful in analyzing containment, breeding, and handling problems. Potential materials include: stainless steel; alloys of niobium, vanadium, or molybdenum; Zircaloy-2; silicon carbide; beryllium oxide; alumina; oxides commonly formed as surface films on metals; tritides, such as $TiT_2$ which may be used in targets for neutron generation (*8*) and in extraction systems; and various grades of steels for steam-generating systems. A small amount of experimental data is available on tritium diffusion in these materials. There is much more information for protium and deuterium (*9, 10, 11, 12*) which is applicable to tritium, if the data are corrected for the effects of isotope mass.

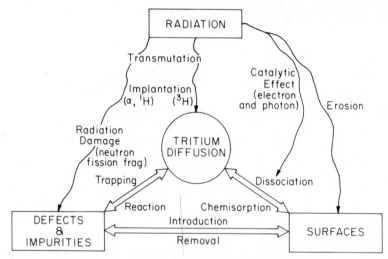

*Figure 1. Diagram of interrelations among tritium diffusion, surface effects, and radiation*

Tritium diffusion occurs in response to gradients in chemical potential and temperature in solids and is influenced by interaction with defects, impurities, and with surfaces. These interactions may involve:

(1) Boundary processes such as dissociation and chemisorption.

(2) External surfaces (such as those between two solids or between a solid and a fluid) and internal surfaces (such as grain boundaries) which act as sources and sinks for defects and impurities.

(3) Surface films, such as oxides on metal surfaces, which may change the rate of tritium transfer between phases.

Potential interrelations among these factors are represented diagrammatically in Figure 1. Several possible effects of radiation on the processes affecting tritium diffusion are indicated, including:

(1) Catalytic effect on dissociation or adsorption.

(2) Erosion of the surface or of surface films.

(3) Increased defect and impurity fluxes to or from the surfaces.

(4) Increased defect and impurity concentrations within the solid.

The effects of these processes on tritium diffusion in solids have not been investigated. Several related problems, however, have been discussed or are under investigation (13). For example, the effect of oxide films on tritium release and permeation in stainless steel (14), trapping effects in niobium (11), and photodesorption (15, 16) have all been reported in some detail. This chapter will focus attention on those aspects of tritium absorption and transport where information is available and where experimental investigation is in progress. We will also try to identify areas where additional investigation is needed.

## Interaction of Tritium Atoms with Defects and Impurities

Tritium diffusion and its ultimate distribution throughout a solid is strongly affected by lattice defects and impurities because of repulsive or attractive interactions such as: trapping of tritium at immobile defects or impurities, diffusion of tritium in response to gradients in the chemical potential of impurities, migration of tritium with moving defects. These processes are related to radiation effects because radiation damage, transmutation, and ion bombardment change the number and distribution of defects and impurities and may implant tritium within the lattice.

**Tritium Solution in Solids.** Dissolved tritium may occupy either octahedral (O) or tetrahedral (T) interstitial positions in metals depending on the crystal lattice of the host structure. Neutron diffraction, ion channeling, and nuclear magnetic resonance techniques have identified the equilibrium interstitial sites in several metals (17, 18, 19, 20). Generally, the T-site is occupied in metals with a body-centered cubic (BCC) lattice, and the O-site is occupied in metals with a face-centered cubic (FCC) or hexagonal close-packed (HCP) lattice. Chromium (BCC) appears to be an exception, as an ion-channeling study indicates that the O-site is occupied (21).

Atomic and molecular solutions of tritium have been observed in nonmetallic solids (22–29), but there is very limited direct evidence on the specific location of tritium in such materials. Generally, interstitial solution is assumed, although substitutional replacement may be anticipated in some cases, as in polymers. Tritium solution and diffusion in silica and silicate glasses is normally molecular (22, 23). However, atomic diffusion of tritium ions has been observed in quartz and fused silica (*24*), and electrolysis apparently introduces atomic hydrogen in vitreous silica (*25*). Tritium solution in polymers is normally molecular as indicated by the direct dependence of permeation rate on the first power of the hydrogen pressure (*26*). In this instance, isotopic exchange of tritium with the hydrogen in the polymer could alter the diffusion kinetics and eventually degrade the polymer by breaking the bonds where replacement tritium decayed. Tritium solution in metal oxides may be atomic in the few cases where it has been studied, e.g., ZnO (*27*) and $TiO_2$ (*28*). However, the pressure dependence of the permeation rate of hydrogen in $ZrO_2$ suggests a molecular solution (*29*).

Tritium diffusivity has been measured in various materials by gas phase charging (*28*), electrolytic charging (*25*), ion bombardment (*24*), and recoil from $^6Li$ (*14*). Results for several types of solid are shown in Figure 2. Generally, tritium diffusivities correlate with protium or deuterium diffusivities by the inverse-square-root-of-mass relation where data exist to make the comparison. Silver and aluminum are apparent exceptions because of effects associated with ion implantation or radiation, as discussed later. Protium and deuterium diffusivities in metals have been summarized previously (*9, 10, 11, 12*). Little work has been done with oxides, but this area is being investigated currently on materials of interest in fusion reactor development (*30*).

**Trapping by Immobile Defects or Impurities.** An attractive interaction between diffusing tritium and lattice defects or impurity atoms is not accounted for in the simple diffusion equation:

$$\partial C/\partial t = \nabla D \nabla C \qquad (1)$$

Therefore, this equation is not applicable under such conditions. More complex formulations are required to incorporate additional processes into the analysis. Formal representation of trapping has been approached in two ways: incorporation of additional terms in the continuum diffusion equation to account for trapping (*31, 32, 33, 34, 35*) and analysis based on the concept of mean-free passage time taken from the theory of stochastic processes (*36*). The former approach has been used extensively but may not be appropriate when considering singularities of atomic dimension. The latter approach takes into account details of the lattice, trap site, and shifting of the saddle point energy caused by the perturbation of the lattice by the defect, but it has not been applied to analysis of hydrogen diffusion experiments.

Trapping sites that have been identified or suspected in metals include: dislocations in type 304L stainless steel, nickel, vanadium, niobium, iron, and molybdenum (*37–42*); interstitial or vacancy clusters

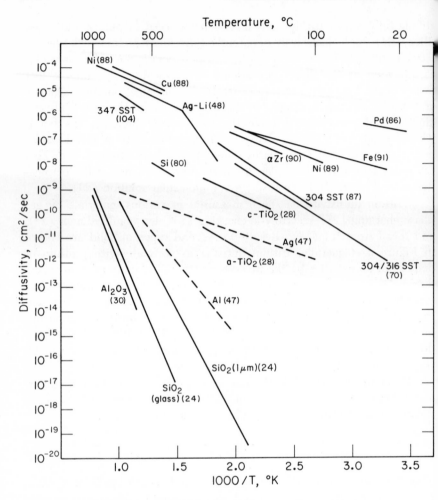

Figure 2.  *Diffusivity of tritium in various materials*

in copper, silver, and gold (*43*); oxygen in niobium and copper (*44, 45*); and voids or gas bubbles in iron, silver–lithium, and aluminum–lithium alloys (*46, 47, 48*). Potential trapping sites in nonmetallic solids include Schottky or Frenkel defects and impurity atoms.

General representation of diffusion with trapping at only one type trap has been developed (*33*) by replacing the simple differential equation with:

$$\partial C/\partial t + N\partial n/\partial t = D\nabla^2 C \tag{2}$$

where $\partial n/\partial t = kC(1-n) + pn$, $N$ is the number of traps per unit volume, $n$ is the fractional occupancy of trap sites, and $k$ and $p$ are capture and release probabilities, respectively. The resulting equations are non-linear and have no general analytic solution. Finite difference methods yielded approximate solutions for boundary and initial conditions corresponding to permeation and evolution from a plane sheet ·(49). Exact solutions have been obtained for certain limiting cases where the equation reduces to a linear form (33, 34, 50). Characteristics of trapping effects on permeation and evolution (Table I) show that: apparent diffusivity is less than the true lattice diffusivity, actual solubility is greater than the lattice solubility, and steady state processes are unaffected.

The significance of trapping on tritium diffusion within a solid depends on the concentrations of both tritium and trapping sites in the solid (33, 34). In the case of a Tokamak fusion reactor with a molten lithium blanket, the effective tritium pressure throughout the blanket region is very low; therefore, tritium concentrations are low. For example, in the University of Wisconsin Tokamak fusion reactor, UWMAK-I (7), the tritium content of the lithium ($10^5$ kg) is estimated at $4.5 \times 10^{-5}$ mole fraction which corresponds to an effective tritium gas pressure of $10^{-4}$ Pa ($10^{-9}$ atm). The quantity of tritium in solution in the large mass ($6 \times 10^6$ kg) of type 316 stainless steel alloy is only $\sim 11000$ cm³. If the void volume arising from radiation damage were 0.1% throughout the structural framework of the blanket, the additional tritium trapped in the voids at $10^{-4}$ Pa ($10^{-9}$ atm) would be $\sim 10^{-3}$ cm³. The effective tritium diffusivity will not be altered significantly by such a small void volume.

**Internal Surfaces.** Internal surfaces, such as grain and subgrain boundaries, may serve as trap sites or act as short circuit diffusion paths. For example, hydrogen solubility in polycrystalline nickel is distinctly greater than in single crystals over a wide temperature range. Further,

**Table I.   Effects of Trapping and Surface Films on Permeation**[a]

|  | Reversible Trapping | Irreversible Trapping | Continuous Film |
|---|---|---|---|
| Steady state permeation rate ($P_\infty$) | $P_\infty = P_\infty$ (ideal) | $P_\infty = P_\infty$ (ideal) | $P_\infty < P_\infty$ (ideal) |
| Apparent diffusivity ($D^*$) | $D^* < D_L$ | $D^* < D_L$ | $D^* < D_L$ |
| Instantaneous permeation rate ($P_t$) | $P_t < P_t$ (ideal) | $P_t < P_t$ (ideal) | $P_t < P_t$ (ideal) |
| Instantaneous evolution rate ($E_t$) | $E_t > E_t$ (ideal) | $E_t = E_t$ (ideal) | $E_t < E_t$ (ideal) |

[a] $D_L =$ true lattice diffusivity.

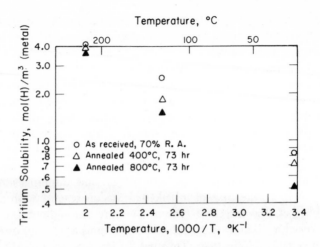

*Figure 3. Effect of metallurgical conditions on the apparent solubility of tritium in nickel*

the solubility in polycrystalline nickel deviates increasingly from the expected temperature dependence as the temperature is lowered below 700 K (*51*). Trapping effects are also observed in tritium solubility measurements for cold-worked and annealed nickel (*52*) between 300 and 500 K (Figure 3).

Tritium has been introduced into silver and aluminum by ion bombardment (*47*) and its diffusivity calculated from release rate measurements. The release rates were slower than expected from calculations based on gas phase charging and evolution measurements. This lowering of apparent diffusivity was attributed to the formation of tritium bubbles. The bubbles were assumed to form at accumulations of point defects or dislocation loops generated during ion bombardment when the lattice is supersaturated with tritium. Release rates were, therefore, controlled by bubble migration at low temperatures or resolution and diffusion at higher temperatures. As shown in Figure 2, apparent tritium diffusivities calculated by this technique are very much lower than for other metals or alloys and are less than calculated from hydrogen diffusivity data (*53*).

Neutron irradiation of Ag–Li and Al–Li alloys generates tritium by an (*n,α*) reaction (*48, 54*). Autoradiographic examination of the aluminum alloys containing 0.4–2.7 wt % lithium revealed gas bubbles along grain boundaries (*54*). These bubbles presumably contained both tritium and helium. Annealing of the irradiated samples apparently led to bubble agglomeration (*54*). After annealing at relatively high temperatures (i.e., 898 K), the Al–2.7 wt % Li alloy was free of bubbles because all the tritium and helium was released during the anneal. The outgassing behavior of the silver alloy showed nonideal diffusion characteristics between 553 and 873 K (*48*). This nonideal diffusion was attributed to trapping of the tritium at point defect clusters and is similar to the results from ion bombardment (*47*).

**Tritium–Dislocation Interaction.** There is much information on the interaction of hydrogen with lattice defects in metals and alloys. The prototypical case is that of α-iron where substantial deviation between the anticipated and measured temperature dependence of the diffusivity is observed at lower temperatures (Figure 4). In this case, the extent of cold work, and hence the dislocation density, substructure, and microvoid formation, appear to play major roles in affecting the measured diffusivity. Steady state permeation rates are unaffected (Ref. 55 and Figure 5).

The behavior exhibited by iron, however, is not unique. Similar features are observed in gold, where the diffusivity differences correlate with heat treatment and are attributed to vacancies, vacancy clusters, or stacking-fault tetrahedra (*53*). Platinum (*53*), silver (*53*), copper (*53*), nickel (*52*), molybdenum (*56*), vanadium (*13*), and niobium (*10*),

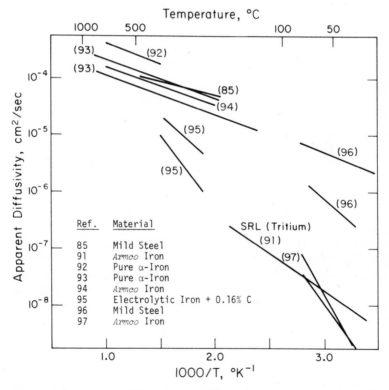

*Figure 4.* *Apparent diffusivity of hydrogen isotopes in α-iron and mild steel. Data of Ref. 91 obtained with tritium. All other data for protium.*

exhibit similar behavior. This behavior is also not unique to dislocation trapping, but occurs with any reversible trap.

Evidence for trapping of hydrogen at dislocations in the refractory metals has been derived from mechanical tests. Variations in the yield points of molybdenum wires saturated with hydrogen at high temperature

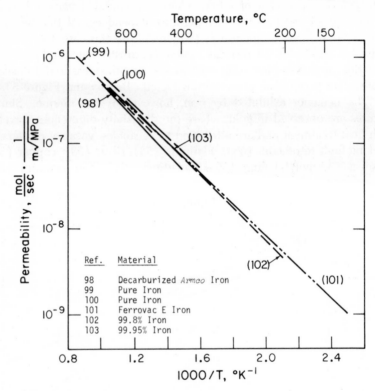

*Figure 5. Permeability of hydrogen in α-iron*

and then quenched were attributed to hydrogen segregation at dislocations (37). Recent deuterium permeation studies and autoradiographic examination of molybdenum wire saturated with tritium at 428 K provided further evidence of trapping but did not identify the trapping site (56). Yield point behavior following hydrogen charging has been taken as evidence of a hydrogen–dislocation interaction in vanadium (38), niobium (39), and nickel (57).

Both external and internal surfaces can act as sources or sinks for dislocations (40). During plastic deformation, a dislocation may move inwardly from the external surface dragging tritium atoms which were picked up at the surface. This has been reported for type 304L stainless steel tensile specimens tested in a 69-mPa tritium atmosphere (40).

Concentration of tritium along slip planes was also observed in the auto-radiographs. A free surface is also a dislocation sink; therefore, tritium may be pulled out of the interior of a metal and rejected to the surrounding atmosphere during plastic deformation. Tritium release from iron, type 304L stainless steel, Inconel 718, nickel, and 5086 aluminum correlated with plastic deformation (*40*).

Temperature is a particularly important parameter in the deformation-enhanced tritium transport phenomenon. Both the binding of the tritium to the dislocation and the ability of the tritium to diffuse with the moving dislocation are temperature dependent. At a high temperature, binding is weak, and little transport occurs because of the small tritium atmosphere. At low temperatures, tritium movement is restricted, and the dislocations break away from the tritium atmosphere and, again, little transport occurs. Practical effects of dislocation transport are restricted, therefore, to an intermediate temperature range.

**Effects of Impurity Fluxes.** Interstitial diffusion of tritium is generally considered in terms of gradients in the chemical potential of tritium and temperature gradients (Soret effect). However, a diffusive flux may also develop in response to gradients in chemical potential of other independent species in solution. The phenomenological equations of irreversible thermodynamics provide general relations among the thermodynamic forces and the conjugate fluxes (*58*). The special case of isothermal diffusion of one interstitial and one substitutional solute illustrates the nature of the cross effects (*59*). The interstitial $J_1$ and substitutional $J_2$ fluxes are related to the concentration gradients $\nabla C_1$ and $\nabla C_2$ through the diffusion coefficients $D_{ij}$:

$$J_1 = -D_{11}\nabla C_1 - D_{12}\nabla C_2 \tag{3}$$

$$J_2 = -D_{21}\nabla C_1 - D_{22}\nabla C_2 \tag{4}$$

For the assumptions made in the derivation, the ratios of the diffusion coefficients in the volume fixed frame of reference were found to be:

$$\frac{D_{12}}{D_{11}} = X_1 \epsilon_{12} - \left(\frac{X_1^2}{(L_{11})_k X_2}\right)\left((L_{22})_k - X_2(L_{33})_k\right) \tag{5}$$

and

$$\frac{D_{21}}{D_{22}} = X_2 \epsilon_{21} + X_2^2\left[\frac{(L_{33})_k}{(L_{22})_k}\right] \tag{6}$$

In Equation 5, $X_i$ is the mole fraction of $i$, $(L_{ii})_k$ are coefficients in the Kirkendall frame of reference, and the interaction coefficients $\epsilon_{ij}$ are from the Taylor expansion of the activity coefficients $\gamma_i$ in the mole fraction:

$$\ln \gamma_i = \ln \gamma_i^{\circ} + \epsilon_{ii} X_i + \epsilon_{ij} X_j \qquad (7)$$

The initial terms in the expression for the ratio of diffusion coefficients are of purely thermodynamic origin; the second terms are kinetic.

The ratios $D_{12}/D_{11}$ and $D_{21}/D_{22}$ cannot be calculated without the necessary thermodynamic and kinetic data. However, their magnitudes may be roughly estimated. This will permit the assessment of the relative contribution of the cross terms to tritium diffusion in dilute solution. The thermodynamic terms $X_1\epsilon_{12}$ and $X_2\epsilon_{21}$ will be less than 1 for dilute solutions because $X_1$ and $X_2 \ll 1.0$, and interaction parameters are estimated to be between $-5$ and $+5$ based on experimental determination of alloying effects on hydrogen solution in liquid iron and cobalt ($60, 61$). Estimates of the kinetic terms indicate that they may be small also. In both expressions, ratios of $L$'s occur, which are expected to be about 1. More important, however, is the fact that these ratios are multiplied by a mole fraction ($\ll 1.0$) or its square, except for the term $[X_1^2/X_2][(L_{22})_k/(L_{11})_k]$ where the ratio $X_1^2/X_2$ could easily vary from $10^2$ to $10^{-2}$ depending on the concentrations.

This type of analysis indicates that the contributions of $D_{12}\nabla C_2$ and $D_{21}\nabla C_1$ to the fluxes $J_1$ and $J_2$ are expected to be smaller than the direct diffusion effects. The contribution need not be insignificant, however, since the gradients may be relatively large under prevailing conditions in fission or fusion reactors. Further investigation of the cross effects is clearly needed to evaluate their possible contribution to tritium diffusion.

### External Surface Films

Oxides are probably the most common form of surface film on metals and also form on readily oxidized materials such as silicon. Control of surface oxidation during fabrication of large engineering structures depends on the alloy. Close control is necessary when fabricating refractory alloys to avoid excessive oxygen pickup which severely degrades mechanical behavior. Thin oxide films will be present on these alloys, however, as well as on structures fabricated from stainless steel, where oxidation control is less important. Furthermore, operating exposure to oxidizing atmospheres at elevated temperature will lead to growth of oxide films, whereas contact with materials such as liquid lithium will remove the oxide film. Oxide–metal interfaces act as sources and sinks for vacancies and oxygen atoms during growth or reduction of an oxide film.

The extent to which tritium transport is affected by an oxide film will depend on the stability of the oxide in the ambient environment, its thickness, its permeability relative to the substrate, and its structural continuity and perfection. Oxides usually have a lower permeability than the base materials and thus restrict tritium uptake and release. Anomolies in permeation behavior of Inconel (*62*), Incoloy 800 (*63*), type 304 stainless steel (*64, 65*), and niobium (*66*) have been observed and attributed to oxide films formed prior to exposure to hydrogen.

**Stability of Oxide Films.** The thermodynamic stability of an oxide on a pure substance may be evaluated from available thermochemical data on dissociation pressure and for possible reactions with species in the surrounding environment that may contact the surface (*67*). For example, oxide reduction by hydrogen gas may occur:

$$2nH_2(g) + M_mO_{2n}(s) \rightleftharpoons 2nH_2O(g) + mM(s) \qquad (8)$$

Possible reactions between oxides of chromium, niobium, iron, and molybdenum with hydrogen gas are readily evaluated by this equation and are illustrated in Figure 6.

The stability of oxides on commercial alloys is more difficult to assess, since the composition of the oxide may vary widely depending on the circumstances surrounding its formation. On stainless steel, for example, the oxide may be nearly pure $Cr_2O_3$, $FeO \cdot Cr_2O_3$ spinel, $MnO \cdot Cr_2O_3$ spinel, or one of the iron oxides (*63, 67, 68*). $Cr_2O_3$ is relatively stable in hydrogen, whereas the iron oxides may be reduced unless the oxygen potential of the gas phase is relatively high.

Aluminum alloys are interesting, because several oxides of the general form $Al_2O_3 \cdot nH_2O$ may form where $n$ is 0, 1, 2, or 3. The hydrated oxides readily form in the presence of relatively low water vapor pressures and are more stable in hydrogen than the pure $\alpha Al_2O_3$ (corundum).

**Effects of Oxide Films.** Few experimental data exist on diffusivity or permeability of tritium (or any hydrogen isotope) in oxides which normally form as surface films on materials. Where data are available, tritium diffusivity and permeability for the oxide are lower than for the base material. This is shown for tritium diffusivity in silicon and silicon dioxide in Figure 2. Tritium diffusivity in aluminum oxide and titanium dioxide (*53*) is also lower than in the corresponding metals. Experiments with oxidized zirconium, niobium, stainless steel (*65, 69, 70*), and titanium (*71*), conform to the expectation of a lower release, absorption, or permeation rate than with "clean" metal surfaces.

Tritium evolution from type 304 stainless steel, Zircaloy-2, and niobium has been analyzed successfully by assuming an oxide–metal laminate (*70*). The oxide films were assumed to be continuous and free of high diffusivity paths. The surface of the type 304 stainless steel specimen had an apparent diffusivity of about $10^{-15}$ $cm^2/sec$, a value comparable with results on type 302 and type 347 stainless steels obtained from hydrogen evolution following cathodic charging (*72*). In both cases, the boundary conditions at the oxide–metal interface are:

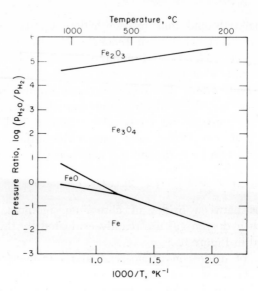

a.   Iron Oxides

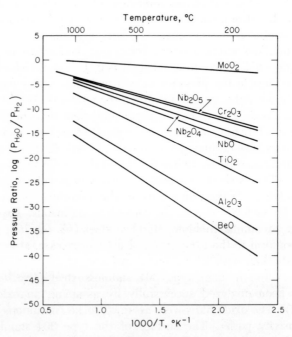

b.   Chromium, Niobium, and Molybdenum Oxides

*Figure 6.   Temperature dependence of the equilibrium pressure
ratio* $(p_{H_2O}/p_{H_2})$

$$D_o\, \partial C_o/\partial x = D_m\, \partial C_m/\partial x, \text{ and } C_o = C_m \qquad (9)$$

This last condition is not true in general and should be replaced by $C_o/C_m = k$, where the distribution coefficient $k$ accounts for differences in the activity coefficients of tritium in the two phases (73). The necessary thermodynamic data are lacking, however, so the simpler condition $C_o = C_m$ may be taken as an expedient approximation until solubility studies on oxides have been made.

Hydrogen pickup by titanium and its alloys is delayed very effectively by the naturally occurring oxide films (71). Experiments on the hydriding of Ti–5Al–2½Sn alloy at low temperatures have been rationalized with a model that represents the oxide film thickness $\delta$ and diffusivity $D_o$ as a film coefficient ($h = D_o/\delta$) in the relation:

$$D_m \nabla C_m = h(C_s - C) \qquad (10)$$

where $C_s$ is the surface concentration of hydrogen in the metal at saturation. In this instance, tritium diffusivity in pure rutile ($TiO_2$) single crystals was known (28). The oxide film thickness was not measured, but was estimated from available studies (74) on titanium oxidation. Further, the large anisotropy in diffusivity observed in $TiO_2$ is a characteristic that may be anticipated in other oxides with non-cubic lattices. This factor could be significant in determining the overall permeation rate through highly textured metals where oxide film growth is epitaxial.

Perfection of an oxide film on a metal base is questionable in light of the available information on the morphology of growing oxides (75). Film formation apparently begins by the nucleation of isolated islands which then grow together. Presumably, this growth pattern could lead to the presence of high diffusivity paths along the oxide grain boundaries. Permeation through stainless steel at low pressures has been analyzed by assuming an impermeable oxide with a small area of pores (69). The model predicts defect-controlled permeation at low pressures and either oxide- or metal-controlled permeation at higher pressures. The apparent thickness dependence of hydrogen diffusivity in oxides on type 302 and type 347 stainless steel alloys may result from changes in film perfection with oxide film thickness (72).

Formation of an oxide film is often accompanied by some solution of oxygen in the metal. Since oxygen dissolves interstitially in many metals (75), for example, $\alpha$-titanium, $\alpha$-zirconium, niobium, and vanadium, an enhanced or diminished tritium concentration could occur from attractive or repulsive oxygen–hydrogen interaction. The only reported observation of this effect is a study of the Ti–O–H ternary, that demonstrates that the hydrogen concentration is lowered as oxygen concentration increases for a fixed hydrogen potential (76).

**Effects Associated with Unstable Films.** Removal or growth of an oxide film during tritium transport may influence diffusion in one of several ways:

(1) Interaction between the interstitial tritium and the vacancy flux caused by changes of oxide thickness.

(2) Interaction between the interstitial tritium and oxygen flux in the metal lattice.

(3) Change of the chemical potential of tritium at the substrate surface upon addition or removal of an oxide film.

Whereas removal of an oxide film may occur by several processes, such as erosion, reduction by tritium gas, dissociation, or chemical reaction, oxide growth is restricted by requiring an adequate oxygen potential in the environment at the interface.

Oxide films may grow by cation diffusion in the substrate to the metal–oxide interface or by anion diffusion through the oxide to this interface (77, 78, 79). In both instances, there will be a vacancy flux in the substrate. Vacancies are generated at the interface and diffuse into the substrate or vacancies in the substrate diffuse to the interface, respectively. Furthermore, vacancy absorption by the substrate may lead to Kirkendall void formation along grain boundaries or at the substrate–oxide interface.

Tritium migrates by interstitial diffusion, but a vacancy flux in the metal lattice could interact with diffusing tritium in the same manner as discussed for an impurity flux. Also, the presence of excess vacancies in the metal lattice or vacancy clusters could serve as trapping sites leading to a higher than normal accumulation of tritium near the surface and a lower apparent diffusivity. High concentrations of tritium in the near-surface regions have been observed in copper (Figure 7), as well as in silicon (80), stainless steel, zirconium, and niobium (70). Interpretation is not unequivocal, however, and a high surface concentration of tritium could be attributed to interaction with dissolved oxygen or with an oxide film directly, rather than an accumulation of vacancies with associated tritium atmospheres.

Reaction between surface oxide films and hydrogen has been observed in permeation experiments. Apparatus constructed of type 304 stainless steel has reacted with deuterium gas at elevated temperatures to produce water vapor which is then available for reaction with other surfaces (81). Thus, oxygen could be transferred from the oxide film on stainless steel to a more reactive metal such as niobium. Oxidation of the niobium would alter its permeability to hydrogen or increase trapping effects (82).

Hydrogen evolving into a vacuum through an oxidized metal surface has reacted with the oxygen on the exit surface to form water vapor (83).

This phenomenon directly affects a permeation measurement, especially at low hydrogen pressures, because all of the evolving hydrogen may not be accounted for unless the water vapor is measured, and this hydrogen is included in the total quantity evolved.

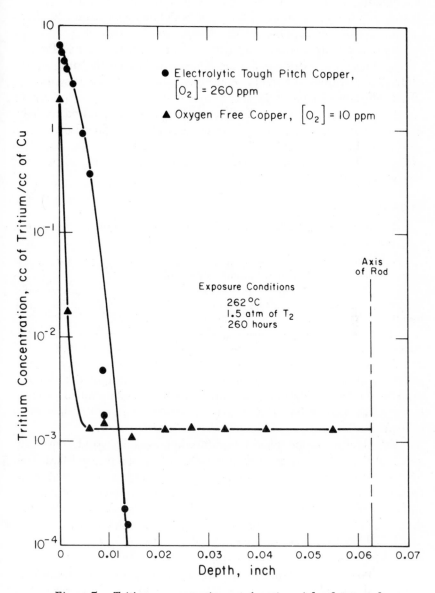

*Figure 7.    Tritium concentration as a function of depth into rods*

## Boundary Processes

The importance of boundary processes, such as dissociation and adsorption, on dissolution of tritium in a solid and its subsequent diffusion is attested to by numerous observations. Carbon dioxide, carbon monoxide, and sulfur dioxide, for example, poison the surface of iron preventing or hindering hydrogen solution (84). Consequently, radiation might be expected to influence tritium diffusion in a solid either by directly affecting adsorption, desorption, or dissociation or by removing adsorbed species that might interfere with tritium diffusion across the surface.

Boundary processes are the mechanisms by which the surface concentration of tritium in a solid is established, and, therefore, they directly affect diffusion. In terms of the formal mathematical description, the boundary processes establish the boundary conditions required for solution of the diffusion equations. Electron- and photon-induced desorption and sputtering are radiation effects that will directly affect the boundary processes. These have been discussed earlier in this volume and will not be covered further.

## Summary

Three general processes may affect tritium diffusion in solids in a radiation environment:

(1) Growth or reduction of surface oxide films may generate a flux of vacancies, oxygen, or metal atoms in the bulk phases.

(2) Rates of boundary processes which control tritium transfer across an interface (e.g., dissociation or adsorption by a solution) may be altered by radiation.

(3) Surfaces act as sources and sinks for defects and impurities generated by radiation which may interact with diffusing tritium.

Tritium diffusion in solids in a radiation field has not been investigated directly, and only a few materials subjected to radiation damage prior to the tritium diffusion measurements have been studied. Consequently, the relative importance of these various processes is difficult to evaluate.

Estimates of trapping and interdiffusion effects discussed above suggest that these two processes should make only a minor perturbation on tritium diffusion and distribution in a fusion reactor system. The complexity of such reactor systems, however, requires a much more detailed analysis before definitive assessment of these effects can be made. Furthermore, there are several important areas where information is currently lacking that must be investigated if tritium diffusion in reactor systems is to be analyzed and controlled. These include:

(1) Pressure and temperature dependence of tritium solubility in oxides and the influence of defect structures and the Fermi energy of the oxide on the tritium solubility.

(2) Thermodynamics of metal–oxygen–hydrogen ternary systems and the possible altered hydrogen solubility in the near-surface regions when oxide films are present.

(3) Tritium flux in the presence of gradients or fluxes of vacancies or other impurities in a material.

(4) Direct measurements of the effects of irradiation on transient and steady state tritium permeation through materials.

*Literature Cited*

1. "The Safety of Nuclear Power Reactors (Light Water Cooled) and Related Facilities," *USAEC* (1973) **WASH-1250**.
2. Compere, E. L., Freid, S. H., Nestor, C. W., "Distribution and Release of Tritium in High-Temperature Gas-Cooled Reactors," *USAEC* (1974) **ORNL-TM-4303**.
3. Hickman, R. G., "Tritium in Fusion Power Reactor Blankets," *USAEC* (1973) **UCRL-74633**.
4. Draley, J. E., Krakowski, R. A., Coultas, T. A., Maroni, V. A., "An Engineering Design Study of a Reference Theta-Pinch Reactor," *USAEC* (1975) **ANL-8019**.
5. Booth, L. A., *Nucl. Eng. Des.* (1973) **34**, 263.
6. Spano, A. H., Belozerov, A., *Nucl. Fusion* (1974) **14**, 281.
7. Badger, B., et al., "Wisconsin Tokamak Reactor Design, Vol. 1, Report **UMFDM-68**, Nuclear Engineering Department, University of Wisconsin, Madison, Wisc., 1974.
8. Nickerson, R. A., "Hydrogen–Isotope Flow in the Rotating Target Neutron Source," *USAEC* (1975) **UCRL-51737**.
9. Stickney, R. E., "The Chemistry of Fusion Technology," Plenum, New York, 1972.
10. Voekl, J., Alefeld, G., "Diffusion in Solids," Academic, New York, 1975.
11. Birnbaum, H. K., Wert, C. A., *Ber. Bunsengese. Phys. Chem.* (1972) **76**, 806.
12. Perkins, W. G., *J. Vac. Sci. Technol.* (1973) **10**, 543.
13. Maroni, V. A., *J. Nucl. Mater.* (1975) **53**, 293.
14. Elleman, T. S., Verghese, K., *J. Nucl. Mater.* (1975) **53**, 299.
15. Lichtman, D., ADV. CHEM. SER. (1976) **158**, 171.
16. Brumback, S., Kaminsky, M., ADV. CHEM. SER. (1976) **158**, 183.
17. Woolan, E. O., Cable, J. W., Koehler, W. C., *J. Phys. Chem. Solids* (1963) **24**, 1141.
18. Skold, K., Nelin, G., *J. Phys. Chem. Solids* (1967) **28**, 236.
19. Carstanjen, H. D., Sizmann, R., *Ber. Bunsengese. Phys. Chem.* (1972) **76**, 1223.
20. Rush, J. J., Flotow, H. E., *J. Chem. Phys.* (1968) **48**, 37.
21. Picraux, S. T., Vook, F. L., *Phys. Rev. Lett.* (1974) **33**, 1216.
22. Weaver, E. A., *J. Chem. Phys.* (1967) **47**, 4891.
23. Laska, H. M., Doremus, R. H., Jorgensen, P. J., *J. Chem. Phys.* (1969) **50**, 135.
24. Matzke, H., *Z. Naturforsch.* (1967) **A22**, 965.
25. Jorgensen, P. J., Norton, F. J., *Phys. Chem. Glasses* (1969) **10**, 23.
26. Crank, J., Park, G. S., "Diffusion in Polymers," Academic, New York, 1968.
27. Thomas, D. G., Lander, J. J., *J. Chem. Phys.* (1956) **25**, 1136.
28. Caskey, Jr., G. R., *Mater. Sci. Eng.* (1974) **14**, 109.
29. Smith, T., "Hydrogen Permeation of Oxide Films on Zirconium," USAEC Report **NAA-SR-6267**, 1962.
30. Elleman, T. S., "Tritium Diffusion in Ceramic CTR Materials," USERDA, Report **ORO-4721-1**, 1975.

31. Gauss, H., Z. *Naturforsch.* (1965) **A20**, 1298.
32. Hurst, D. G., "Diffusion of Fission Gas," Report **CRRP-1124**, Atomic Energy of Canada, Ltd., 1962.
33. McNabb, A., Foster, P. K., *AIME Trans.* (1963) **227**, 618.
34. Oriani, R. A., *Acta Metall.* (1970) **18**, 147.
35. Sy Ong, A., Elleman, T. S., *Nucl. Instrum. Methods* (1970) **86**, 117.
36. Koiwa, M., *Acta Metall.* (1974) **22**, 1259.
37. Lawley, A., Liebmann, W., Maddin, R., *Acta Metall.* (1961) **9**, 841.
38. Eustace, A. L., Carlson, O. N., *AIME Trans.* (1961) **221**, 238.
39. Wilcox, B. A., Huggins, R. A., *J. Less-Common Met.* (1960) **2**, 292.
40. Louthan, Jr., M. R., Caskey, Jr., G. R. Donovan, J. A., Rawl, D. E., *Mater. Sci. Eng.* (1972) **10**, 357.
41. Baker, C., Birnbaum, H. K., *Scripta Metall.* (1972) **6**, 851.
42. Blakemore, J. S., *Metall. Trans.* (1970) **1**, 145.
45. Harper, S., Callcut, J. A., Townsend, D. W., Eborall, P., *J. Instrum. Methods* (1961) **90**, 414.
44. Baker, C., Birnbaum, H. K., *Acta Metall.* (1973) **21**, 865.
45. Harper, S., Sallcut, J. A., Townsend, D. W., Eborall, P., *J. Instrum. Methods* (1961) **90**, 414.
46. Ellerbrock, H. G., Vibrans, G., Stüwe, H. P., *Acta Metall.* (1972) **20**, 53.
47. Matzke, H., Z. *Metallkd.* (1967) **58**, 572.
48. Migge, H., *Phys. Status Solidi* (1967) **35**, 673.
49. Caskey, Jr., G. R., Pillinger, W. L., *Metall. Trans.* (1975) **6A**, 467.
50. Foster, P. K., McNabb, A., Payne, C. M., *AIME Trans.* (1965) **233**, 1022.
51. Stafford, S. W., McLellan, R. B., *Acta Metall.* (1974) **22**, 1463.
52. Louthan, Jr., M. R., Donovan, J. A., Caskey, Jr., G. R., *Acta Metall.* (1975) **23**, 745.
53. Caskey, Jr., G. R., Derrick, R. G., presented at the Spring Meeting of the Metallurgical Society, Pittsburgh, PA, May 1974.
54. Shiraishi, K., Nagasaki, R., *J. Nucl. Sci. Technol.* (1965) **2**, 499.
55. Gonzales, O. D., *AIME Trans.* (1969) **245**, 607.
56. Caskey, Jr., G. R., Louthan, Jr., M. R., Derrick, R. G., *J. Nucl. Mater.* (1975) **55**, 279.
57. Smith, G. C., "Hydrogen in Metals," I. M. Bernstein and A. W. Thompson, Eds., American Society of Metals, Metals Park, Ohio, 1974.
58. Howard, R. E., Lidiard, A. B., *Rep. Prog. Phys.* (1964) **XXVII**, 161.
59. Kirkaldy, J. S., *Adv. Mater. Res.* (1970) **4**, 55.
60. Weinstein, M., Elliott, J. F., *AIME Trans.* (1963) **227**, 382.
61. Wooley, F. E., Pehlke, R. D., *AIME Trans.* (1965) **233**, 1454.
62. Rudd, D. W., Vose, D. W., Johnson, S., *J. Phys. Chem.* (1961) **65**, 1018.
63. Savage, H. C., Strehlow, R. A., *Trans. Am. Nucl. Soc.* (1973) **17**, 152.
64. Austin, J. H., Elleman, T. S., Verghese, K., *J. Nucl. Mater.* (1973) **48**, 307.
65. Louthan, Jr., M. R., Derrick, R. G., *Corros. Sci.* (1975) **15**, 565.
66. Pennington, C. W., Verghese, K., Elleman, T. S., *Trans. Am. Nucl. Soc.* (1973) **17**, 151.
67. Dushman, S., "Scientific Foundation of Vacuum Technique," 2nd ed., John Wiley, New York, 1962.
68. Swansiger, W. A., Musket, R. G., Weirick, L. J., Bauer, W., *J. Nucl. Mater.* (1974) **53**, 307.
69. Strehlow, R. A., Savage, H. C., *J. Nucl. Mater.* (1974) **53**, 323.
70. Austin, J. H., Elleman, T. S., *J. Nucl. Mater.* (1972) **43**, 119.
71. Caskey, Jr., G. R., "Hydrogen in Metals," I. M. Bernstein and A. W. Thompson, Eds., American Society of Metals, Metals Park, Ohio, 1974.
72. Piggot, M. R., Siarkowski, A. C., *J. Iron Steel Inst.* (1972) **210**, 901.
73. Ash, R., Barrer, R. M., Palmer, D. G., *Br. J. Appl. Phys.* (1965) **16**, 873.
74. Johnson, D. L., Bashra, N. M., Tato, L. C., *Mater. Sci. Eng.* (1971) **8**, 175.

75. Kofstad, P., "High Temperature Oxidation of Metals," John Wiley, New York, 1966.
76. Hepworth, M. T., Schumann, R., *AIME Trans.* (1962) **224**, 928.
77. Hales, R., *J. Phys. Chem. Solids* (1971) **32**, 1417.
78. Hales, R., Dobson, P. S., Smallman, R. E., *Acta Metall.* (1969) **17**, 1323.
79. Reimann, D. K., Stark, J. P., *Acta Metall.* (1970) **18**, 63.
80. Ichimiya, T., Furichi, A., *Int. J. Appl. Radiat. Isot.* (1968) **19**, 575.
81. Beavis, L. C., *J. Vac. Sci. Technol.* (1973) **10**, 386.
82. Birnbaum, H. K., *Scripta Metall.* (1973) **7**, 925.
83. Begeal, D. R., *J. Vac. Sci. Technol.* (1975) **12**, 405.
84. Srikrishnan, V., Liu, H. W., Ficalora, P. J., *Scripta Metall.* (1975) **9**, 663.
85. Sykes, C., Burton, H. H., Gregg, C. C., *J. Iron Steel Inst.* (1947) **156**, 155.
86. Sicking, G., Buchold, H., *Z. Naturforsch.* (1971) **A26**, 1973.
87. Chaney, K. F., Powell, G. W., *Metall. Trans.* (1970) **1**, 2356.
88. Katz, L., Guinan, M., Borg, R. J., *Phys. Rev.* (1971) **B4**, 330.
89. Louthan, Jr., M. R., Donovan, J. A., Caskey, Jr., G. R., *Scripta Metall.* (1974) **8**, 643.
90. Cupp, C. R., Flubacher, P., *J. Nucl. Mater.* (1962) **6**, 213.
91. Louthan, Jr., M. R., Caskey, Jr., G. R., presented at International Conference on "Effect of Hydrogen on Behavior of Materials," Moran, Wyoming, Sept. 7–11, 1975.
92. Geller, W., Tak-Ho-Sun, *Arch. Eisenhuettenwes.* (1950) **21**, 423.
93. Stross, T. M., Tompkins, F. C., *J. Chem. Soc.* (1956) 230.
94. Eichenauer, W., Künzig, H., Pebler, A., *Z. Metallkd.* (1958) **49**, 220.
95. Hill, M. L., Johnson, E. E., *AIME Trans.* (1959) **215**, 717.
96. Frank, R. C., Swets, D. E., Fry, D. L., *J. Appl. Phys.* (1958) **29**, 892.
97. Beck, W., O'M Bockris, J., McBreen, J., Nanis, L., *Proc. Roy. Soc.* (1966) **A290**, 220.
98. Chang, P. L., Bennett, W. D. G., *J. Iron Steel Inst.* (1952) **170**, 205.
99. Schenck, H., Taxhet, H., *Arch. Eisenhuettenwes.* (1959) **30**, 661.
100. Gorman, J. K., Nardella, W. R., *Vacuum* (1962) **12**, 19.
101. Bryan, W. L., Dodge, B. F., *J. AIChE* (1963) **9**, 223.
102. Wagner, R., Sizman, R., *Z. Angew. Physik* (1964) **18**, 193.
103. Gonzales, O. D., *AIME Trans.* (1967) **239**, 929.
104. Randall, D., Salmon, O. N., "Diffusion Studies I. Permeability of Type 347 Stainless Steel to Hydrogen and to Tritium," *USAEC* (1953) **KAPL-904**.

RECEIVED January 5, 1976. This paper was prepared in connection with work under Contract No. AT(07-2)-1 with the U.S. Energy Research and Development Administration.

# INDEX [1]

[1] In this index, many of the subjects are listed as being on the first page of the chapters. In most of these cases, information about the entries must be found within that chapter.

*The text of this book is set in 10 point Caledonia with two points of leading. The chapter numerals are set in 30 point Garamond; the chapter titles are set in 18 point Garamond Bold.*

*The book is printed offset on Text White Opaque 50-pound. The cover is Joanna Book Binding blue linen.*

*Jacket design by Linda Mattingly. Editing and production by Virginia Orr.*

*The book was composed by Service Composition Co., Baltimore, Md., printed and bound by The Maple Press Co., York, Pa.*

| DATE DUE | | | |
|---|---|---|---|
| Chemistry Dept | | | |
| | | | |
| | | | |
| | | | |
| | | | |
| | | | |
| | | | |
| | | | |
| | | | |
| | | | |
| | | | |

Radiation                158165